An Introduction to Solar Energy for Scientists and Engineers

An Introduction to Solar Energy for Scientists and Engineers

SOL WIEDER
Fairleigh Dickinson University

JOHN WILEY & SONS
New York Chichester Brisbane Toronto Singapore

Library of Congress Cataloging in Publication Data:

Wieder, Sol.
An introduction to solar energy for scientists
and engineers.

Includes index.
1. Solar energy. I. Title.
TJ810.W5 621.47 81-13014
ISBN 0-471-06048-8 AACR2

Printed in the United States of America

10 9 8 7 6 5 4 3 2 1

to my wife Suzanne

A woman of valour who can find. . .
Her children rise up and call her blessed;
Her husband also and he praiseth her:
Many daughters have done valiantly,
But thou excellest them all.

—*Proverbs*

PREFACE

As a result of events of the past decade, there has been an increased awareness of societal problems related to energy resources. Colleges and universities have begun to expand their curricula to include the subject of energy and, where appropriate, to incorporate energy-related material into existing courses. Solar energy courses are especially attractive to undergraduate science and engineering majors. Much of the material is sufficiently basic so that the fundamentals acquired in their introductory courses can be applied quite easily. This text in fact evolved from lecture notes for a one-semester junior level course in solar energy offered to physics, chemistry, and engineering students, but which also included those majoring in mathematics, computer science, and environmental science. The prerequisites were one year each of general physics and calculus.

Although a number of solar energy books were available, no single work seemed appropriate—some were too qualitative and elementary whereas others were too technical—for the intended audience. Still others offered a largely phenomenological treatment. This text deals with solar energy on a quantitative level but nevertheless is introductory and intended to prepare the student for more advanced work in solar energy.

Even though written from a physicist's perspective, the book includes many numerical examples in order to appeal to engineering students. Rather than being a "do-it-yourself" manual for solar construction projects or a technical handbook of hard experimental data for industrial design applications, this text is geared to students who desire a deeper understanding of the fundamental principles behind solar energy technology. Whereas emphasis is placed on the physics and engineering of solar energy, material covering its history, economics, and architecture may be found in the references cited at the end of each chapter.

The book is divided into four parts. The first three chapters, dealing with the sun, its apparent motion across the sky, and the insolation it provides at the earth's surface, give the student an insight into the nature of terrestrial insolution. Chapters 4 and 5 survey the principles of heat transfer and optics and present the background material for the following two chapters. Chapters 6 and 7 cover solar heaters with applications to space heating and hot water supply systems and include topics such as flat plate collectors, arrays, thermal storage, and solar-assisted systems. The final part, Chapters 8 and 9, discusses thermodynamic and photovoltaic conversion of solar energy to useful work. In addition, the book has two appendices: one summarizing the formulas for the sun's motion and the other deriving approximate equations for diffuse solar flux. The material can be covered in one semester with time available to include supplementary subjects.

I am grateful to my colleagues and students for the many interesting discussions throughout the development of this work. My thanks to Professors Ralph Hautau and Vincent P. Tomaselli for their critical comments on the manuscript, to Edmond Jaoudi for his help with the drawings, and to Phyllis Rind for her assistance in typing.

For their tolerance and understanding during those times that I devoted to the text rather than to them, I extend special thanks to my sons Ari, Jonah, and Jeremy. Most of all, I am grateful to my wife Suzanne to whom this work is dedicated. She patiently typed and retyped the manuscript, contributing many valuable suggestions. Her inspiration and loyalty helped make this book possible.

Sol Wieder

CONTENTS

Chapter 4
Elements of Heat Transfer

Chapter 5
The Optics of Collectors

Chapter 6
Solar Heating Panels

An Introduction to Solar Energy for Scientists and Engineers

CHAPTER 1

The Solar Constant

The Model Sun

The sun is the ultimate origin of most of the energy presently available on earth. This includes the energy for direct heating, as well as wind energy, hydroelectric power, and energy derived from fossil fuels. Fossil fuels exist today as a consequence of photosynthesis, the process through which plants convert solar energy to chemical energy. A complete understanding of solar energy technology is only possible with a thorough analysis of solar radiation.

The sun, our closest star, provides the energy to maintain life on earth and produces the necessary gravitational attraction to keep our planet in a nearly circular orbit. It has a mass of $M_\odot = 1.99 \times 10^{30}$ kg ($\simeq 3.3 \times 10^5$ earth masses) and a radius of $R_\odot = 6.96 \times 10^8$ m ($\simeq 109$ earth radii).[1] The earth–sun distance varies from 1.0167 AU (aphelion, ~July 4) to 0.983 AU (perihelion, ~January 4) and has an average value of 1 AU (1 AU = 1 astronomical unit $\simeq 1.5 \times 10^{11}$ m).

The interior of the sun is inaccessible to us for direct experimentation. However, based on observations of the solar surface and

[1]The subscript $\odot$ is a symbol used to represent solar quantities.

theoretical considerations, it is believed that the interior temperature is about 15 million kelvins. The chemical composition of the sun is mainly hydrogen with a lesser amount of helium. These two elements, which account for 96 to 99 percent of the sun's mass, are under enormous pressure and only the large gravitational pull of the sun keeps this mass together. Energy is generated in the interior through the nuclear fusion of hydrogen into helium. This energy finds its way to the surface and is eventually emitted into space primarily in the form of electromagnetic radiation. The surface of the sun, the *photosphere*, is actually a transition region in which the density falls off rapidly. As we move from the interior of the sun to the outer part of the photosphere, we pass from an optically opaque medium to a relatively transparent one. Furthermore, the temperature falls to approximately 6000 K. Above the photosphere is the sun's atmosphere, which is called the *chromosphere* because it selectively absorbs certain colors of the radiation emitted from the pho-

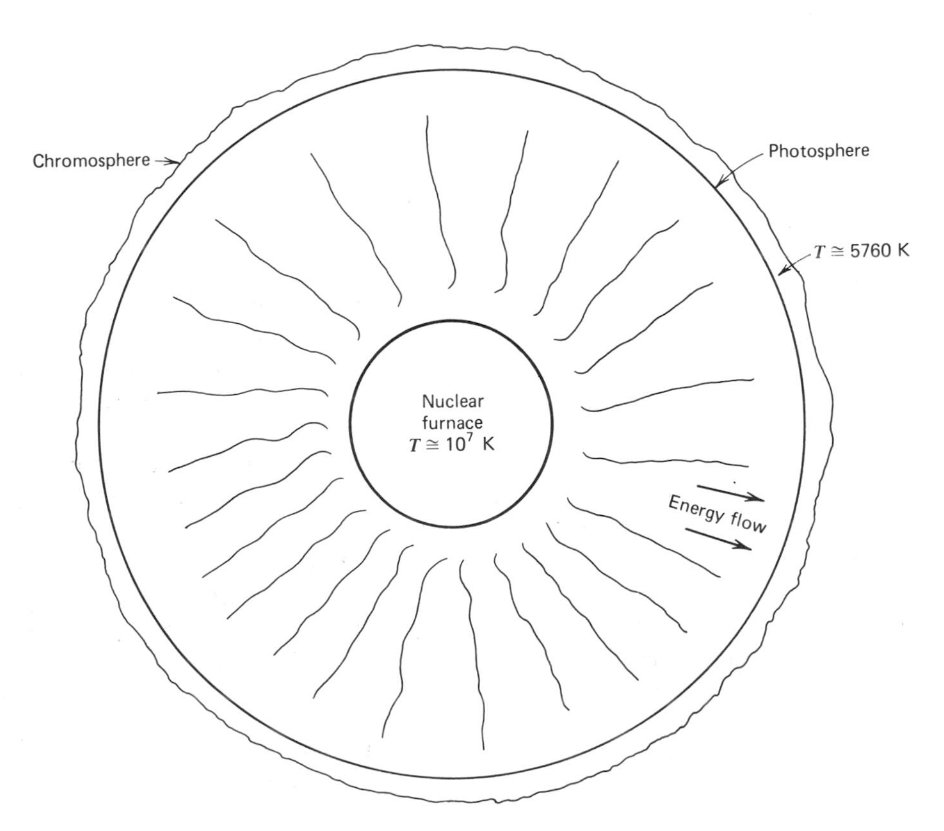

FIGURE 1.1 *A simplified model of the sun.*

tosphere. Because it is relatively transparent, we will ignore its effects on the emitted solar radiation.

Most of the radiation reaching us emanates from the photosphere so that the solar spectrum is determined by the optical and thermal properties of the solar surface. The simple model being used here assumes that the sun behaves as a black body whose surface is maintained at $T \simeq 6000$ K. This surface temperature is kept constant by a source of energy located in the interior. As a result of this elevated temperature, the surface glows and electromagnetic radiation is emitted in all directions of space (Figure 1.1).

Black Body Emission

Electromagnetic radiation is composed of waves of oscillating electric and magnetic fields. Each wave is characterized by a wavelength λ and a frequency ν. In free space all the waves travel at the same speed, $c = 2.9979 \times 10^8$ m/sec. The frequency, wavelength, and speed of each wave are related by

$$\nu\lambda = c$$

The higher the frequency is, the shorter the wavelength and vice versa. The entire electromagnetic spectrum is shown in Figure 1.2. Only a very narrow band of wavelengths, those in the range $400 \text{ nm} < \lambda < 700 \text{ nm}$, are visible to the human eye.[2] Those wavelengths bordering the visible on the violet end ($\lambda < 400$ nm) are called ultraviolet and are invisible. Those wavelengths bordering on the red ($\lambda > 700$ nm) are the infrared and are also invisible. As we will see, approximately half of the solar radiation is in the infrared; the visible components make up less than 40 percent of the solar energy.

When electromagnetic radiation is incident on the surface of a body, it can either be transmitted, reflected, or absorbed. If the body is opaque, no transmission is possible. The radiant energy per unit time per unit area per unit wavelength incident on the surface is called the incident *spectral flux*, $F_\lambda^{(i)}$. Similarly, the absorbed and reflected spectral fluxes are denoted by $F_\lambda^{(a)}$ and $F_\lambda^{(r)}$, respectively. The subscript λ indicates that we are dealing with a single component

[2]A nanometer (1 nm) $= 10^{-9}$ m $= 10^{-7}$ cm. It is often convenient to measure wavelengths in microns (μm), where 1 μm $= 10^{-6}$ m $= 10^{-4}$ cm so that 1000 nm $=$ 1 μm. The spectral range of radiation detected by the eye varies among individuals. We will define the visible range to include those wavelengths between 0.4 μm and 0.7 μm.

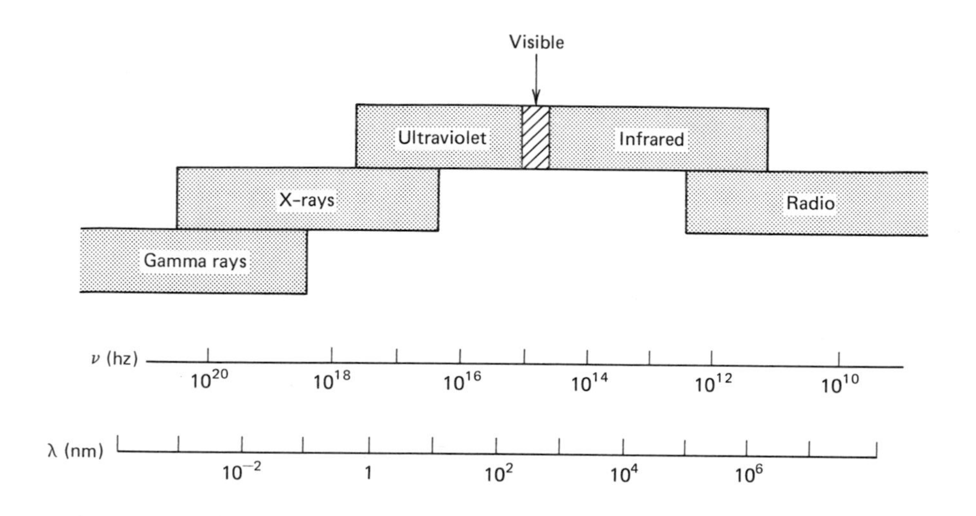

FIGURE 1.2 *The electromagnetic spectrum.*

wavelength. The total flux in the distribution is

$$F = \int_0^\infty F_\lambda \, d\lambda$$

We will define the spectral absorptivity a_λ and the spectral reflectivity r_λ of a body's surface by

$$a_\lambda = \frac{F_\lambda^{(a)}}{F_\lambda^{(i)}} \qquad \text{and} \qquad r_\lambda = \frac{F_\lambda^{(r)}}{F_\lambda^{(i)}} \tag{1.1}$$

When the body is opaque, what is not reflected from the surface must be absorbed and we may write

$$a_\lambda + r_\lambda = 1 \tag{1.2}$$

Actually, a_λ and r_λ, for a real surface, depend on the wavelength of the incident flux and on the direction of incidence of the radiation. For example, many surfaces absorb radiation well at normal incidence, yet do not absorb efficiently when the radiation is incident at glancing angles. We will neglect this dependence on direction of incidence and assume, for simplicity, that the surface is an *isotropic absorber*. However, the spectral reflectivity and absorptivity do vary appreciably with the wavelength of the incident radiant flux. Many pigments that appear white to the eye because they reflect well in the visible spectrum may in fact be excellent absorbers of infrared radiation.

It is useful to define the following idealizations of real surfaces.

Black Body Any body whose surface absorbs *all* components of incident electromagnetic radiation regardless of the wavelength or the direction of incidence is called a *black body*. For such bodies we have

$$a_\lambda = 1 \qquad (r_\lambda = 0) \text{ for all } \lambda$$

White Body Any body whose surface reflects *all* components of incident electromagnetic radiation regardless of the wavelength and the direction of incidence is called a *white body* (or perfect reflector). For such bodies we have

$$a_\lambda = 0 \qquad (r_\lambda = 1) \text{ for all } \lambda$$

Gray Body Any body whose surface absorptivity is between that of a black body and that of a white body but is *independent* of the wavelength and direction of incidence of the incident radiation is called an (isotropic) *gray body*. For such bodies we have

$$a_\lambda = a \qquad (\text{for all } \lambda) \qquad \text{where } 0 < a < 1$$

No real surfaces are perfectly black or white. For solar radiation, black matte lacquer has a mean absorptivity of $a = 0.97$. Polished silver, which is highly reflective, has $a = 0.07$.

It is an experimental fact of nature that when any opaque body is maintained at a constant temperature its surface emits characteristic electromagnetic radiation called *thermal radiation*. This radiation is generally emitted in all directions and contains all of the wavelengths of the electromagnetic spectrum. The thermal flux leaving the body depends both on the surface characteristics of the body as well as on its Kelvin temperature T. For isotropically absorbing surfaces the emitted thermal flux is isotropic and its spectral distribution is given by

$$\boxed{F_\lambda = \epsilon_\lambda B_\lambda(T)} \qquad (1.3a)$$

where ϵ_λ is a characteristic of the surface called the *spectral emissivity* and where $B_\lambda(T)$ is called *Planck's function*. This universal function of

λ and T is given by

$$\boxed{B_\lambda(T) = \frac{a}{\lambda^5(e^{b/\lambda T} - 1)}} \tag{1.3b}$$

The constants in the function are

$$a = 2\pi hc^2 = 3.7405 \times 10^{-16}\ \mathrm{W - m^2}$$

and

$$b = \frac{hc}{k} = 1.4388 \times 10^{-2}\ \mathrm{m - K}$$

where

h (Planck's constant) $= 6.6252 \times 10^{-34}$ J-s (joule-sec)
c (speed of light) $= 2.9979 \times 10^{8}$ m/sec
k (Boltzmann's constant) $= 1.3806 \times 10^{-23}$ J/K (joule/K)

The total radiant flux emitted by the surface is

$$F = \int_0^\infty \epsilon_\lambda B_\lambda(T)\, d\lambda \tag{1.4}$$

It can be shown from the laws of thermodynamic equilibrium that the thermal emissivity and the optical absorptivity are in fact related. This relationship is established by Kirchhoff's law.[3]

The spectral emissivity of an isotropic surface is equal to its spectral absorptivity or

$$\boxed{\epsilon_\lambda = a_\lambda} \tag{1.5}$$

It follows from Equation 1.5 that the black body for which $a_\lambda = 1$ is the most efficient radiator with $\epsilon_\lambda = 1$ for all wavelengths. Thus for a black body Equation 1.3a gives

$$F_{\lambda\,\mathrm{black}} = B_\lambda(T)$$

[3]This version of Kirchhoff's law is applicable only to idealized isotropic surfaces such as black, white, and gray bodies. The more general form implies that when the thermal radiation leaving a surface is not isotropic, the thermal emissivity in a given direction is equal to the absorptivity of radiation incident from that direction.

so that the function describing the spectral flux emitted by a black surface at a Kelvin temperature T is simply the Planck function. It also follows from Equation 1.5 that a white body ($a = \epsilon = 0$) emits *no* thermal radiation and that a gray body emits radiation according to

$$F_{\lambda\,\text{gray}} = \epsilon B_\lambda(T) \qquad (0 < \epsilon < 1) \tag{1.6}$$

The emission spectrum of a black, gray, and real body at 6000 K is plotted in Figure 1.3.

Note that the spectral function for a black body is equal to the Planck function, whereas the spectral function for a gray body has the same shape but is reduced by a factor ϵ. It is therefore important that we consider the mathematical properties of $B_\lambda(T)$.

Figure 1.4 shows the Planck function as a function of wavelength for various Kelvin temperatures. Each curve has a finite area under it and each has a wavelength, λ_{max}, at which $B_\lambda(T)$ is a maximum. Therefore, at any finite temperature, the energy carried by those components whose wavelengths are either very short or very long is negligibly small. Furthermore, the most energy is carried by those wavelengths in the region where $B_\lambda(T)$ is largest.

It can be shown that the following mathematical properties of $B_\lambda(T)$ are valid.

$$\boxed{\lambda_{\text{max}} = \frac{\alpha}{T} \qquad \text{(displacement law)}} \tag{1.7a}$$

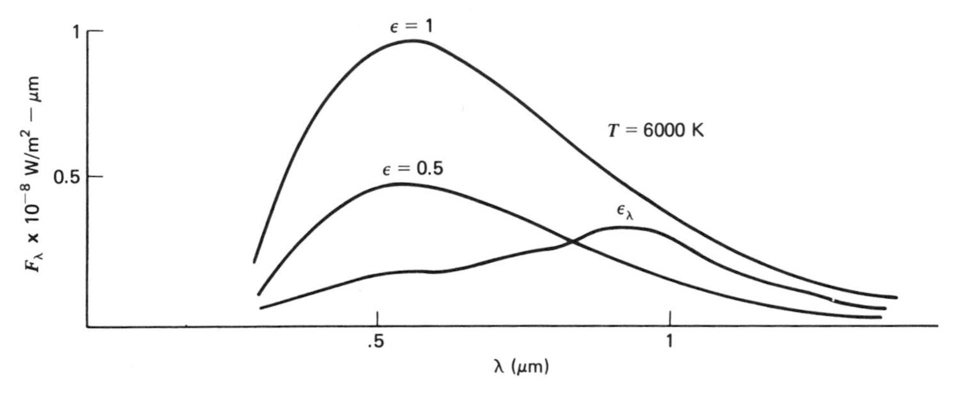

FIGURE 1.3 *The thermal emission spectrum of a black ($\epsilon = 1$), gray ($\epsilon = 0.5$), and real (ϵ_λ) body, each at 6000 K. The real body represented here is a more efficient emitter of infrared radiation than it is of visible radiation.*

and

$$\int_0^\infty B_\lambda(T)\,d\lambda = \sigma T^4 \qquad \text{(Stefan–Boltzmann law)} \tag{1.7b}$$

where α and σ are universal constants.

From Equation 1.7a we observe that the wavelength at which $B_\lambda(T)$ is a maximum varies inversely with the Kelvin temperature. From Equation 1.7b we see that the total area under $B_\lambda(T)$ and consequently the total thermal flux emitted by a black body is proportional to the fourth power of the Kelvin temperature.

In the following presentation we will continue to express all quantities in the MKS (meter-kilogram-second) system. Energy is expressed in joules (J), power (energy/time) in watts (W), and area in m^2. In these units the total flux is in watts/m^2 and the wavelength is conveniently expressed either in meters, microns ($1\,\mu m = 10^{-6}$ m), or nanometers ($1\,nm = 10^{-9}$ m). The constants in Equation 1.7 are

$$\alpha = 2898\ \mu m - K$$

$$\sigma = \frac{5.6696 \times 10^{-8}\ W}{m^2 - K^4} \qquad \text{(Stefan–Boltzmann constant)}$$

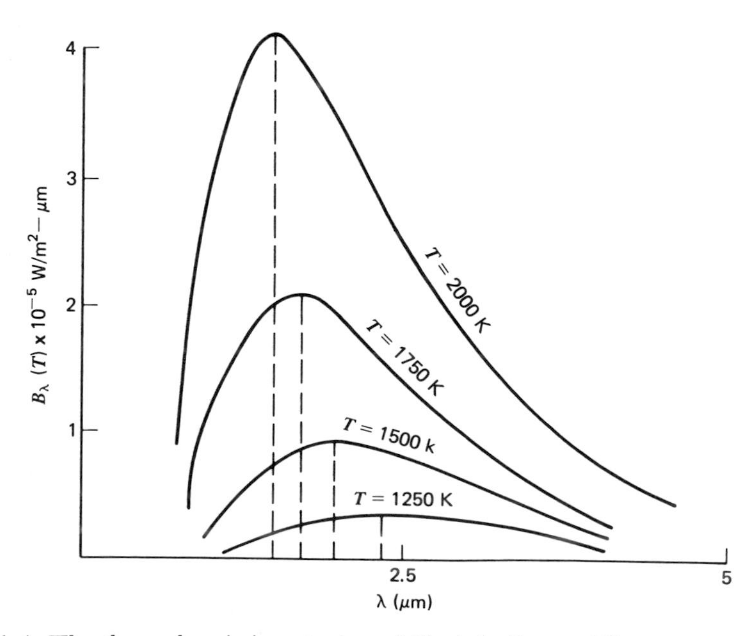

FIGURE 1.4 *The thermal emission spectra of black bodies at different temperatures. The dashed lines mark the values of* λ_{max}.

The λ_{max} of a given Planck distribution is often said to represent the characteristic "color," although it is not necessarily the color detected by the human eye. Nonetheless, the fact that λ_{max} decreases with increasing T explains why a black body appears red hot at some temperature and becomes white hot as the temperature is raised. The whiteness indicates the presence of bluish components. Using Equation 1.7a, we find that a black body at a temperature of 7000 K has a characteristic wavelength $\lambda_{max} = 0.414\ \mu m$ (blue), whereas one with $T = 5800$ K has $\lambda_{max} = 0.500\ \mu m$ (green). If a black body is only at room temperature ($T = 300$ K), then $\lambda_{max} = 966\ \mu m$ (infrared) and the body appears black to the eye.

The total flux emitted by a black body is derived from the Stefan–Boltzmann law Equation 1.7b as

$$F_{black} = \sigma T^4 = (5.67 \times 10^{-8}) T^4 \tag{1.8}$$

Thus black objects whose surfaces are maintained at 7000, 5800, and 300 K, respectively, emit fluxes of 1.36×10^8, 6.4×10^7, and 4.59×10^2 W/m^2, respectively. Note the dramatic increase in black-body emission as we raise the Kelvin temperature. Since the emission varies as T^4, a twofold increase in T produces a sixteen fold increase in F_{black}. The gray body generalization of Equation 1.8 is

$$F_{gray} = \epsilon \sigma T^4 \qquad (0 < \epsilon < 1) \tag{1.9}$$

Radiative Emission from the Sun

If we now take the model of the sun to be a black body at a steady-state temperature T, then the radiant flux emitted at the solar surface can be represented by a Planck distribution. The observed spectral distribution of the sun differs slightly from $B_\lambda(T)$ because the sun is neither in radiative equilibrium nor even in steady state. Nevertheless, a good approximation to the solar spectrum is a black-body curve corresponding to a temperature of $T_\odot \simeq 5800$ K, as can be seen from Figure 1.5. We will use this black-body approximation subsequently.

Using Equation 1.7a, we see that the characteristic wavelength of the solar spectrum is

$$\lambda_{max} = \frac{2.9 \times 10^3\ \mu m - K}{5800\ K} = 0.500\ \mu m = 500\ nm$$

which corresponds to green light. From Equation 1.7b we find the

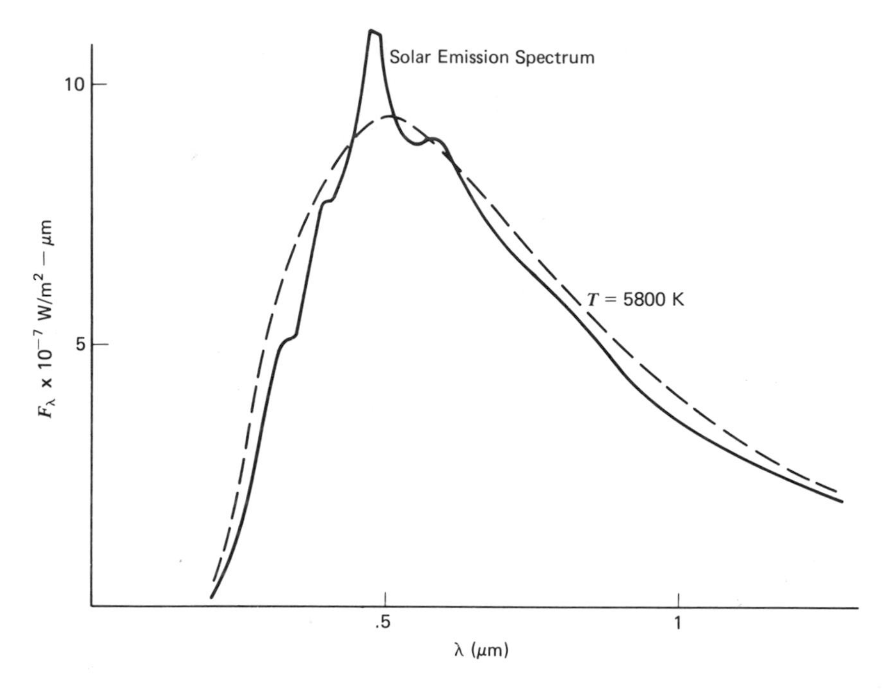

FIGURE 1.5 *The spectral distribution of the flux emitted from the sun's surface. The dashed line is the emission spectrum of a black body at* 5800 *K.*

total flux leaving the surface of the sun to be

$$F_\odot = \sigma T_\odot^4 = \left(\frac{5.670 \times 10^{-8}\ \mathrm{W}}{\mathrm{m}^2 - \mathrm{K}^4}\right)(5800\ \mathrm{K})^4 = 6.416 \times 10^7\ \mathrm{W/m^2}$$

This radiation is *diffuse* (traveling in all directions) when it leaves the solar surface. The total radiant power emitted from the sun is obtained by multiplying the flux above by the surface area of the sun. We find

$$\begin{aligned} P_\odot &= F_\odot 4\pi R_\odot^2 \\ &\simeq (6.42 \times 10^7\ \mathrm{W/m^2})4\pi(6.96 \times 10^8\ \mathrm{m})^2 \\ &\simeq 3.91 \times 10^{26}\ \mathrm{W} \end{aligned}$$

If the sun emits radiation isotropically, this enormous power, called *luminosity* by astronomers, is emitted equally in all directions of space. As the distance from the sun increases, this power is spread over spherical surfaces of increasing area. Consequently, the intensity varies inversely as the square of the distance from the center of the sun. At a distance r the surface area is $4\pi r^2$ so that the radiant flux crossing such a surface is

$$F = \frac{P_\odot}{4\pi r^2} = \frac{4\pi R_\odot^2 F_\odot}{4\pi r^2}$$

or

$$F = R_\odot^2 F_\odot / r^2$$
$$\simeq \frac{3.11 \times 10^{25}}{r^2} \text{W/m}^2 \quad (1.10)$$

Because the earth's distance from the sun varies throughout the year, the total flux reaching the earth also changes. At the mean earth–sun distance of $r = 1.5 \times 10^{11}$ m, the flux is

$$S = F \simeq \frac{3.11 \times 10^{25}}{(1.50 \times 10^{11})^2} = 1382 \text{ W/m}^2 \quad (1.11)$$

The value of the flux is called the *solar constant*, which, as already mentioned, is not actually a constant but varies with season and somewhat with solar activity. Note also that the numerical value in Equation 1.11 has been obtained by assuming the solar spectrum to be that of a black body at ~5800 K. If we were to change this temperature to 5762 K, an accurate computation would show that the solar constant would drop to ~1352 W/m^2. The value of the solar constant has been measured by various investigators to range from 1350 to 1382 W/m^2. The discrepancy amounts to approximately 2 percent. We will arbitrarily take 1352 W/m^2 as the value of the solar constant, taking the spectrum temperature to be ~5760 K.

Equation 1.10 is also valid for the spectral distribution and we may write

$$S_\lambda = \frac{R_\odot^2}{r^2} F_{\odot\lambda} = \frac{R_\odot^2}{r^2} B_\lambda(5760 \text{ K})$$
$$\simeq 2.165 \times 10^{-5} B_\lambda(5760 \text{ K})(\text{in W/m}^2\text{-m}) \quad (1.12)$$

with

$$S = \int_0^\infty S_\lambda \, d\lambda = 1352 \text{ W/m}^2$$

Consequently, the spectral distribution of the flux as it arrives at the

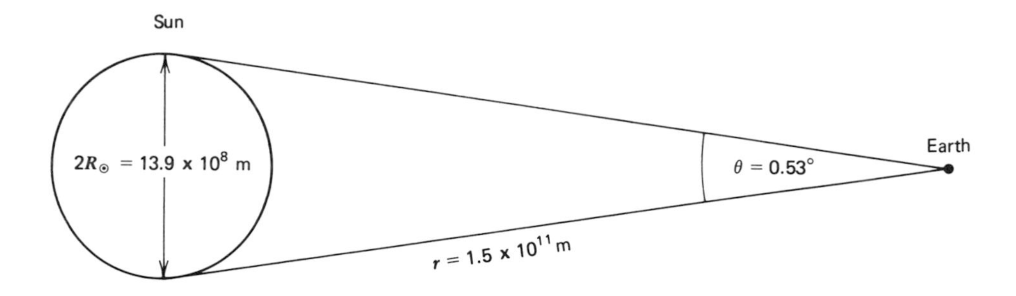

FIGURE 1.6 *The geometry for determining the divergence angle of the solar constant.*

top of the earth's atmosphere is essentially the same as that emitted by the sun. However, each spectral component has been attenuated equally during transit.

Although the emitted flux leaving the solar surface is diffuse, it becomes almost monodirectional or beamlike by the time it reaches our atmosphere. As we get farther from the sun, the solar disc appears to shrink in size and all energy appears to come from a well-defined direction. Because the sun is not quite a point, but appears as a disc, the radiation is not perfectly beamlike and diverges slightly. From Figure 1.6 we can see that the divergence is determined by the ratio of the solar diameter to the earth–sun distance. The angular divergence is therefore

$$\Delta\theta = \frac{2R_\odot}{r} = \frac{(2)(6.96 \times 10^8\ \text{m})}{(1.5 \times 10^{11}\ \text{m})} = 9.28 \times 10^{-3}\ \text{rad} = 0.53°$$

This is slightly more than $\frac{1}{2}°$ so that the flux making up the solar constant can be approximated as monodirectional radiation.

The Spectral Composition of the Solar Constant

As noted, the spectrum of the solar constant is given by Equation 1.12, which has the form

$$S_\lambda = \text{constant} \times B_\lambda(T)$$

The fraction of the energy f transmitted by those wavelengths between 0 and λ is proportional to the area under the black-body curve between these limits. It can be determined by evaluating the integrals in the expression

$$f_\lambda(T) = \frac{\int_0^\lambda B_\lambda(T)\,d\lambda}{\int_0^\infty B_\lambda(T)\,d\lambda} = \frac{\int_0^\lambda a\,d\lambda/[\lambda^5(e^{b/\lambda T}-1)]}{\int_0^\infty a\,d\lambda/[\lambda^5(e^{b/\lambda T}-1)]} \tag{1.13}$$

It would appear from Equation 1.13 that the integrals must be evaluated for each and every temperature. This is not the case as can be seen by transforming the integral using the substitution $x = \lambda T$. Equation 1.13 can be written, using Equation 1.7b, as

$$f_\lambda(T) = f(\lambda T) = f(x) = \int_0^x \frac{a\,dx}{\sigma x^5(e^{b/x}-1)} \tag{1.14}$$

Thus if this integral can be evaluated and $f(x)$ tabulated, then the fraction of the energy between 0 and λ can be determined for black-body curves at any temperature. Equation 1.14 has been tabulated in Table 1.1.

We may use Table 1.1 to find the energy fraction for the wavelengths between 0 and $\lambda_1 = 0.4\ \mu$m for a black-body spectrum of 5760 K. The value of the fraction for

$$x_1 = \lambda_1 T = (0.4\ \mu\text{m})(5760\ \text{K}) = 2304\ \mu\text{m-K}$$

is, from interpolation in Table 1.1,

$$f(x_1) = f(2304\ \mu\text{m-K}) = 0.121 = 12.1\%$$

{*Note.* The fraction for the same wavelengths at a reduced temperature of 3000 K [$x_1 = (0.4)(3000) = 1200\ \mu$m-K] would only be $f(1200\ \mu\text{m-K}) = 0.002 = 0.2\%$.}

Table 1.1 can also be used to determine the energy fraction for wavelengths between λ_1 and λ_2. We set

$$f_{\lambda_1,\lambda_2} = f_{\lambda_2} - f_{\lambda_1}$$

For example, to find the energy fraction for the wavelengths between $\lambda_1 = 0.4$ and $\lambda_2 = 0.7\ \mu$m for $T = 5760$ K, we use $f(4032\ \mu\text{m-K}) - f(2304\ \mu\text{m-K}) = 0.488 - 0.121 = 0.367 \simeq 37\%$. The remaining fraction carried by those wavelengths longer than $\lambda_2 = 0.7\ \mu$m is obtained using

$$1 - f_{\lambda_2} = 1 - 0.488 \simeq 0.51 = 51\%$$

TABLE 1.1 *The function $f(x)$ defined by Equation 1.14.*

x(μm-K)	$f(x)$	x(μm-K)	$f(x)$	x(μm-K)	$f(x)$
1100	0.001	4600	0.580	8100	0.860
1200	0.002	4700	0.594	8200	0.864
1300	0.004	4800	0.608	8300	0.868
1400	0.008	4900	0.621	8400	0.871
1500	0.013	5000	0.634	8500	0.875
1600	0.020	5100	0.646	8600	0.878
1700	0.029	5200	0.658	8700	0.881
1800	0.040	5300	0.669	8800	0.884
1900	0.052	5400	0.680	8900	0.887
2000	0.067	5500	0.691	9000	0.890
2100	0.083	5600	0.701	9100	0.893
2200	0.101	5700	0.711	9200	0.895
2300	0.120	5800	0.720	9300	0.898
2400	0.140	5900	0.730	9400	0.901
2500	0.161	6000	0.738	9500	0.903
2600	0.183	6100	0.746	9600	0.905
2700	0.205	6200	0.754	9700	0.908
2800	0.228	6300	0.762	9800	0.910
2900	0.251	6400	0.770	9900	0.912
3000	0.273	6500	0.776	10000	0.914
3100	0.296	6600	0.783	11000	0.932
3200	0.318	6700	0.790	12000	0.945
3300	0.340	6800	0.796	13000	0.955
3400	0.362	6900	0.802	14000	0.963
3500	0.383	7000	0.808	15000	0.969
3600	0.404	7100	0.814	16000	0.974
3700	0.424	7200	0.819	17000	0.978
3800	0.443	7300	0.824	18000	0.981
3900	0.462	7400	0.830	19000	0.983
4000	0.483	7500	0.834	20000	0.986
4100	0.499	7600	0.840	30000	0.995
4200	0.516	7700	0.844	40000	0.998
4300	0.533	7800	0.848	50000	0.999
4400	0.549	7900	0.852		
4500	0.564	8000	0.856		

If we approximate the solar spectrum by a black-body distribution at 5760 K, then as previously seen, approximately 12% of the energy is transmitted by wavelengths shorter than 0.4 μm. This is mainly in the form of ultraviolet radiation. The visible portion of the solar spectrum contains 37% of the energy whereas those wavelengths longer than 0.7 μm (primarily infrared) contain 51 percent. Therefore nearly two-thirds of the energy arriving from the sun is invisible to the human eye. Of this, the overwhelming fraction is in the infrared.

In summary, the solar flux arriving at the top of the earth's atmosphere is primarily electromagnetic in character. Its spectral distribution closely resembles one emitted by a black surface at 5760 K. Approximately one-half of the energy arrives in infrared rays whereas one-third is in the visible spectrum. The flux is essentially beamlike or monodirectional radiation with a divergence of $\sim\frac{1}{2}°$. The (seasonally averaged) total flux crossing a surface oriented toward the sun is called the solar constant and is approximately equal to

$$S = 1352 \text{ W/m}^2 = \frac{1.94 \text{ Ly}}{\text{min}} = \frac{429 \text{ Btu}}{\text{hr-ft}^2}$$

where

$$1 \text{ British thermal unit (Btu)} = 252 \text{ cal} = 2.929 \times 10^{-4} \text{ kwhr}$$

and

$$1 \text{ Langley (Ly)} = 1 \text{ cal/cm}^2$$

We will show in Chapter 3 that the flux reaching the earth's surface is reduced considerably after passing through the atmosphere. To make any predictions about the abundance and availability of terrestial solar energy, we need to consider the apparent motion of the sun across the celestial sphere. Both the position of the sun throughout the day and the length of the day itself determine the amount of solar energy available to solar collectors. In Chapter 2 we will present those elements of solar astronomy necessary to explain the apparent motion of the sun.

PROBLEMS

1-1. If the sun were to cool down to 5500 K, estimate the new value of the earth's solar constant and the characteristic color of the spectrum. Sketch the spectral distribution.

1-2. Find the solar constants for Venus and Mars at their mean distances from the sun. Take $r_{Venus} = 1.08 \times 10^8$ km and $r_{Mars} = 2.28 \times 10^8$ km.

1-3. Two orbiting probes are being launched, one to orbit Mercury and the other to orbit Jupiter. Each probe's electronic equipment is to be powered by silicon photovoltaic cells that are 10 percent efficient. If a total of 1 kw of electrical power is required, find the area of the photocells necessary in each case. Is solar power feasible for Jupiter missions? (Take $r_{Mercury} = 0.58 \times 10^8$ km and $r_{Jupiter} = 7.78 \times 10^8$ km.)

1-4. (a) A planet's albedo is defined as its average reflectivity over the solar spectrum. Show that if a planet is considered to be at a uniform temperature, and if the thermal emissivity of its surface is assumed to be unity, its steady-state (effective) temperature would be

$$T_{eff} = \left[\frac{(1 - \text{albedo})S}{4\sigma}\right]^{1/4}$$

where S is the solar constant for the planet and σ is the Stefan–Boltzmann constant. (Explain how a planet can have a nonzero reflectivity and yet have an emissivity of approximately unity.)

(b) If the earth's albedo is 0.33, find the effective temperature of the earth and the characteristic wavelength of its emission spectrum.

1-5. The tungsten filament of a clear 150-W bulb operates at a temperature of 2700°C. The filament's average emissivity at this temperature is $\epsilon = 0.4$.

(a) Plot the spectral flux emitted by the filament and determine λ_{max}.

(b) Estimate the surface area of the filament if it is assumed that all the energy leaves via radiation.

(c) A 60-W bulb also operates at 2700°C. Repeat (a) and (b) for this bulb and compare the results with those of the 150-W bulb.

1-6. A wall dimmer reduces the electrical power supplied to the 150-W bulb in Problem 1-5 so that it consumes a total of 60 W.

(a) Find the operating temperature of the filament.

(b) Find the characteristic wavelength, λ_{max}, for the emitted spectrum and compare it with that of the 60-W bulb operating at full power.

(c) Using Table 1.1, find the fraction of the radiant energy emitted by the 60-W bulb (operating at full power) in the visible range, $0.4\ \mu m \leq \lambda \leq 0.7\ \mu m$.

(d) Find the same fraction emitted by a 150-W bulb operating at 60 W. Which is a more efficient way of producing illumination?

1-7. Show that only 25 percent of the total radiant energy of a black-body spectrum at *any* temperature is contained by wavelengths shorter than the characteristic wavelength λ_{max}.

1-8. A solar panel consists of an absorber plate placed under a glazing. The glazing only transmits wavelengths greater than $0.3\ \mu$. The absorber absorbs all wavelengths except those in the interval $0.48\ \mu \leq \lambda \leq 0.52$, which it reflects. If the collector is $2\ m^2$ in area and is oriented toward the sun above the atmosphere, find the heating power produced in the absorber due to absorption of solar radiation. (*Hint.* Use Table 1.1.)

1-9. Find the angular divergence of the flux in the solar constants for Mercury and Jupiter.

1-10. (a) Using the property of Equation 1.14, $f(x) \xrightarrow[x\to\infty]{} 1$, and the fact that $\int_0^\infty u^3\, du/(e^u - 1) = \pi^4/15$, show that the Stefan–Boltzmann constant is

$$\sigma = \frac{\pi^4}{15}(a/b^4)$$

(*Hint.* Set $u = b/x$.)

(b) Evaluate σ from the values of a and b given in Equation 1.3b.

REFERENCES

1. Abell, G., *Exploration of the Universe*, Holt, Rinehart and Winston, New York (1964), Chapter 23.
2. Bartky, W., *Highlights of Astronomy*, University of Chicago Press, Chicago (1961), Chapter 6.
3. Duffie, J. A. and W. A. Beckman, *Solar Energy Thermal Processes*, Wiley-Interscience, New York (1974), Chapters 1, 4, and 5.
4. McDaniels, D., *The Sun*, Wiley, New York (1979), Chapter V.

5. Reif, F., *Fundamentals of Statistical and Thermal Physics,* McGraw-Hill, New York (1965), Section 9.15.
6. Wyatt, S., *Principles of Astronomy,* 2nd ed., Allyn & Bacon, Boston (1971), Chapter 12.
7. Zemansky, M. W., *Heat and Thermodynamics,* 5th ed., McGraw-Hill, New York (1968), Chapters 4 and 13.

CHAPTER 2

Solar Astronomy

The availability of solar energy at the earth's surface depends primarily on the optical state of the atmosphere and the apparent daily motion of the sun across the celestial sphere. The sun's motion is important because its trajectory determines the degree to which solar energy is attenuated by the atmosphere. The seasonal variation of the sun's path also causes the number of daylight hours to vary. To explain the sun's apparent motion about an observer on earth, we need to study both the revolution of the earth about the sun and the rotation of the earth on its axis.

The Earth's Orbit

The earth's motion about the sun is affected primarily by the gravitational attraction between the earth and the sun. Although the moon and the planets do influence the earth's orbit somewhat, they produce only small perturbations. We will ignore the latter and assume that the earth is the only celestial object orbiting about the sun. Furthermore, we will assume that the sun and the earth are tiny spheres or at least behave as such because of the large distance separating them. In fact, the earth–sun separation is more than 200 times the sun's radius and more than 20,000 times the earth's radius.

Because the sun is far more massive than the planets, we may assume that the sun remains approximately stationary as the earth moves in its orbit. Using the classical laws of Newtonian dynamics along with Newton's law of gravitation, we can show that (1) the earth moves in a fixed plane about the sun and (2) the earth's orbit is an ellipse, with the sun situated at a point known as a *focus* (Kepler's First Law). The fixed plane containing the earth's orbit is called the *ecliptic plane* because only when the moon passes through this plane can an eclipse occur. The orbit of the earth, although elliptical, is very nearly circular. The sun is at a point offcenter along the major (long) axis at the focus of the ellipse.

The earth's orbit is best expressed in plane polar coordinates as

$$r = \frac{a(1-\epsilon^2)}{1+\epsilon \cos\theta} \tag{2.1}$$

where the *semimajor axis* (the mean orbital distance) is $a = 1.497 \times 10^{11}$ m ($\simeq$93,000,000 mi) and the *eccentricity* is $\epsilon = 0.0167$. The eccentricity is a measure of the deviation from a circle; it is zero for a circle and approaches one as the ellipse becomes flat. The elliptical orbit of the earth is shown somewhat exaggerated in Figure 2.1.

From Equation 2.1 we determine that the smallest value of r (called *perihelion*) occurs when $\theta = 0$ and is given by

$$r_p = a(1-\epsilon) = 1.497 \times 10^{11}(1-0.0167) = 1.471 \times 10^{11} \text{ m}$$

Similarly, the maximum value of r (called *aphelion*) occurs when

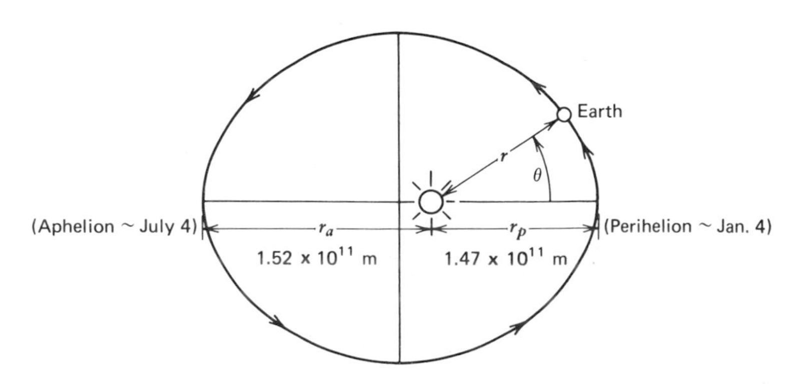

FIGURE 2.1 *The earth's orbit shown with an exaggerated eccentricity.*

$\theta = 180°$ and is

$$r_a = a(1+\epsilon) = 1.497 \times 10^{11}(1+0.0167) = 1.521 \times 10^{11} \text{ m}$$

Note that the perihelion and aphelion distances differ by less than 2 percent from the mean value. Consequently, the earth's orbit is almost a perfect circle. The earth actually reaches perihelion and aphelion on approximately January 4 and July 4, respectively. The earth is therefore closer to the sun in the Northern Hemisphere's winter than in the summer.

Seasons are a consequence of the inclination of the earth's axis of rotation to a line perpendicular to the ecliptic plane. The angle of inclination is 23.5° and remains constant throughout the year. The rate of rotation is also constant and equal to one rotation every 23.93 hr. This time interval, called a *sidereal day*, is equal to the time the stars appear to make one revolution about an observer on earth. The angle between the earth's axis and the line connecting the earth and the sun does, however, vary seasonally. At one point in the orbit the earth's axis of rotation is tilted toward the sun making an angle of 66.5° with this line. This occurs on about June 21 and is known as the *summer solstice*. The time between two successive summer solstices is called a *tropical year* and is, for all practical purposes, equal to the time between two successive perihelions or a *sidereal year*. Each year is approximately (365.25) (24) = 8766 hr. One-quarter of a tropical year after the summer solstice the earth's axis makes an angle of 90° with the earth–sun line. This is the *autumnal equinox*, which occurs on about September 21. The same situation takes place one-quarter of a year before the summer solstice at the *vernal equinox* (~March 21). At the equinoxes, day and night are of equal duration to all observers on the earth. One-half of a tropical year after the summer solstice we arrive at the *winter solstice* (~December 21), at which time the earth's axis of rotation is tilted away

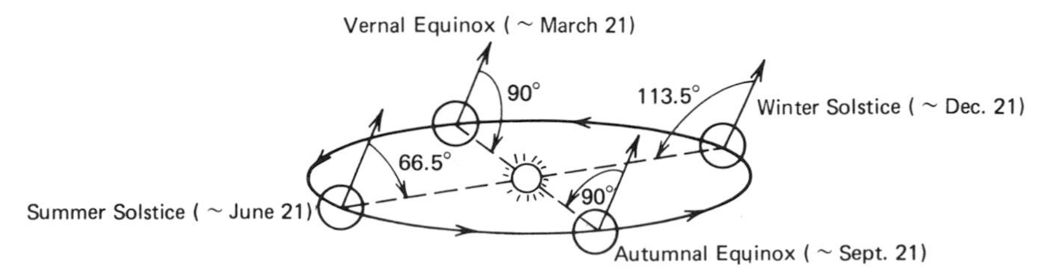

FIGURE 2.2 *The seasonal variation of the angle between the earth's polar axis and the earth–sun line.*

from the sun, making an angle of 113.5° (Figure 2.2) with the earth–sun line.

The Solar Day

The sun's apparent motion to an observer situated at the center of the earth is a consequence of two separate motions. They are the earth's revolution about the sun and the earth's rotation about its axis. The sun appears to revolve about the earth in a time interval called a *solar day*. The solar day is not constant but varies slightly during the year in a somewhat irregular manner. To understand this, let us suppose that the earth rotated on its axis but did not revolve about the sun. The solar day would then be due entirely to the earth's rotation and would be equal to the sidereal day. The sun would appear to revolve about the earth's polar axis with a uniform frequency of

$$f_{\text{rot}} = \frac{1}{T_{\text{sidereal}}} = \frac{1 \text{ rev}}{(23.93)(60)^2 \text{ sec}} = 1.16 \times 10^{-5} \text{ rev/sec} \qquad (2.2)$$

The motion would be such that the sun would rise in the east and set in the west.

Now let us consider the effect of the revolution of the earth about the sun. Suppose for simplicity that the polar axis of the earth were perpendicular to the ecliptic and that the earth's orbit were perfectly circular. If the earth were not rotating at all, the sun would appear to revolve about the earth once every year rising in the *west* and setting in the *east*. The frequency would be equal to

$$f_{\text{rev}} = \frac{1 \text{ rev}}{(365.25)(24)(60)^2 \text{ sec}} = 3.17 \times 10^{-8} \text{ rev/sec} \qquad (2.3)$$

It is evident from Equations 2.2 and 2.3 that f_{rev} is much smaller than f_{rot}. Because the effects of rotation and revolution produce opposite apparent motions of the sun, we have, from Equations 2.2 and 2.3,

$$\begin{aligned} f_{\text{net}} = f_{\text{rot}} - f_{\text{rev}} &= (1.16 - 0.00317)10^{-5} \text{ rev/sec} \\ &= 1.157 \times 10^{-5} \text{ rev/sec} \end{aligned} \qquad (2.4)$$

The apparent motion of the sun would be east to west with an effective period

$$T = \frac{1}{f_{\text{net}}} \simeq 8.64 \times 10^4 \text{ sec} = 24 \text{ hr} \qquad (2.5)$$

This time is called the *solar day*. We now show why the solar day is not quite constant throughout the year.

The value of f_{rev} in Equation 2.3 was obtained assuming that the annual drift of the sun is due east. Actually, because the spin axis of the earth is tilted, the drift is due east only at the solstices. However, at the equinoxes, the sun is traveling either northeast or southeast as it crosses the equator. Therefore its effective easterly drift is smaller at the equinoxes than at the solstices. At the equinoxes the sun crosses the equator at an angle of 23.5° so that we have

$$f_{rev}^{effective} = f_{rev} \cos 23.5° = 0.917\, f_{rev}$$

Thus $f_{rev}^{effective}$ varies from a maximum of f_{rev} at the solstices to a minimum of $0.917\, f_{rev}$ at the equinoxes. It follows from Equation 2.4 that even if the earth's orbit were perfectly circular, we would have the longest solar day at the solstices and the shortest solar day at the equinoxes. The solar day would vary slightly in length in a regular manner completing two cycles in a tropical year.

The elliptical orbit of the earth also contributes to the irregularity of the solar day. The earth's orbital velocity varies with the distance from the sun so that f_{rev} is largest when the earth is closest to the sun and smallest when it is farthest. This effect tends to make the solar day longest at perihelion and shortest at aphelion. The complete cycle occurs once a year.

Because the earth's axis is tilted to the ecliptic and its orbit is elliptical, the solar day varies throughout the year in a somewhat irregular manner. The mean value of the solar day taken over the year is exactly

$$T_{\text{mean solar day}} = 24 \text{ hr} \qquad (2.6)$$

The reason for this exactness is that Equation 2.6 *defines* the *mean solar hour* to be $\frac{1}{24}$ of the mean solar day. Until 1967, our measurement of all time was based on this interval.[1] The variation of the solar day about its mean value is shown in Figure 2.3.

The solar day has a maximum length of 24 hr + 30 sec near the winter solstice and a minimum length of 24 hr − 19 sec shortly before the autumnal equinox. A secondary maximum occurs near the summer solstice, whereas a secondary minimum occurs near the vernal equinox. Mean values of the solar day occur on about February 11, May 14, July 26, and November 3. The difference between the longest and shortest solar day is less than 1 min or 0.07 percent of the

[1]Since 1967 an atomic clock has been used as the international standard. The second is defined as 9,192,631,770 periods of a certain radiation emitted by the ^{133}Cs atom.

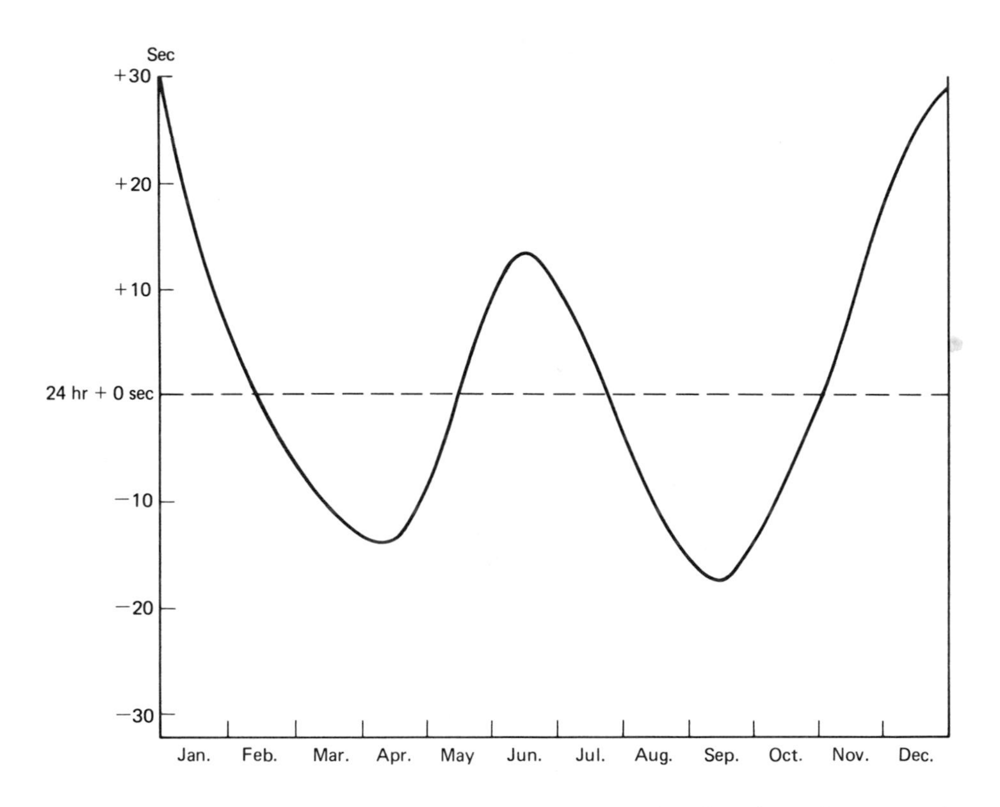

FIGURE 2.3 *The seasonal deviation of the apparent solar day about the mean solar day.*

mean value. Thus for most purposes the solar day can be regarded as being nearly constant and equal to 24 mean solar hours throughout the year.

The Equation of Time

Meridian transit of the sun or *solar noon* occurs when the sun crosses an imaginary line in the sky extending from the North Pole to the South Pole and passing directly over an observer's head. As already noted, the time between two successive solar noons (a solar day) is very nearly constant throughout the year. This means that if an observer observes solar noon at, say 11:55 AM local time, he will observe the next solar noon at no more than 19 sec before or 30 sec after 11:55 AM on the following day. However, over a period of a month or more, this effect is cumulative and solar noon

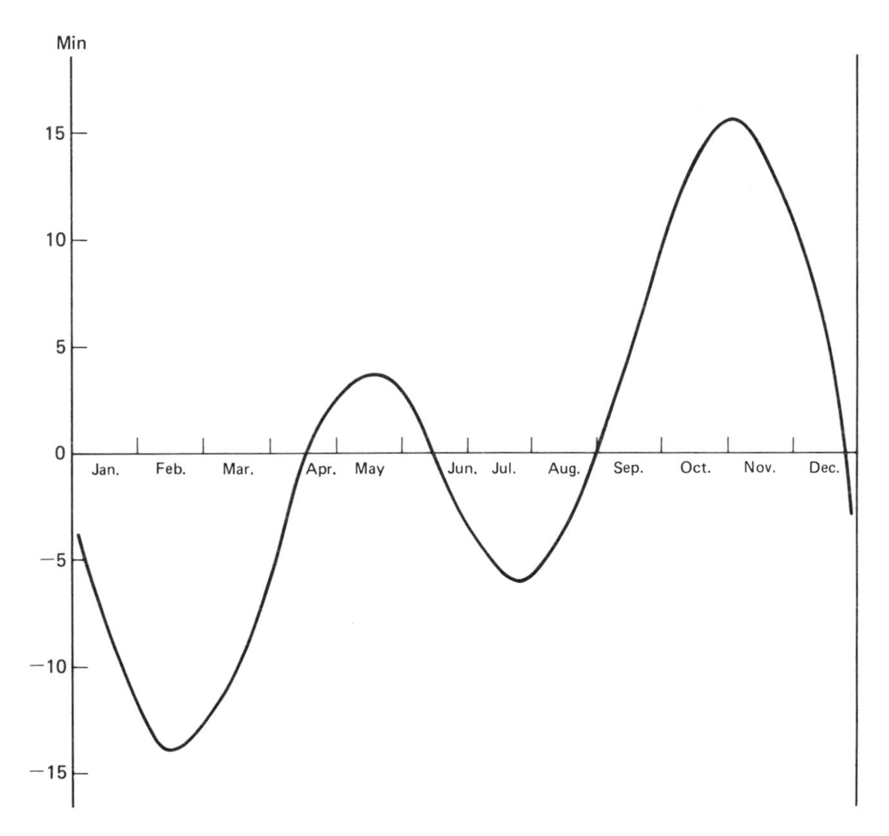

FIGURE 2.4 *The equation of time (EOT).*

may occur many minutes before or after 11:55 AM. In many solar applications it is important to know exactly when solar noon occurs local time on any given day.

The curve that gives the cumulative effect of the variation of the solar day is called the equation of time (EOT). It is plotted in Figure 2.4 and tabulated in Table 2.1. Mathematically, Figure 2.3 is the negative derivative of Figure 2.4; conversely, Figure 2.4 is the negative of the integral of Figure 2.3. As with all integrals, a constant can be added without affecting the result. This constant affects the zeros of the EOT. In practice, the EOT is centered so that its average value is zero, indicating that the net cumulative effect over a year vanishes. This fixes the zeros of the EOT so that they occur on approximately April 16, June 14, September 1, and December 25. Greenwich time is set so that solar noon at the Greenwich observatory in England coincides with 12:00 noon on approximately these dates. It is therefore simple to verify that the local standard time at which solar noon

TABLE 2.1 *The Equation of Time in Minutes*

Day of Month	January	February	March	April	May	June	July	August	September	October	November	December
1	−4	−14	−13	−4	+3	+2	−3	−6	0	+10	+16	+11
4	−5	−14	−12	−3	+3	+2	−4	−6	+1	+11	+16	+10
7	−6	−14	−11	−2	+3	+2	−5	−6	+2	+12	+16	+9
10	−8	−14	−10	−1	+4	+1	−5	−5	+3	+13	+16	+7
13	−9	−14	−10	−1	+4	0	−6	−5	+4	+14	+16	+6
16	−10	−14	−9	0	+4	0	−6	−4	+5	+14	+15	+4
19	−11	−14	−8	+1	+4	−1	−6	−4	+6	+15	+15	+3
22	−12	−14	−7	+1	+4	−2	−6	−3	+7	+15	+14	+2
25	−12	−13	−6	+2	+3	−2	−6	−2	+8	+16	+13	0
28	−13	−13	−5	+2	+3	−3	−6	−1	+9	+16	+12	−2

occurs for any observer is given by the relation

$$\text{Solar noon (in local standard time)} = 12{:}00 - 4(\text{Long}_{st} - \text{Long}_{loc}) - \text{EOT}. \qquad (2.7a)$$

where the last two terms on the right are in minutes. Long_{loc} is the observer's local meridian, whereas Long_{st} is the standard meridian for the observer's time zone. The standard meridians for the continental United States are: Eastern, 75°W; Central, 90°W; Mountain, 105°W; and Pacific, 120°W. The factor of four in Equation 2.7a arises because the sun traverses 1° of longitude every 4 min.

For example, suppose we wish to determine the approximate time at which solar noon occurs in New York City (74°W) on February 4. For Eastern Standard Time (EST), we take $\text{Long}_{st} = 75°\text{W}$. The EOT (Table 2.1) gives −14 min for February 4. Using Equation 2.7a, we find

$$\begin{aligned}\text{solar noon} &= 12{:}00 - 4(75 - 74) - (-14) \\ &= 12{:}00 - 4 + 14 = 12{:}00 + 10\text{ min} = 12{:}10\text{ EST}\end{aligned}$$

Solar time on any given day is defined as local time shifted so that solar noon occurs at 12:00 noon, solar time. We therefore have the general rule.

$$\text{Solar time} = \text{standard time} + 4(\text{Long}_{st} - \text{Long}_{loc}) + \text{EOT} \qquad (2.7b)$$

Thus, for the preceding example, solar time lags 10 min behind local time. Throughout this text all time will be in terms of local solar time unless otherwise indicated.

Solar Coordinates

The apparent motion of the sun across the celestial sphere depends in part on the observer's position on the earth's surface. We will begin the mathematical description of the sun's motion by considering a coordinate system for an observer at the center of the earth. This is called a *geocentric* coordinate system and generally uses the north–south axis as the z direction. The remaining two mutually perpendicular (x and y) axes are situated in the earth's equatorial plane (Figure 2.5). The x axis is oriented toward some local meridian;

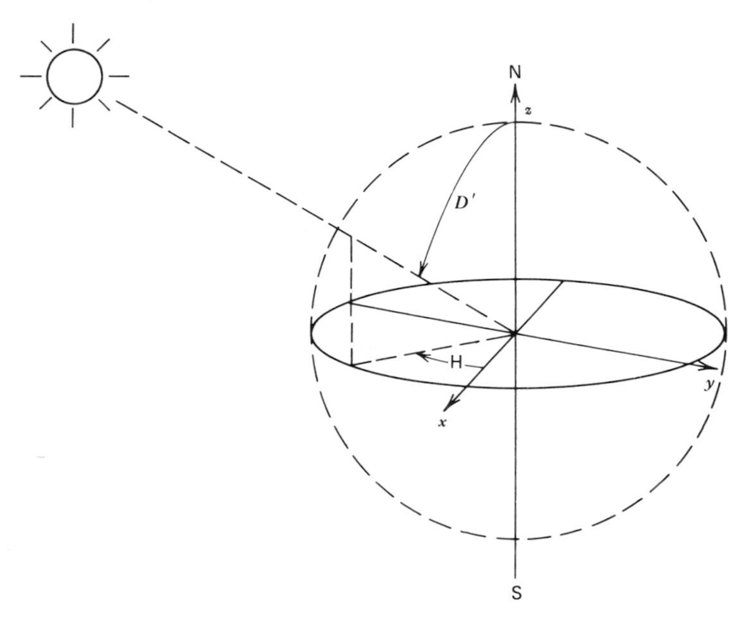

FIGURE 2.5 *The geocentric solar coordinates—the codeclination D' and the hour angle H.*

the y axis is 90° to the east in the equatorial plane. In this coordinate system the sun's position can be fixed by two angles. The first is the *codeclination* D' and the second is the *hour angle* H. The codeclination (complement of the *declination*[2] D) is the angle between the sun's rays and the North Pole. As we will see, this angle varies seasonally from 66.5° at the summer solstice to 113.5° at the winter solstice, passing through 90° at each of the equinoxes. The hour angle is the angle between the projection of the sun's rays in the equatorial ($x - y$) plane and the local meridian or x axis. Its hourly variation takes it from negative values before solar noon to positive values after solar noon; it passes through zero at solar noon.

The codeclination is actually the angle between the North Pole and the line in the ecliptic plane joining the earth and sun. Note that although the angle between the earth's axis and the line per-

[2]In many texts the declination D is used to give the position of the sun. The angle D is measured from the equatorial plane and varies from +23.5° at the summer solstice to −23.5° at the winter solstice. In this text we will use the codeclination D' measured from the North Pole. This angle varies from 66.5° at the summer solstice to 113.5° at the winter solstice.

pendicular to the ecliptic plane remains constant at 23.5°, the angle D' varies with the earth's position. Some elementary vector algebra shows this angle to be given by

$$\cos D' = \sin 23.5° \sin \alpha$$

where α is the angle between the earth–sun line on the day of the year in question and this line at the vernal equinox.[3] Because the earth's orbit is approximately circular, we may set

$$\alpha = \frac{360°}{365.25 \text{ days}} \times n$$

and obtain

$$\boxed{\cos D' = \sin 23.5° \sin \frac{360° \times n}{365.25 \text{ days}}} \tag{2.8}$$

where n is the number of days after the vernal equinox. Thus the cosine of the codeclination oscillates with a period of one year ranging from 66.5° at the summer solstice to 113.5° at the winter solstice (Figure 2.6).

The hour angle is simply related to the observer's local solar time by

$$\boxed{H = \pm\frac{360°}{24 \text{ hr}} t} \tag{2.9}$$

where t is the number of hours before or after solar noon. The negative sign refers to time before solar noon.

EXAMPLE

Find the solar coordinates D' and H at 3:00 PM EST on October 10 for an observer at New York City (74°W). To find D', we note that October 10 is approximately 201 days after the vernal

[3] The declination D ($D = 90° - D'$) can be expressed as $\sin D = \sin 23.5° \sin \alpha$. This expression is often approximated by the formula $D \simeq 23.5° \sin \alpha$. The approximation introduces an error of no more than 2% for any day of the year.

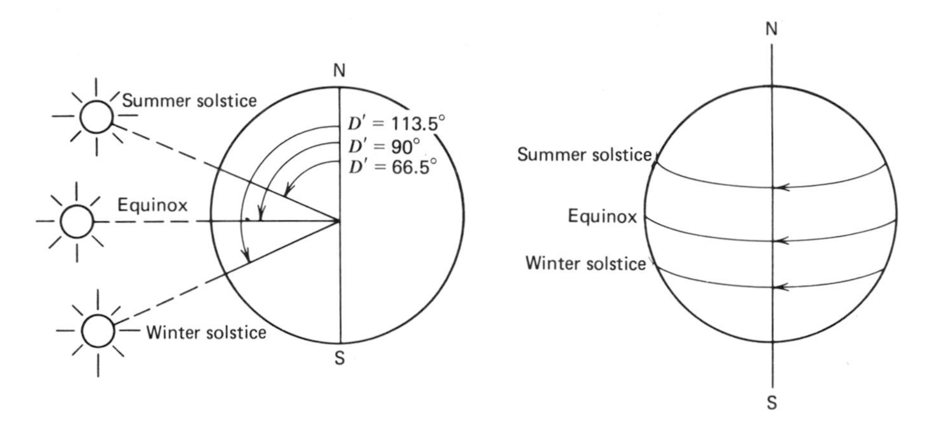

FIGURE 2.6 *The seasonal variation of the codeclination.*

equinox, and from Equation 2.8

$$\begin{aligned} D' &= \cos^{-1}(\sin 23.5°)\left(\sin \frac{360°}{365.25} \times 201\right) \\ &= \cos^{-1}(0.399)(-0.311) \\ &= 97.1° \end{aligned}$$

To find H, we first determine the local solar time corresponding to 3:00 PM. From the EOT, Table 2.1, we find a value of +13 min for October 10 so that

$$\begin{aligned} \text{solar noon} &= 12{:}00 - 4(75 - 74) - (13) \\ &= 11{:}43 \text{ EST} \end{aligned}$$

Thus 3:00 PM EST is 3:17 PM local solar time, which is 3.28 hr after solar noon. Using Equation 2.9, we find the hour angle to be

$$H = \frac{360°}{24 \text{ hr}}(3.28) = 49°$$

Local Solar Coordinates

The codeclination D' and the hour angle H are defined with respect to the earth's center and are not particularly suitable coordinates for an observer situated on the surface of the earth. For such an observer, it is far more convenient to use the *zenith angle* Z and the *azimuth angle* A as solar coordinates. The coordinate system used here is situated on the earth's surface at a latitude L and aligns the z direction with the vertical. Due south is taken as the x axis and due east as the y axis. The angle Z is the angle between the sun's rays and the vertical (Figure 2.7). Its complement is called the *altitude* of the sun. The azimuth angle A is the angle between the projection of the sun's rays in the horizontal $(x-y)$ plane and the direction due south (x axis). Like the hour angle, the azimuth angle is negative before, zero at, and positive after solar noon. Unlike H, the azimuth angle A does not follow solar time in a simple linear manner as given by Equation 2.9.

It is possible to relate the local coordinates Z and A to the geocentric coordinates D' and H using some basic vector relations. These give the following results.

$$\boxed{\cos Z = \cos D' \cos L' + \sin D' \sin L' \cos H} \tag{2.10a}$$

and

$$\boxed{\tan A = \sin D' \sin H/(\sin D' \cos L' \cos H - \cos D' \sin L')} \tag{2.10b}$$

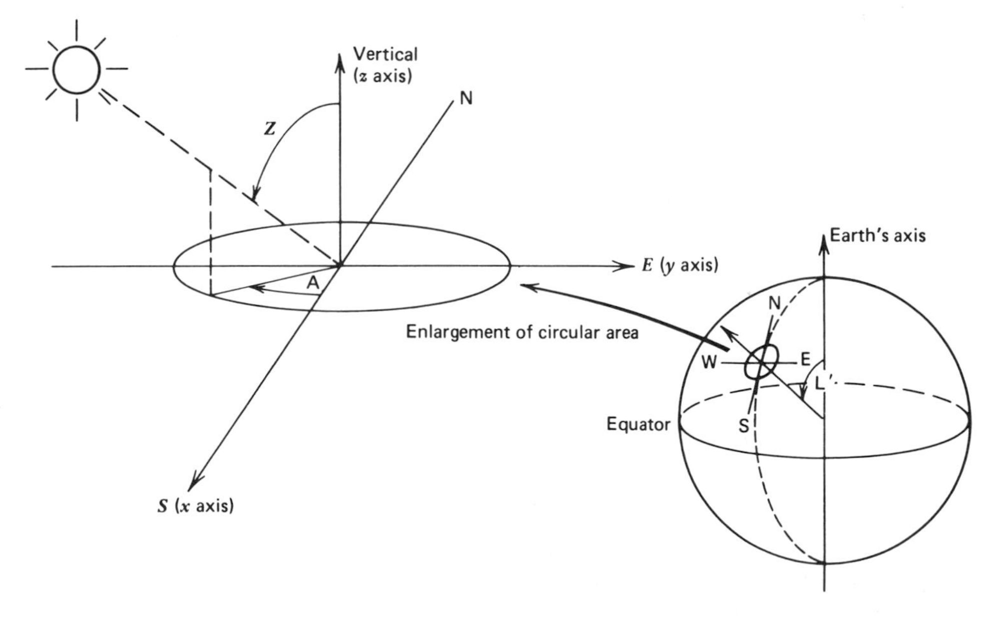

FIGURE 2.7 *Local solar coordinates—the zenith angle Z and the azimuth angle A.*

where L' is the colatitude (complement of the latitude L) of the observer.

Whereas the zenith angle, Z, gives the obliquity of the sun's rays to the earth's surface, the azimuth angle, A, gives the geographical coordinates of the sun's position (e.g., south of west, north of east, etc). Both angles vary hourly (through H) and seasonally (through D'). Furthermore, this variation differs for observers at different latitudes.

EXAMPLE

Find the local coordinates Z and A at 3:00 PM EST on October 10 for an observer at New York City (Long = 74°W, $L = 41°\mathrm{N}$).

From the previous example, we have $D' = 97.1°$ and $H = 49.3°$. Using Equation 2.10a, we have (for $L' = 90° - 41° = 49°$)

$$\cos Z = \cos 97.1° \cos 49° + \sin 97.1° \sin 49° \cos 49.3°$$
$$= 0.407$$

or

$$Z = 66° \text{ (to the vertical).}$$

Using Equation 2.10b, we find

$$\tan A = \sin 97.1° \sin 49.3°/(\sin 97.1° \cos 49° \cos 49.3° - \cos 97.1° \cos 49°)$$
$$= 1.49$$

or

$$A = 56° \text{ (west of south)}$$

Sunrise, Sunset, and the Number of Daylight Hours

Equations 2.8, 2.9, and 2.10 provide much information about the apparent motion of the sun relative to an observer on the earth's surface. Note that on any given day the sun is highest in the sky at solar noon at which time $H_{noon} = A_{noon} = 0$ and Z_{noon} is a minimum.

Setting $H = 0$ in Equation 2.10a we find, after using the trigonometric identity, $\cos(\theta_1 - \theta_2) = \cos\theta_1 \cos\theta_2 + \sin\theta_1 \sin\theta_2$, that

$$\cos Z_{\text{noon}} = \cos(D' - L')$$

or

$$\boxed{Z_{\text{noon}} = |D' - L'|.} \tag{2.11}$$

Hence the zenith angle at solar noon is always equal to the difference between the solar codeclination and the observer's colatitude. At New York City, for example, the noon time zenith angle varies from $113.5° - 49° = 64.5°$ at the winter solstice to $66.5° - 49° = 17.5°$ at the summer solstice.

As we move away from solar noon, Z increases steadily. When Z reaches 90°, the sun's rays are parallel to the earth's surface and we approach sunrise or sunset. The time of sunrise and sunset depends on the terrain and on which point of the solar disc we take as characteristic of the sun. It is sufficient for our purposes to *define* sunrise and sunset to be those times at which the zenith angle is 90°. Setting $Z = 90°$ in Equation 2.10a and solving for H, we obtain the sunset (sunrise) hour angle as

$$\cos H_s = -\cot D' \cot L'$$

or

$$\boxed{H_s = \pm\cos^{-1}(-\cot D' \cot L')} \tag{2.12}$$

where the negative sign refers to sunrise. The sunset (sunrise) hour angle H_s is directly related to the number of hours from solar noon to sunset (sunrise) by Equation 2.9. The sunset hour in solar time is obtained from Equation 2.12, using Equation 2.9 or

$$t_s = \frac{24\text{ hr}}{360°} H_s \tag{2.13}$$

The number of daylight hours is simply equal to twice the result in Equation 2.13 or

$$\boxed{T_{\text{daylight}} = 2t_s = \frac{24\text{ hr}}{180°} H_s} \tag{2.14}$$

The value of the azimuth angle at sunset is obtained by substituting the sunset hour angle, Equation 2.12, into Equation 2.10b. After some simplification we obtain

$$\cos A_s = \frac{-\cos D'}{\sin L'}$$

or

$$\boxed{A_s = \pm\cos^{-1}(-\cos D'/\sin L')} \tag{2.15}$$

where the negative sign again refers to sunrise.

EXAMPLE

Find the time of sunset and sunrise, the number of daylight hours, the noontime zenith angle, and the sunset and sunrise azimuth angles of the sun on October 10 for an observer at New York City.

From the results of the preceding examples, we have $D' = 97.1°$ and $L' = 49°$. The noontime zenith angle is, from Equation 2.11,

$$Z_n = D' - L' = 97.1° - 49° = 48.1°$$

The sunset (or sunrise) hour angle is, from Equation 2.12,

$$H_s = \pm\cos^{-1}(-\cot 97.1° \cot 49°) = \pm 83.8°$$

Using Equation 2.13, we find the number of hours from solar noon to sunset to be

$$t_s = \frac{24}{360} \times 83.8° = 5.59 \text{ hr} = 5{:}35 \text{ PM solar time}$$

Using the EOT for October 10, we can show that sunset corresponds to 5:18 PM EST. A similar calculation indicates that sunrise occurs at 6:26 AM solar time, which corresponds to 6:08 AM EST. The total number of daylight hours is

$$T_{\text{daylight}} = 2t_s = 2(5.59) = 11.18 \text{ hr} = 11 \text{ hr } 11 \text{ min}$$

The sunset (or sunrise) azimuth is found, from Equation 2.15, as

$$A_s = \pm\cos^{-1}(-\cos 97.1°/\sin 49°)$$
$$= \pm 80.6°$$

Thus the sun rises at 80.6° east of south (9.4° south of east) and sets 80.6° west of south (9.4° south of west) on October 10 at New York City.

The following rules apply to latitudes in the Northern Hemisphere. During the summer the noontime zenith angle is smallest, the day has more than 12 hr of daylight, and the sun sets north of west. At the equinoxes the noontime zenith angle is equal to the latitude angle of the observer, the day has exactly 12 hr of daylight, and the sun sets due west. During the winter the noontime zenith angle is largest, the day has less than 12 hr of daylight, and the sun sets south of west.

Obliquity of the Sun's Rays to Tilted Surfaces

As we will see in Chapter 3, the amount of direct solar energy intercepted by a fixed surface is determined, in part, by the angle θ between the perpendicular to the surface and the sun's rays. The function $\cos\theta$ is called the *obliquity factor*. This factor depends both on the local solar coordinates of the sun (Z and A) and on the tilt coordinates of the surface (Δ and ψ). The tilt angle of the surface, Δ, is the angle between the perpendicular line to the surface (the surface normal) and the vertical (Figure 2.8). The azimuth angle of the surface, ψ, is the angle between the projection of the surface normal in the horizontal plane and the direction of due south. The angles Δ and ψ fix the direction of the normal to the surface in the same manner in which Z and A fix the direction of the sun's rays. The obliquity of the sun's rays to an inclined surface is given by

$$\boxed{\cos\theta = \cos Z \cos\Delta + \sin Z \sin\Delta \cos(A - \psi)} \qquad (2.16)$$

The hourly variation of $\cos\theta$ is determined by the dependence of Z and A on the hour angle, as given by Equation 2.10. Note that for a horizontal surface, that is, one for which $\Delta = 0$, we have

$$\cos\theta = \cos Z$$

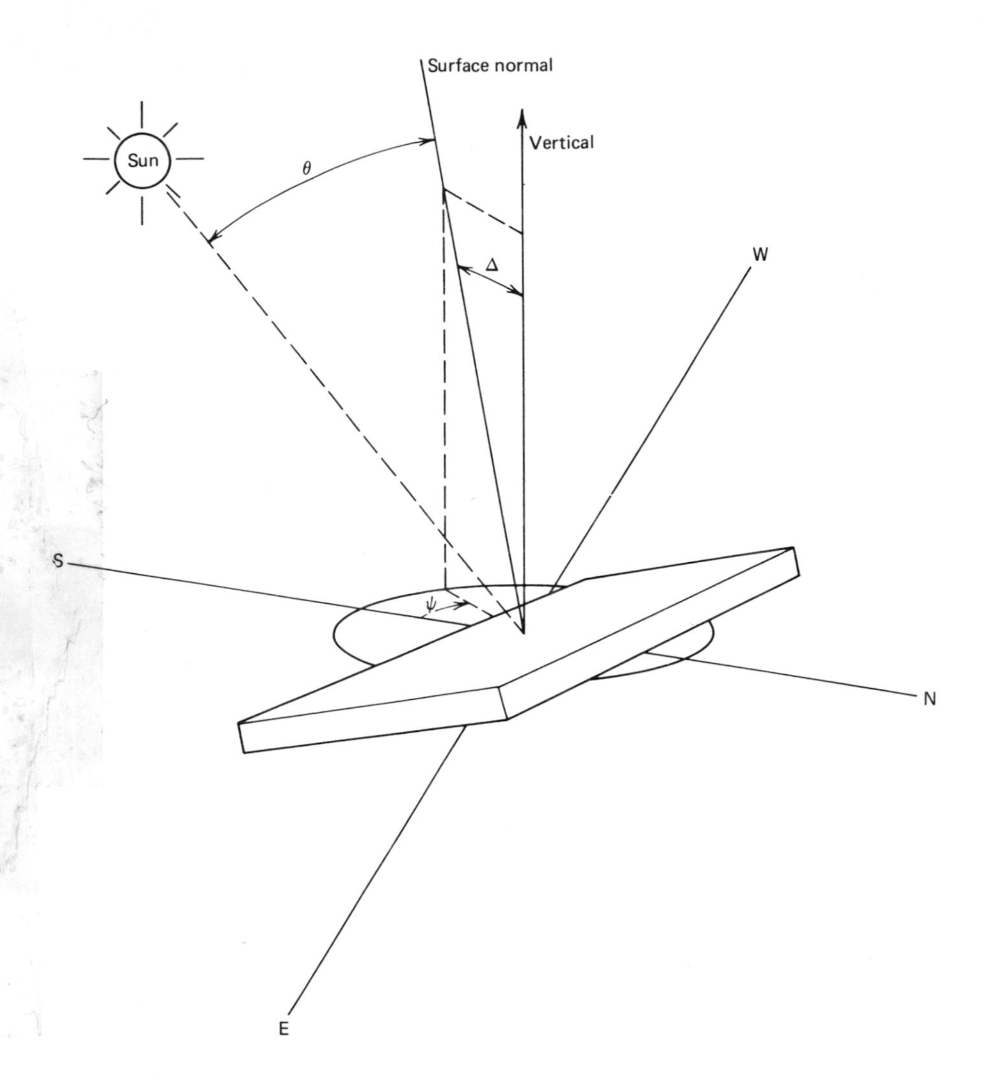

FIGURE 2.8 *A tilted surface showing the tilt Δ, the azimuth ψ, and the obliquity angle to the sun's rays θ.*

or

$$\theta = Z$$

as expected. On the other hand, for a tracking collector for which $Z = \Delta$ and $A = \psi$, we find

$$\cos\theta = \cos^2 Z + \sin^2 Z = 1$$

or

$$\theta = 0°$$

For a collector facing due south ($\psi = 0$), the noontime ($A_n = 0$) obliquity, θ_n, becomes

$$\cos \theta_n = \cos Z_n \cos \Delta + \sin Z_n \sin \Delta$$
$$= \cos(Z_n - \Delta)$$

or

$$\theta_n = |Z_n - \Delta|$$

Equation 2.16 shows that under certain conditions the obliquity factor can become negative (i.e., $\theta > 90°$). This means that the sun's rays fall behind the surface. If the surface is a solar collector designed to collect direct sunlight only on its front surface, collection will not occur when $\cos \theta$ is negative. A clear example of this situation is a vertical- ($\Delta = 90°$) collecting surface with a westerly tilt ($\psi = 90°$). Equation 2.16 gives

$$\cos \theta = \sin Z \cos(A - 90°) = \sin Z \sin A$$

Before solar noon A is negative so that $\cos \theta$ must also be negative. Thus, as expected, a west-facing vertical collector intercepts no direct solar energy before solar noon. It can be shown that a similar phenomenon occurs during the summer for a vertical surface with a southerly tilt ($\psi = 0$). Since the sun rises north of east and sets north of west in the summer, the surface will not begin collecting until sometime after sunrise and will cease collecting sometime before sunset. Since Equation 2.16 determines the obliquity of the sun's rays to any flat surface, it plays an important role both in the installation of solar collector panels as well as in the architectural design of homes intended to be energy efficient. The equations used in this chapter to determine the solar coordinates and the obliquity of the sun's rays are summarized in Appendix 1.

Tracking Systems

It is interesting to study the mathematical aspects of the apparatus required to keep a surface oriented toward the sun throughout the day. Whether one uses the local coordinates Z and A or the geocentric coordinates D' and H, it is evident that two coordinates are required to fix the position of the sun. Therefore the tracking apparatus must contain two mutually perpendicular axes about which to rotate the surface.

The two axes for tracking in local coordinates Z and A are shown in Figure 2.9*a*. Axis A is aligned with the vertical, whereas axis Z is situated in the horizontal plane. Axis A actually turns axis Z in the horizontal plane and produces the tracking of the azimuth angle. The surface is mounted on axis Z; this axis rotates the surface so that it tracks the sun's zenith angle. As seen from Equation 2.10, the daily variation of A and Z is not uniform and servomotors, controlled by

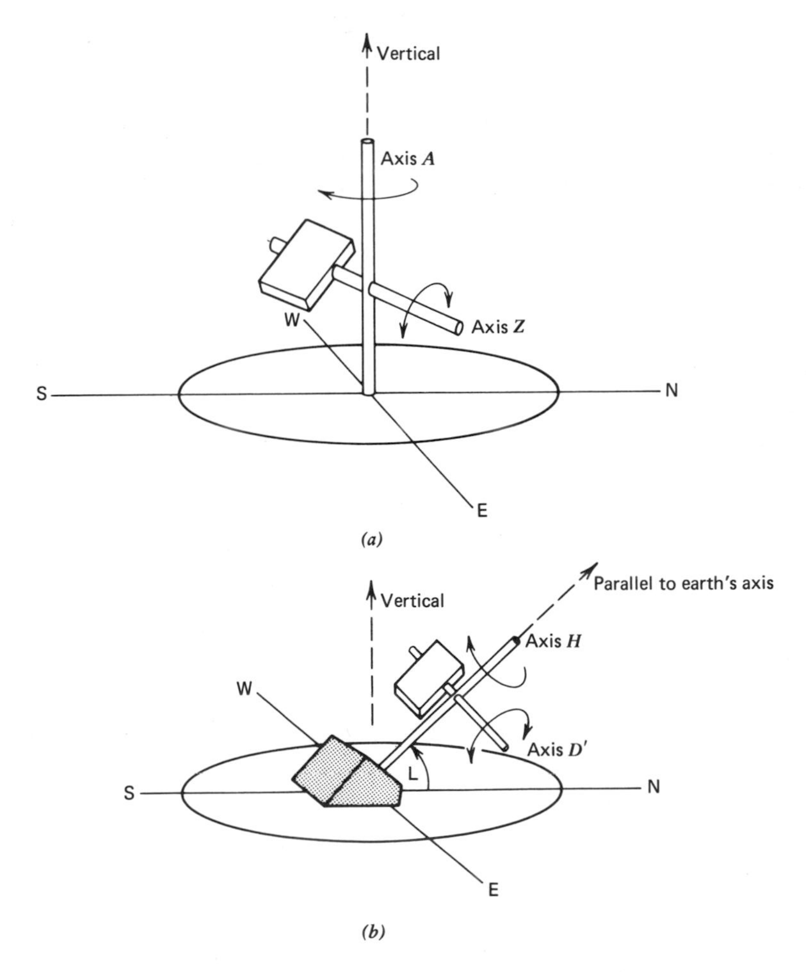

FIGURE 2.9 (*a*) *Azimuthal tracking about axes* Z *and* A. (*b*) *Equatorial tracking about axes* D' *and* H.

thermal or optical sensors, are normally required for tracking. This type of tracking is called *azimuthal.*

It is also possible to arrange the apparatus so that it tracks the solar coordinates D' and H. The required perpendicular axes are shown in Figure 2.9*b*. Because the variation of the solar hour angle is produced by the sun's apparent motion about the rotation axis of the earth, the axis H must be fixed so that it is parallel to the earth's polar axis. This is accomplished by aligning axis H along a north–south line and raising the north end until the axis makes an angle with the ground equal to the latitude L of the apparatus. Axis H actually turns axis D' so that it follows the hour angle of the sun. The surface is mounted on and rotated about axis D' in order to follow the codeclination of the sun. This type of tracking is called *equatorial*; it has two distinct advantages: (1) Because it varies seasonally, D' may be held fixed during a given day and adjusted periodically depending on the accuracy required. (2) Because the hour angle varies almost uniformly during the day, a 24 hr clockdrive may be used to rotate axis H. However, the apparatus must be reset every morning so that its angular position about the axis H coincides with the sun's hour angle. The required periodic adjustments are obtained from the EOT.

Insolation at Zero Air Mass

The insolation level at ground is primarily determined by two factors—the optical state of the atmosphere and the solar coordinates. In order to establish the dependence on the latter, we will assume *zero air mass* conditions. This means that the atmosphere is taken to be totally transparent so that it does not affect the solar energy in any way.

If a surface's tilt and azimuth are Δ and ψ and the sun's direction is characterized by Z and A, then the flux intercepted is

$$F(t) = S \cos \theta(t) \tag{2.17}$$

where S is the solar constant, θ is the obliquity angle between the sun's rays and the surface normal, and t is the solar time. Using Equations 2.8, 2.9, 2.10, and 2.16, we can establish the time dependence of $\cos \theta$. (The computations are best done with a computer or a programmable calculator.) Some results obtained from Equation 2.17 have been plotted in Figure 2.10 for an observer at a colatitude of $L' = 49°$ during the summer and winter solstices. The total intercepted energy/area for the day is given by the area under each curve.

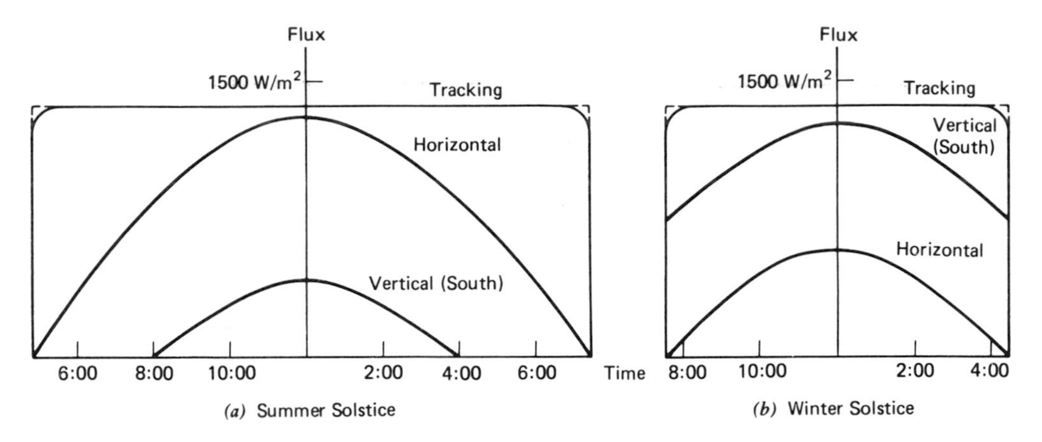

FIGURE 2.10 *The hourly flux on a horizontal, on a south-facing vertical, and on a tracking surface situated at a colatitude of $L' = 49°$ under zero air mass conditions. Case (a) is the summer solstice. Case (b) is the winter solstice.*

The results in Figure 2.10 are hypothetical because even on a relatively clear day the atmosphere will considerably reduce the insolation level at ground. The changes in the amount and nature of the insolation produced by the atmosphere is the subject of Chapter 3.

PROBLEMS

2-1. (a) Show that the angle between the earth–sun line at the vernal equinox and the same line at perihelion is given approximately by

$$\theta = \frac{360° \times n}{365.25}$$

where n is the number of days between these times.

(b) Using Equation 2.1, estimate the earth–sun distance at the vernal equinox.

(c) Find the ratio of the solar constant at the vernal equinox to that at perihelion.

2-2. (a) Find the local time (CST) at which solar noon occurs in Chicago (42°N, 88°W) on May 4.

(b) What is the noontime zenith angle?

(c) An observer measures the hour angle of the sun to be $H = 30°$. What time is it CST?

(d) When will the sun set—in solar time? in CST?

(e) How many degrees north of west will it set?

2-3. An observer is situated just at the Tropic of Cancer ($L' = 66.5°$).
(a) Find the length of the day at the equinoxes and at the solstices.
(b) Find the noontime zenith angle and the sunset azimuth angle at these times.
(c) Repeat (a) and (b) for an observer at the equator ($L' = 90°$).
(d) Repeat (a) and (b) for an observer just below the Arctic Circle at a colatitude $L' = 24°$. Explain your answers.

2-4. A solar heating panel is installed at a location in New York City (41°N, 74°W). It is surrounded by an obstructing wall in such a way that no sun can reach it on February 22 until the sun's azimuth is $A = -20°$.
(a) Find the solar time and the local time at which the collector begins to function.
(b) Find the zenith angle of the sun at this time.
(c) If the collector is on the ground 7 m from the wall, find the height of the wall.

2-5. (a) A south-facing wall of a home has a window 2 m high. The home is situated in Philadelphia (40°N, 75°W). Find the minimum extension length, *s*, of an awning required to obscure the noontime sun after April 21.
(b) On what date after the summer solstice will noontime sunlight begin to enter the window?

2-6. On January 6 in Minnesota (46°N, 94°W), a sportsman is ice fishing at the center of a circular lake. The tree line around the lake forms a circle 2 mi in radius and 0.25 mi in height. Find the number of effective daylight hours (i.e., when the sun is visible) that the man can fish.

2-7. Show, using Equations 2.10 and 2.16, that for a surface with a southerly tilt ($\psi = 0$) the obliquity factor is given by $\cos\theta = \cos(L' + \Delta)\cos D' + \sin(L' + \Delta)\sin D' \cos H$.

2-8. A collector 3 m^2 in area is mounted on a vertical south-facing wall of a home near Phoenix (34°N, 112°W) on October 20.
(a) What is the obliquity of the sun's rays at 3:00 PM solar time?
(b) If the direct solar flux is 1000 W/m^2, what heating power is intercepted by the collector at this time?

2-9. A traveler owns a camper trailer with a solar panel mounted on the roof. He parks his vehicle in Baltimore (39°N, 76°W) on November 30. He orients the vehicle due west and tilts the panel south so that it is directed toward the sun at solar noon.

How fast must he drive in order to keep the sun always perpendicular to the panel? If he is not to exceed 55 mph, how far north should he be to attempt this feat?

2-10. A system for tracking D' and H, as described in Figure 2.9 is set up on December 25 so that at solar noon the receiving surface is pointed toward the sun. The system is driven about the H axis by a timing motor at a rate of 1 rev/24 hr. If the change in solar declination over a day is ignored, find the obliquity between the surface normal and the sun's rays on the following day at solar noon. (*Note.* The solar day on December 25 is ~24 hr + 30 sec.)

2-11. A vertical south-facing collector is operating on a clear day at the summer solstice near Boston (42°N, 71°W).
(a) Find the total number of daylight hours.
(b) Find the total time that the collector is collecting solar energy.

2-12. (a) Estimate the time it takes for the solar disc to drop below the horizon for an observer at the equator during the equinox.
(b) Repeat for an observer at a colatitude of $L' = 49°$. (*Hint.* Find the time it takes for the solar zenith angle to change from 89.74 to 90.26°, the difference being approximately equal to the angle subtended by the solar disc.)
(c) Show that for northern latitudes well below the Arctic Circle that this time (in minutes) is given approximately by

$$\Delta t = \frac{(4)(0.53)}{\sqrt{\sin^2 D' \sin^2 L' - \cos^2 D' \cos^2 L'}}$$

REFERENCES

1. Abell, G., *Exploration of the Universe,* Holt, Rinehart and Winston, New York (1964), Chapters 3–6.
2. Bartky, W., *Highlights of Astronomy,* University of Chicago Press, Chicago (1961), Chapter 2.
3. Brinkworth, B. J., *Solar Energy for Man,* Wiley, New York (1972).
4. Duffie, J. A. and W. A. Beckman, *Solar Engineering of Thermal Processes,* Wiley-Interscience, New York (1980), Chapter 1.
5. Haymes, R. C., *Introduction to Space Science,* Wiley (1971), Chapters 1 and 2.

CHAPTER 3

Terrestrial Insolation

The amount of solar energy available on the earth's surface is considerably smaller than that arriving at the top of the atmosphere. The degree to which solar energy is reduced as it arrives at the earth's surface is determined primarily by the optical state of our atmosphere. As we will see, atmospheric constituents affect solar radiation by two processes—absorption and scattering. The amount of absorption and scattering that occurs for a given component of the solar spectrum depends on the composition of the atmosphere as well as on the wavelength of that component. In certain regions of the spectrum, solar energy is predominantly scattered, whereas in others it is mainly absorbed. The spectral composition of terrestrial insolation is therefore markedly different from the 5760 K black-body curve characteristic of the solar constant.

Equally important, terrestrial insolation can no longer be approximated by a monodirectional beam, as was true of the radiation incident at the top of the atmosphere. Some of the radiation scattered by the atmosphere finds its way to ground as *diffuse* radiation. The diffuse radiation consists of components traveling in many directions. Thus the total solar radiation at the earth's surface consists of a *direct* or monodirectional component, which

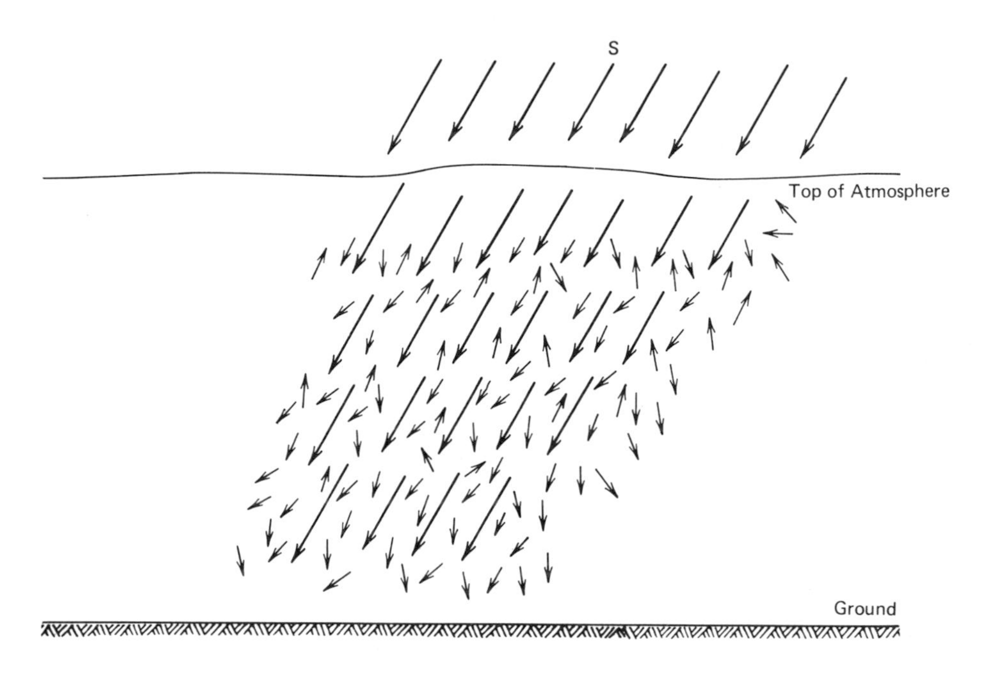

FIGURE 3.1 *The atmospheric solar flux as it penetrates the atmosphere.*

has survived both scattering and absorption, as well as a diffuse component produced by atmospheric scattering (Figure 3.1).

To understand, quantitatively, the way in which solar energy is modified as it traverses our atmosphere, we will present some basic elements of atmospheric physics.

An Atmospheric Model

The state of the atmosphere can be characterized, in part, by such thermodynamic variables as temperature T, density ρ, pressure P, and chemical composition. These parameters vary spatially in the atmosphere and with time. Because this variation is rather unpredictable, theoretical predictions regarding terrestrial insolation are exceedingly difficult to make. In order to draw some theoretical conclusions, it becomes necessary to make some simplifying approximations regarding atmospheric structure.

We first assume that the atmosphere is sufficiently thin when compared with the earth's radius so that it can be regarded as being flat. As we will see, the atmosphere has an effective height of $\simeq 8$ km, which is quite small when compared with the earth's radius ($R =$

6371 km). This approximation is therefore a very good one except possibly near sunrise and sunset when terrestrial insolation is negligibly small. Thus the curvature of the atmosphere is of little consequence in most solar energy applications.

The second approximation used here assumes that the atmospheric parameters vary only with a single coordinate—the altitude z. That is, we may express all the atmospheric parameters in terms of vertical profiles such as $T = T(z)$, $\rho = \rho(z)$, and $P = P(z)$. This approximation is certainly questionable especially when patchy clouds are present. A flat atmosphere whose composition varies only with altitude is said to be *plane stratified.*

The Temperature Profile

The temperature profile of the atmosphere, $T(z)$, is of interest in a variety of problems in atmospheric science and meteorology. Its behavior serves to partition the atmosphere into approximately four layers. The lowest portion, the *troposphere,* extends from sea level to

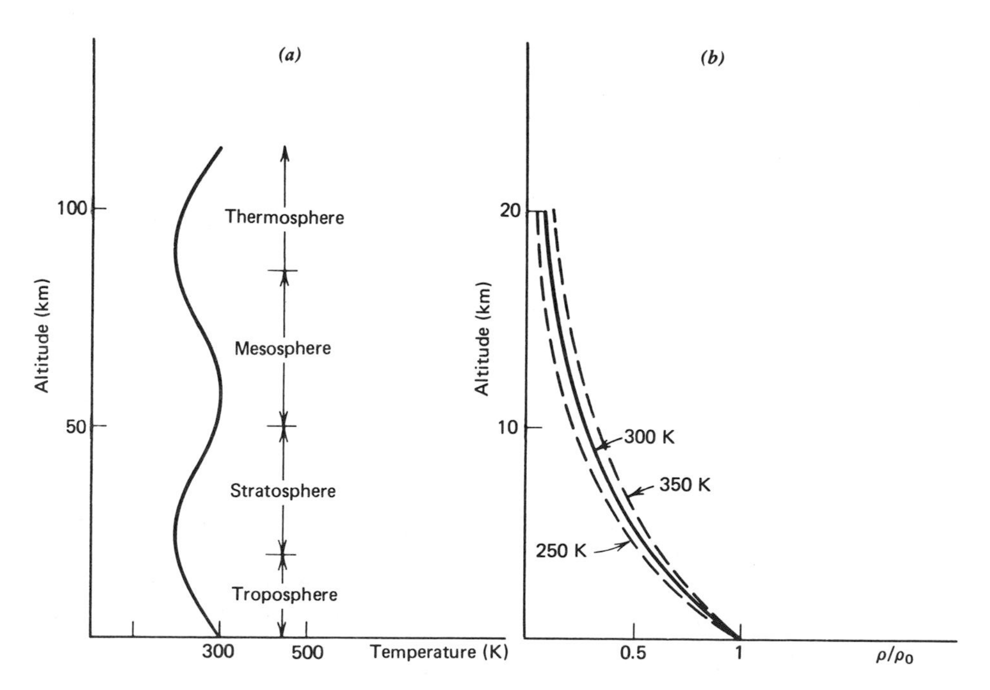

FIGURE 3.2 *(a) The temperature profile of the atmosphere. (b) The density profiles of isothermal atmospheres at different temperatures.*

approximately 12 km above sea level. In this region the temperature decreases with altitude at a rate approximately equal to $\Gamma \simeq 6$ K/km (Figure 3.2*a*). This decrease is called the *thermal lapse rate* and is preserved in part by the rapid expansion and cooling of rising air masses. Extending from 12 to 50 km above sea level is the *stratosphere* in which the temperature increases with altitude. From 50 to 85 km above sea level is the *mesosphere*; above 85 km is the *thermosphere.* Although both the mesosphere and thermosphere are of great interest to atmospheric scientists, they play a secondary role in determining terrestrial insolation. The troposphere, which contains 80 percent of the total atmospheric mass, is primarily responsible for the abundance or scarcity of solar energy at the earth's surface. Except for the absorption of ultraviolet radiation by stratospheric ozone, the major attenuation of solar radiation is produced by the constituents of the troposphere. It is therefore to this lowest layer that we direct our attention. We will characterize the temperature profile of the troposphere by

$$T(z) = T_0 - \Gamma z \tag{3.1}$$

where T_0 is the ground temperature and $\Gamma \simeq 6$ K/km is the thermal lapse rate previously introduced.

The Density Profile

The degree to which solar energy is attenuated is determined to a great extent by the density profile $\rho(z)$. To establish this profile, we will assume that the tropospheric air is well mixed and behaves as an ideal gas. The ideal gas law can be written

$$P = \rho \frac{RT}{\bar{M}} \qquad \text{(ideal gas law)} \tag{3.2}$$

where T is the Kelvin temperature; P, the pressure; $\bar{M}$, the mean molecular mass of air ($\bar{M} = 29$ amu), and R, the ideal gas constant ($R = 8317$ J/kg mole-K).

Because air is a fluid assumed to be in hydrostatic equilibrium, we set the negative pressure gradient supporting the air equal to its weight density and find

$$\frac{dP}{dz} = -\rho g \qquad \text{(hydrostatic equation)} \tag{3.3}$$

where $g = 9.8 \text{ m/sec}^2$ is the acceleration of gravity. Solving for ρ in Equation 3.2 and substituting the result into Equation 3.3, we obtain, after integrating, the pressure profile

$$P(z) = P_0 \exp\left[-\left(\frac{\bar{M}g}{R}\int_0^z \frac{dz}{T(z)}\right)\right] \tag{3.4}$$

Substituting this back into Equation 3.2, we obtain the density profile as

$$\rho(z) = \rho_0 \frac{T_0}{T(z)} \exp\left[-\left(\frac{\bar{M}g}{R}\int_0^z \frac{dz}{T(z)}\right)\right] \tag{3.5}$$

where ρ_0 and T_0 are sea level values.

According to Equation 3.5, the density profile can be obtained once the temperature profile is specified. For simplicity, we will ignore the thermal lapse rate in the troposphere and assume that the temperature is everywhere equal to the sea level value T_0. (The error involved in ignoring the lapse rate is considered in Problem 3-1.) For such an isothermal atmosphere, Equation 3.5 can be integrated to give

$$\boxed{\rho(z) = \rho_0 e^{-z/H}} \tag{3.6}$$

where

$$\boxed{H = \frac{RT_0}{\bar{M}g}}$$

and

$$\boxed{\rho_0 = \frac{P_0\bar{M}}{RT_0}}$$

An atmosphere whose density profile is given by Equation 3.6 is said to be an *exponential* atmosphere; the constant H is called its *scale* height. For a sea level temperature of $T_0 = 273$ K and pressure of $P_0 = 1.01 \times 10^5 \text{ N/m}^2$, we find, using Equation 3.6, that

$$H = \frac{(8317)(273)}{(29)(9.8)} = 8000 \text{ m} = 8 \text{ km}$$

and

$$\rho_0 = \frac{(1.01 \times 10^5)(29)}{(8317)(273)} = 1.29 \text{ kg/m}^3$$

The scale height represents the altitude at which the density of an exponential atmosphere falls to $1/e \simeq 1/2.7$ of its sea level value. It is possible to show that if the entire atmosphere were redistributed with a uniform density equal to its sea level value, it would extend only as high as the scale height (see Problem 3-2). Figure 3.2*b* shows Equation 3.6 plotted for different temperatures.

In obtaining Equation 3.6, we made two fundamental assumptions. The first was that the atmosphere was isothermal. Although this approximation introduces some error in the atmospheric temperature profile, it nevertheless leads to a reasonably accurate description of the density profile and simplifies the analysis to be presented on thermal sky radiation. The second assumption was that the air was well mixed and could be regarded as being composed of molecules with a mean mass $\bar{M} = 29$ amu. The mean mass is obtained by assuming a $\frac{4}{5}$ fraction of nitrogen, N_2, (28 amu) and a $\frac{1}{5}$ fraction of oxygen, O_2, (32 amu). Other constituents such as carbon dioxide, CO_2, and water vapor, H_2O, make up a very small fraction of the atmospheric mass. If well mixed with N_2 and O_2, they do not affect the value of $\bar{M}$; consequently, CO_2 and H_2O conform to the same profile as given by Equation 3.6. If, however, the CO_2 and H_2O molecules were allowed to settle out and establish *diffusive* equilibrium, then each component would have its own distinct profile and scale height according to its molecular mass. Constant mixing in the troposphere prevents diffusive equilibrium.

There are other factors that cause certain molecular species to deviate from the exponential profile. Ozone, O_3, exists primarily in the stratosphere because it is produced there by a photochemical reaction between oxygen and ultraviolet radiation. Both H_2O and CO_2 exist in higher concentrations in the lower troposphere—H_2O because of evaporation from the oceans; CO_2 due to its production by plants and animals.

Absorption and Scattering of Solar Radiation by Atmospheric Components

Atmospheric constituents—whether molecules such as N_2, O_2, CO_2, H_2O, or O_3 or larger particles such as fog droplets, soot, or dust, can affect radiation by either absorption or scattering. In absorption the radiant energy is converted into some other form, usually heat. The fraction absorbed is determined, in part, by the *mass absorption cross section*, $\sigma^a(\lambda)$, of the constituent. This parameter varies from one molecule to another and also depends on the wavelength of the incident radiation. As we will see, molecules of O_2 and N_2 do not absorb appreciably in the solar spectrum. On the other hand, CO_2 and H_2O absorb heavily in selected ranges of the infrared portion of the solar spectrum. Such regions are called characteristic absorption bands (Figure 3.3). In the ultraviolet regions of the solar spectrum, absorption bands are produced by ozone in the stratosphere.

Scattering is a more complicated process than absorption. As in absorption, a fraction of the energy is removed from the incident beam of radiation. This amount is determined by the *mass scattering cross section*, $\sigma^s(\lambda)$, of the constituent. Unlike absorption, scattering does not convert radiant energy into heat but redirects this energy into other directions of space. Atmospheric scattering of solar energy on a clear day is produced primarily by oxygen and nitrogen. Theory suggests that the scattering of solar energy by air molecules varies smoothly with wavelength according to Rayleigh's law

$$\sigma^s_{\text{air}}(\lambda) = \frac{C}{\lambda^4} \qquad \text{(Rayleigh's Law)} \tag{3.7}$$

where C is a parameter with a slight dependence on wavelength.

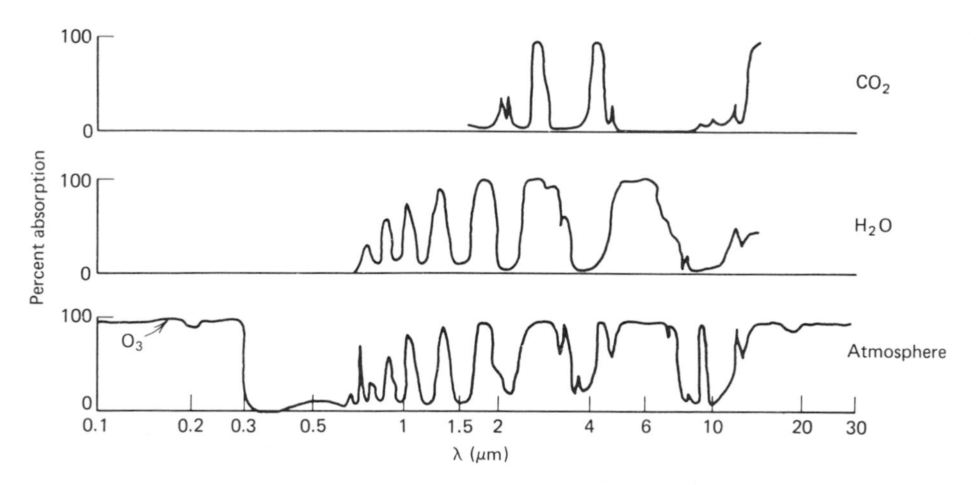

FIGURE 3.3 *The percent absorption of solar radiation by a clear atmosphere and its absorbing constituents CO_2 and H_2O.*

According to Rayleigh's law, short wavelengths such as ultraviolet, violet, and blue are scattered more effectively than are red and infrared. Hence ordinary air produces substantial scattering in the visible spectrum, particularly for blue-violet components, accounting for the bluish color of the sky. Certain particles produce scattering that favors the forward and backward directions (e.g., Rayleigh scattering), whereas others scatter radiation in a more isotropic manner. Particulate matter in the atmosphere, such as dust, soot, and haze, scatters radiation in a still more complicated manner than that predicted by Rayleigh's law. Reddish sunsets result from the scattering of radiation by dust particles near the earth's surface.

The radiation that has survived scattering and absorption is called the *direct* or attenuated component. For a plane-stratified atmosphere, this component is relatively simple to compute and will be shown to be determined by a *single,* wavelength-dependent atmospheric parameter known as the *optical thickness,* τ_λ. This parameter and the way in which it affects the direction radiation will be considered in detail.

Direct Solar Radiation

We assume that our atmosphere has uniformly mixed constituents and a density profile $\rho = \rho(z)$. Until now we have characterized a point in the atmosphere by an altitude or height variable z that ranges from zero at ground to infinity at the "top" of the atmosphere. Because solar radiation arrives at the top of the atmosphere and propagates toward the earth's surface, it is more convenient to introduce a variable, s, which measures the depth of a point in the atmosphere from the top. This variable ranges from $s = 0$ at the top to $s = \infty$ at ground. Increments of altitude and depth are related by $dz = -ds$.

Imagine that a beam of parallel solar rays of spectral intensity I_λ is incident on an infinitesimal layer of the atmosphere of thickness ds situated at a depth s (Figure 3.4). The beam is incident at some zenith angle Z. The fractional change in intensity when the beam emerges from the bottom of the layer can be expressed as

$$\frac{dI_\lambda}{I_\lambda} = -\rho(s)\sigma(\lambda)\, dl \tag{3.8}$$

where $\rho(s)$ and $\sigma(\lambda)$ are the density and the total $[\sigma^a(\lambda) + \sigma^s(\lambda)]$ mass attenuation cross section of the layer, respectively. The slant thickness

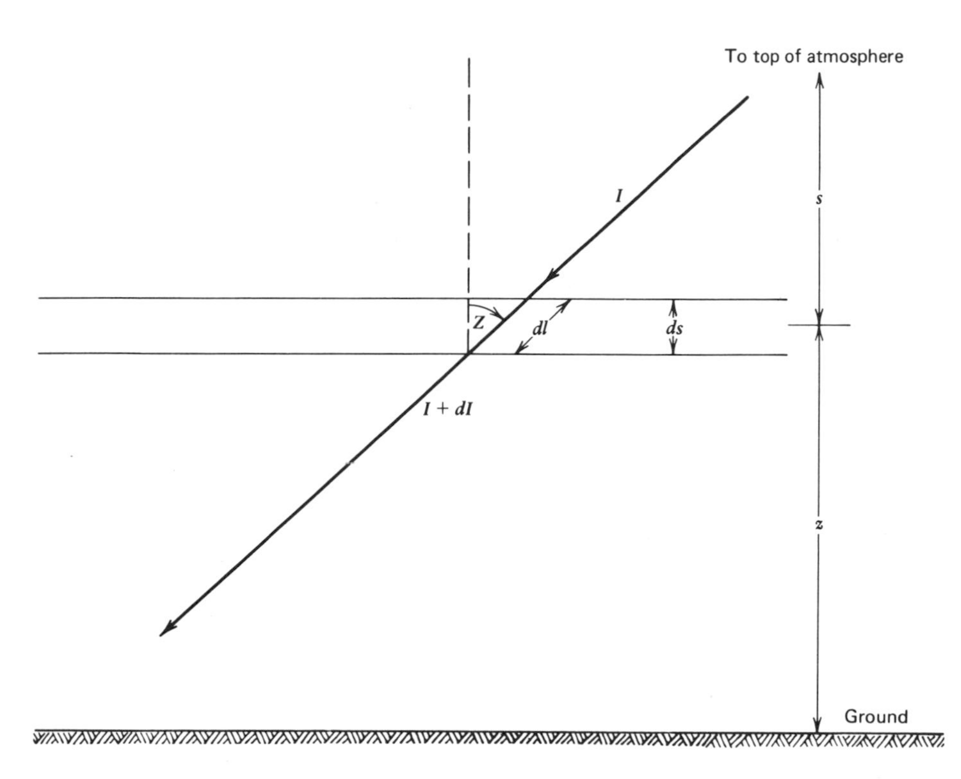

FIGURE 3.4 *The attenuation of a direct solar ray of intensity I by a differential layer of the atmosphere. The ray is incident at zenith angle Z.*

dl is defined by

$$dl = \frac{ds}{\mu_0}$$

where $\mu_0 = \cos Z$. It is convenient to simplify Equation 3.8 and write

$$\frac{dI_\lambda}{I_\lambda} = -k_\lambda(s)\frac{ds}{\mu_0} \tag{3.9}$$

where $k_\lambda(s) = \rho(s)\sigma(\lambda)$ is called the spectral *attenuation* or *extinction* coefficient. This coefficient varies with the altitude of the layer and with the wavelength of the incident radiation.

To find the attenuation produced by the entire atmosphere, we integrate Equation 3.9 from $s = 0$ to $s = \infty$, using the fact that at $s = 0$ (the top of the atmosphere), the spectral distribution is that of the solar constant. We thus find the direct or attenuated spectral intensity at ground to be

$$\boxed{I_\lambda^{(\text{dir})} = S_\lambda e^{-\tau_\lambda/\mu_0}} \tag{3.10}$$

where S_λ refers to the spectral intensity of the solar constant and $\tau_\lambda = \int_0^\infty k_\lambda(s)\, ds$ is called the *spectral optical thickness* of the atmosphere for beam radiation of wavelength λ.

This direct intensity remains essentially monodirectional or beam-like as it traverses the atmosphere so that the spectral flux intercepted by a flat panel situated at the earth's surface but inclined with an obliquity angle σ to the beam is

$$\boxed{F_\lambda^{(\text{dir})} = \mu I_\lambda^{(\text{dir})} = \mu S_\lambda e^{-\tau_\lambda/\mu_0}} \tag{3.11}$$

where $\mu = \cos\theta$. It follows from Equation 3.11 that the direct spectral solar flux falling on a flat surface at ground is determined by three variable factors.

1. The obliquity of the panel to the sun's rays, $\mu = \cos\theta$.
2. The cosine of the solar zenith angle, $\mu_0 = \cos Z$.
3. The spectral optical thickness of the atmosphere, τ_λ, for radiation of wavelength λ.

The obliquity factor applies quite generally to all cases in which beam radiation is incident on an inclined surface. The amount of flux intercepted by a surface decreases as we increase its obliquity to the beam. $F_\lambda^{(\text{dir})}$ decreases as the solar zenith angle increases because the path traversed by the beam through the atmosphere increases. The exponential dependence on cos Z suggests that the available solar flux decreases markedly as we approach sunrise or sunset.

It is sometimes convenient to write Equation 3.10 as

$$I_\lambda^{(\text{dir})} = S_\lambda T_\lambda^{\,m}$$

where $T_\lambda = e^{-\tau_\lambda}$ is called the *spectral optical transparency* of the atmosphere at one *air mass* and $m = 1/\cos Z$ is called the number of air masses traversed by the beam. When the sun is directly overhead ($Z = 0$), the radiation is traversing one air mass. Similarly, $Z = 60°$ and $Z = 70.5°$ correspond to two and three air masses, respectively. If the atmosphere has, for example, $T = 0.9$ (i.e., it is 90 percent transmissive) at one air mass, it will have $T^2 = 0.81$ and $T^3 = 0.73$ at two and three air masses, respectively.

The spectral optical thickness for a given atmospheric state consists of a part due to absorption, $\tau_\lambda^{(a)}$, and a part due to scattering, $\tau_\lambda^{(s)}$. On a clear day the attenuation by absorption is generally due to carbon dioxide and water vapor in the troposphere and to ozone in the stratosphere. Carbon dioxide and water vapor absorb strongly in selected bands in the infrared, whereas ozone absorbs well in selected bands in the ultraviolet.

Attenuation by molecular scattering in a clear atmosphere is primarily due to O_2 and N_2 and varies with wavelength in a more continuous manner than does absorption. Using Rayleigh's law, Equation 3.7, we can derive the following relationship for scattering in an exponential atmosphere (see Ref. 6 p. 238).

$$\tau_\lambda^{(s)} = \left[\frac{32}{3}\pi^3 \frac{(n-1)\bar{M}H}{N_0\rho_0}\right]\frac{1}{\lambda^4}$$

where n is the index of refraction of air and $N_0 = 6.02 \times 10^{26}$ mol/kg mole is Avogadro's number. Using the values of $(n-1) \simeq 2.8 \times 10^{-4}$, $\bar{M} = 29$ amu, $H = 8000$ m, and $\rho_0 = 1.29$ kg/m^3, we obtain

$$\tau_\lambda^{(s)} \simeq \frac{0.0076}{\lambda^4}$$

where λ is expressed in μm. This formula gives values of $\tau_\lambda^s = 0.27$, 0.12, and 0.04 for violet (0.410 μm), green (0.500 μm), and red (0.650 μm) light, respectively. Attenuation by molecular scattering increases markedly as we approach the ultraviolet limit of the visible spectrum and is almost negligible for infrared radiation. A plot of the direct solar intensity at sea level for one air mass is shown in Figure 3.5.

The total direct flux intercepted by a surface at sea level is obtained by integrating over all wavelengths giving

$$F^{(\mathrm{dir})} = \int_0^\infty \mu S_\lambda e^{-\tau_\lambda/\mu_0}\, d\lambda \tag{3.12}$$

In many solar energy applications the quantity of interest is the total flux rather than the spectral flux. For such applications it is sufficient to use an average optical thickness for the solar spectrum, $\bar{\tau}_\lambda = \tau$, and write Equation 3.12 as

$$F^{(\mathrm{dir})} = \mu e^{-\tau/\mu_0}\int_0^\infty S_\lambda\, d\lambda$$

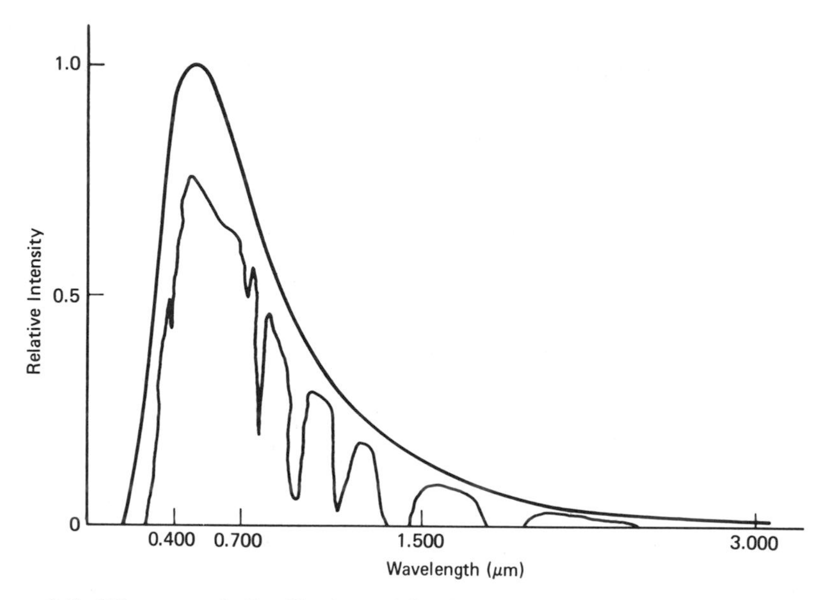

FIGURE 3.5 *The spectral distribution of the direct solar beam at one air mass (lower curve). The spectral distribution of the solar constant (upper curve).*

or

$$\boxed{F^{(\mathrm{dir})} = \mu e^{-\tau/\mu_0} S} \tag{3.13}$$

where $S = 1352\ \mathrm{W/m^2}$ is the solar constant. We will use this simple relation to approximate the direct solar flux on surfaces situated at sea level. The atmosphere is relatively clear when $\tau < 0.3$ and relatively opaque when $\tau > 1$. Note that on a given day when the atmospheric state is fixed, the hourly variation of the direct flux is due to the dependence of $\mu = \cos\theta$ and $\mu_0 = \cos Z$ on the hour angle. This dependence is established using Equations 2.10 and 2.16. Plots of $F^{(\mathrm{dir})}$ versus solar time obtained from Equation 3.13 are given in Figure 3.6.

EXAMPLE

A solar heating panel is situated at a colatitude of $L' = 43°$ and tilted with coordinates $\Delta = 45°$, $\psi = 0$. If the optical thickness of the atmosphere is $\tau = 0.3$, find the direct flux incident on the panel at the winter solstice ($D' = 113.5°$) at 2:00 PM solar time.

Using Equation 2.9, we find the hour angle of the sun to be

$$H = \frac{2}{24} \times 360 = 30°$$

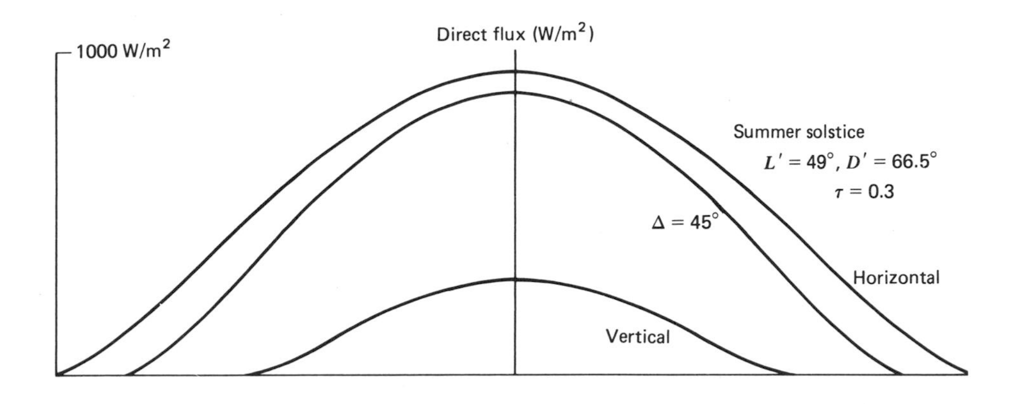

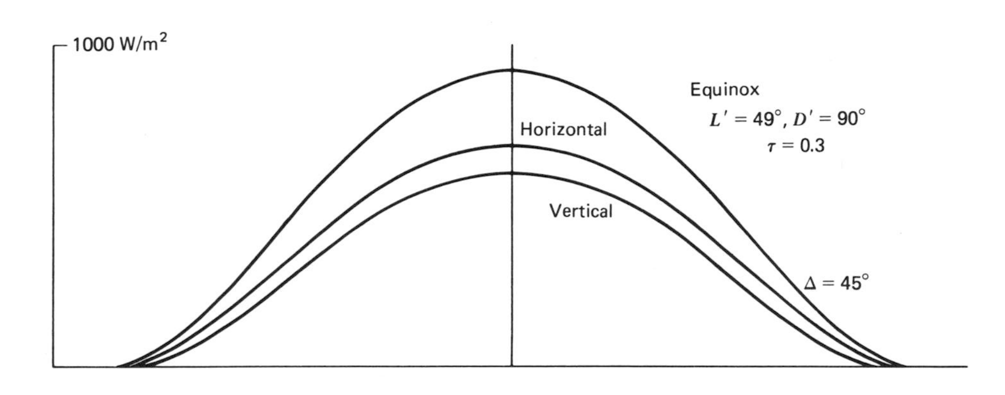

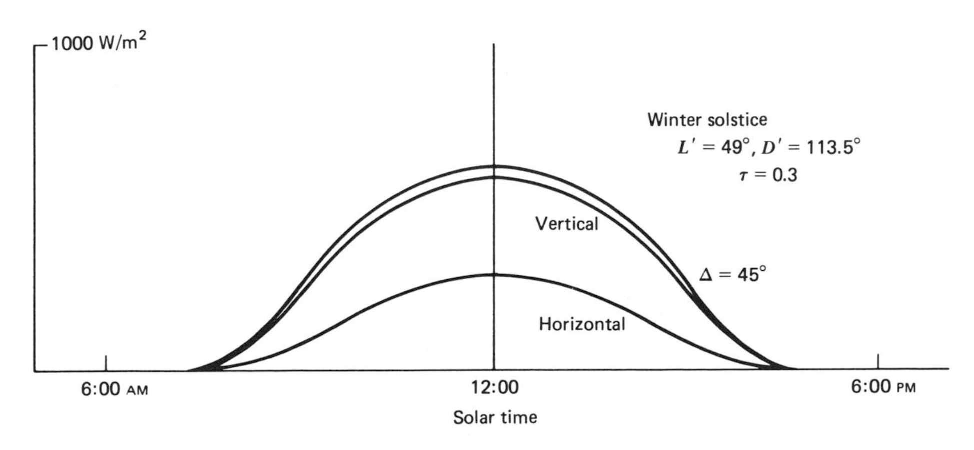

FIGURE 3.6 *The hourly direct flux on surfaces of various tilts at different seasons. All the cases are for an observer at a colatitude of* $L' = 49°$ *and for an atmospheric optical thickness of* $\tau = 0.3$.

From Equation 2.10 we have

$$\mu_0 = \cos Z = \cos 113.5° \cos 43° + \sin 113.5° \sin 43° \cos 30° = 0.25$$

and

$$\tan A = \sin 113.5° \sin 30°/(\sin 113.5° \cos 43° \cos 30° - \cos 113.5° \sin 43°)$$

Solving, we find

$$Z = 75.5° \qquad \text{and} \qquad A = 28.3°$$

From Equation 2.16 we have

$$\mu = \cos\theta = \cos 75.5° \cos 45° + \sin 75.5° \sin 45° \cos 28.3° = 0.78$$

or

$$\theta = 38.7°$$

Substituting these values into Equation 3.13, we obtain

$$F^{(\mathrm{dir})} = (1352)(\cos 38.7°)e^{-0.3/\cos 75.5°} = 318 \text{ W/m}^2$$

Diffuse Flux

The total solar flux falling on a surface at sea level consists of a monodirectional beam of direct flux plus a component of diffuse flux. The diffuse component is composed of radiation that has been scattered or reflected from atmospheric constituents as well as radiation reflected from the underlying terrain. Even for a plane-stratified atmosphere, the diffuse component is exceedingly difficult to treat and no simple formula analogous to Equation 3.13 exists for it. Instead of attempting to obtain exact solutions for the diffuse flux, we will consider a rather semiquantitative analysis in order to estimate it and also to show how environmental factors affect it.

Until now we have not been particularly careful to differentiate between the terms intensity and flux. Although this distinction is not essential for monodirectional radiation, it *is* necessary for diffuse radiation. The term intensity is used to characterize the directional distribution of the radiation. Specifically, it provides information as to how much radiant power per unit area is propagating into the various directions about some point in space. We label a given direction by the unit vector $\hat{\Omega}$. For a plane-stratified atmosphere, we may write the spectral intensity as $I_\lambda(z, \hat{\Omega})$, suggesting that this function depends on both altitude and the direction in question. The intensity function is defined as

$$I_\lambda(z, \hat{\Omega}) = \frac{\text{radiant power}}{\text{area-solid angle-wavelength}} = \frac{dP}{dA - d\Omega - d\lambda}$$

where $d\Omega$ is the differential solid angle about the direction $\hat{\Omega}$.

If $I_\lambda = I_\lambda(z)$, that is, if the intensity does not depend on $\hat{\Omega}$, then we say that the radiation is *isotropic*. Radiant energy is then propagating equally in all directions. If, on the other hand, the function $I_\lambda(z, \hat{\Omega})$ is very intense along a single direction, say, $\hat{\Omega}_0$, and zero along all others, we say that the radiation is *beamlike* or *monodirectional*. Although the direct solar component is essentially monodirectional, the diffuse component is *not* composed of purely isotropic radiation.

Note that in describing the intensity function, no mention is made of any intercepting surface. The concept of flux, on the other hand, is meaningful only in relation to a specific surface. Consider a small flat surface whose orientation is characterized by its *inward* normal, that is, by a unit vector $\hat{\boldsymbol{n}}$ oriented perpendicularly *into* the surface. The spectral flux intercepted by this surface is given by

$$\boxed{F_\lambda(z, \hat{\boldsymbol{n}}) = \int_{(\hat{\boldsymbol{n}} \cdot \hat{\Omega} \geq 0)} I_\lambda(z, \hat{\Omega}) \hat{\boldsymbol{n}} \cdot \hat{\Omega} \, d\Omega} \tag{3.14}$$

where $\hat{\boldsymbol{n}} \cdot \hat{\Omega}$ is the cosine of the angle between $\hat{\boldsymbol{n}}$ and $\hat{\Omega}$. The subscript on the integral reminds us that the integration is over those solid angles for which $\hat{\boldsymbol{n}} \cdot \hat{\Omega} \geq 0$. This assures that only radiation being transported in the hemisphere *toward* the surface is included in Equation 3.14. From Equation 3.14 it follows that the flux intercepted by a surface, even for diffuse radiation, depends in general on the orientation of the surface as well as on its position in the atmosphere.

For beamlike radiation it can be shown that Equation 3.14 reduces to

$$F_\lambda^{(\text{beam})} = F_\lambda^{(\text{normal})} \hat{\boldsymbol{n}} \cdot \hat{\Omega}_0 = F_\lambda^{(\text{normal})} \cos\theta \tag{3.14a}$$

where θ is the obliquity angle between the beam and the surface normal, and where $F_\lambda^{(\text{normal})}$ is the beam flux falling on a surface oriented perpendicularly to the beam.

For isotropic radiation, Equation 3.14 can be integrated to give

$$F_\lambda^{(\text{isotropic})} = I_\lambda^{(\text{isotropic})} \int_{(\hat{n}\cdot\hat{\Omega}\geq 0)} n \cdot \hat{\Omega}\, d\Omega = \pi I_\lambda^{(\text{isotropic})} \tag{3.14b}$$

In this case the intercepted flux is *independent* of the orientation of the surface, as expected. Diffuse radiation is not, in general, isotropic so that the intercepted flux depends somewhat on the orientation of the surface. This dependence is not nearly as marked as that for the direct flux.

Approximate Equations for Total Solar Flux

A thorough treatment of the total intensity (direct plus diffuse) for terrestrial insolation requires some of the most advanced and complicated procedures known to mathematical physicists. To find the intensity function, we need to solve an integrodifferential equation known as the *equation of radiative transfer*. Even for a plane-stratified atmosphere solutions are exceedingly difficult to develop. The difficulty increases as the molecular-scattering process becomes more complex. The solution for the intensity function, $I(z, \hat{\Omega})$, is further complicated by reflections from the underlying terrain.

Even superficial treatment of the equation of radiative transfer is beyond the level of this text. To proceed, we will make some simplifying assumptions: Let us assume that (1) the diffuse component is much smaller than the direct flux; (2) the downward diffuse component falling on a horizontal surface is isotropic; and (3) the underlying terrain is *Lambertian*; that is, it reflects radiation according to Lambert's law. A Lambertian surface is one for which the reflected radiation is always isotropic regardless of the nature of the incident radiation. The upward atmospheric flux is therefore assumed to be totally diffuse or isotropic. Using these assumptions, we see that the diffuse flux is composed of an isotropic downward flux $F^{\downarrow(\text{diff})}$ and an isotropic upward flux $F^{\uparrow(\text{diff})}$. Furthermore, the total diffuse flux intercepted by an inclined surface depends on its tilt angle from the vertical, Δ, but *not* on its azimuth angle, ψ. It can be shown that for isotropic upward and downward fluxes, the total diffuse flux intercepted by an inclined surface is

$$F^{(\text{diff})} = \left[\frac{1+\cos\Delta}{2}\right] F^{\downarrow(\text{diff})} + \left[\frac{1-\cos\Delta}{2}\right] F^{\uparrow(\text{diff})}$$

so that the total (direct plus diffuse) intercepted flux is

$$F = F^{(\mathrm{dir})} + F^{(\mathrm{diff})}$$

or

$$\boxed{F = S\mu e^{-\tau/\mu_0} + \left[\frac{1+\cos\Delta}{2}\right]F^{\downarrow(\mathrm{diff})} + \left[\frac{1-\cos\Delta}{2}\right]F^{\uparrow(\mathrm{diff})}} \tag{3.15}$$

For a horizontal surface ($\Delta = 0$), the diffuse flux is entirely $F^{\downarrow(\mathrm{diff})}$ whereas for a vertical surface ($\Delta = 90°$), this flux is $F^{(\mathrm{diff})} = \frac{1}{2}(F^{\downarrow(\mathrm{diff})} + F^{\uparrow(\mathrm{diff})})$.

The total flux falling on an inclined surface can be computed from Equation 3.15 once the two fluxes $F^{\downarrow(\mathrm{diff})}$ and $F^{\uparrow(\mathrm{diff})}$ have been determined. Using some simplifying approximations (see Appendix 2), we are able to transform the equation of radiative transfer into a pair of simple ordinary differential equations for the upward and downward fluxes. These equations involve three environmental parameters—the optical thickness, τ, of the atmosphere; the *single scattering albedo*, $\tilde{\omega}_0$, of the atmosphere; and the reflectivity, R, of the underlying terrain. The last two parameters are defined as

$$\tilde{\omega}_0 = \frac{\tau^{(s)}}{\tau} \quad \text{and} \quad R = \frac{F^{\uparrow}}{F^{\downarrow}}$$

where $\tau^{(s)}$ is the optical thickness due to scattering and $F^{\downarrow}$ and $F^{\uparrow}$ are, respectively, the downward and upward fluxes at ground. For a totally scattering atmosphere (i.e., no absorption), $\tilde{\omega}_0 = 1$, whereas for a totally absorbing atmosphere (i.e., no scattering), $\tilde{\omega}_0 = 0$. For a terrain of dark vegetation $R < 0.2$, whereas for a snow-covered terrain $R \sim 0.5$. The parameters τ, $\tilde{\omega}_0$, and R of course depend on wavelength, but we will use values averaged over the solar spectrum. In terms of these parameters, the downward and upward diffuse components at ground are shown in Appendix 2 to be (see Ref. 6 p. 203).

$$F^{\downarrow(\mathrm{diff})} = S\mu_0\left[\frac{1}{1+G}(Ge^{\gamma^+\tau} + e^{\gamma^+\tau}) - e^{-\tau/\mu_0}\right] \tag{3.16a}$$

$$F^{\uparrow(\mathrm{diff})} = RS\mu_0\left[\frac{1}{1+G}(Ge^{\gamma^+\tau} + e^{\gamma^-\tau})\right] \tag{3.16b}$$

where

$$G = -\left[\frac{\gamma^- + A - BR}{\gamma^+ + A - BR}\right]e^{(\gamma^- - \gamma^+)\tau}$$

$$\gamma^{\pm} = \tfrac{1}{2}(C - A) \pm \tfrac{1}{2}[(C + A)^2 - 4BD]^{1/2}$$

and where

$$A = \frac{(2 - \tilde{\omega}_0)}{2\mu_0} \qquad B = \tilde{\omega}_0$$

$$C = (2 - \tilde{\omega}_0) \qquad D = \frac{\tilde{\omega}_0}{(2\mu_0)} \tag{3.17}$$

It is straightforward to verify that when scattering is absent ($\tilde{\omega}_0 \to 0$), the downward diffuse flux vanishes (i.e., $F^{\downarrow(\text{diff})} \to 0$). Furthermore, it is evident that when the reflectivity of the terrain vanishes ($R \to 0$), the upward flux also vanishes (i.e., $F^{\uparrow(\text{diff})} \to 0$). Note, however, that the downward diffuse flux *does* depend on R (through the function G) because reflected radiation can be rescattered back to ground by the atmosphere.

EXAMPLE

Find the direct and total fluxes intercepted by a horizontal surface when the solar zenith angle is $Z = 53°$ and when the environmental parameters are:

$$\tau = 0.4 \qquad \tilde{\omega}_0 = 0.5 \qquad \text{and} \qquad R = 0.2$$

Since $\mu = \mu_0$ for a horizontal collector, the direct flux is obtained as

$$F^{(\text{dir})} = S\mu_0 e^{-\tau/\mu_0} = (1352)(\cos 53°)e^{-0.4/\cos(53°)} = 419\ \text{W/m}^2$$

The constants for the diffuse fluxes for this case are

$$A = 1.25 \qquad B = 0.5 \qquad C = 1.5 \qquad D = 0.417$$

so that $\gamma^+ = 1.42$, $\gamma^- = -1.17$, and $G = 0.003$.

Substituting these values into Equation 3.16, we find

$$F^{\downarrow(\text{diff})} = 94\ \text{W/m}^2$$

and

$$F^{\uparrow(\text{diff})} = 103\ \text{W/m}^2$$

Because the collector is horizontal, Equation 3.15 gives

$$F = 419 + 94 = 513\ \text{W/m}^2$$

Thus approximately 18 percent of the flux is diffuse.

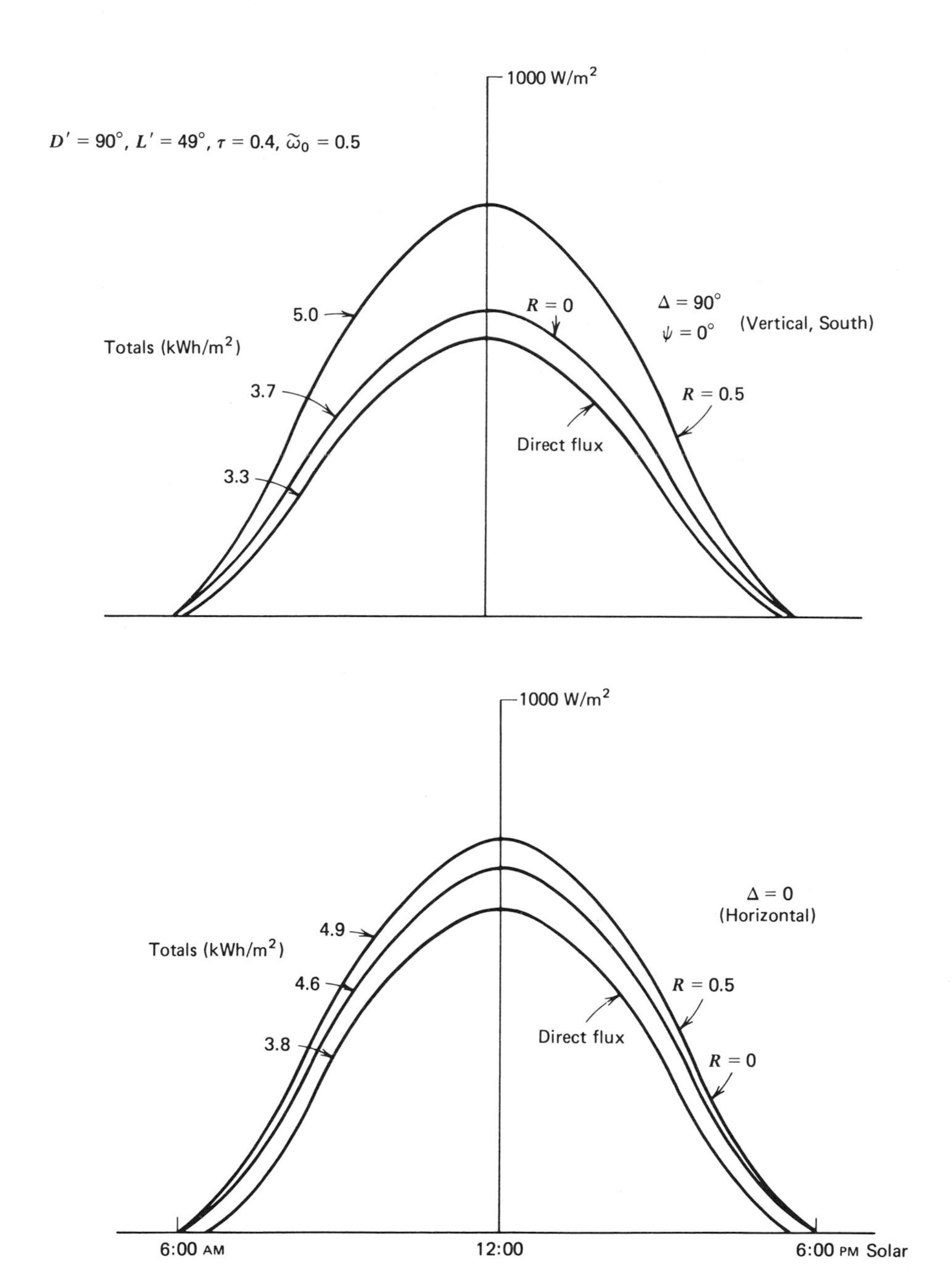

FIGURE 3.7 *The hourly direct and total fluxes on a horizontal and vertical south-facing surface at the equinox. Results for terrain reflectivities of $R = 0$ and $R = 0.5$ are shown. The cases are for an observer at $L' = 49°$ with atmospheric parameters $\tau = 0.4$ and $\tilde{\omega}_0 = 0.5$. Approximate daily totals are shown for each case.*

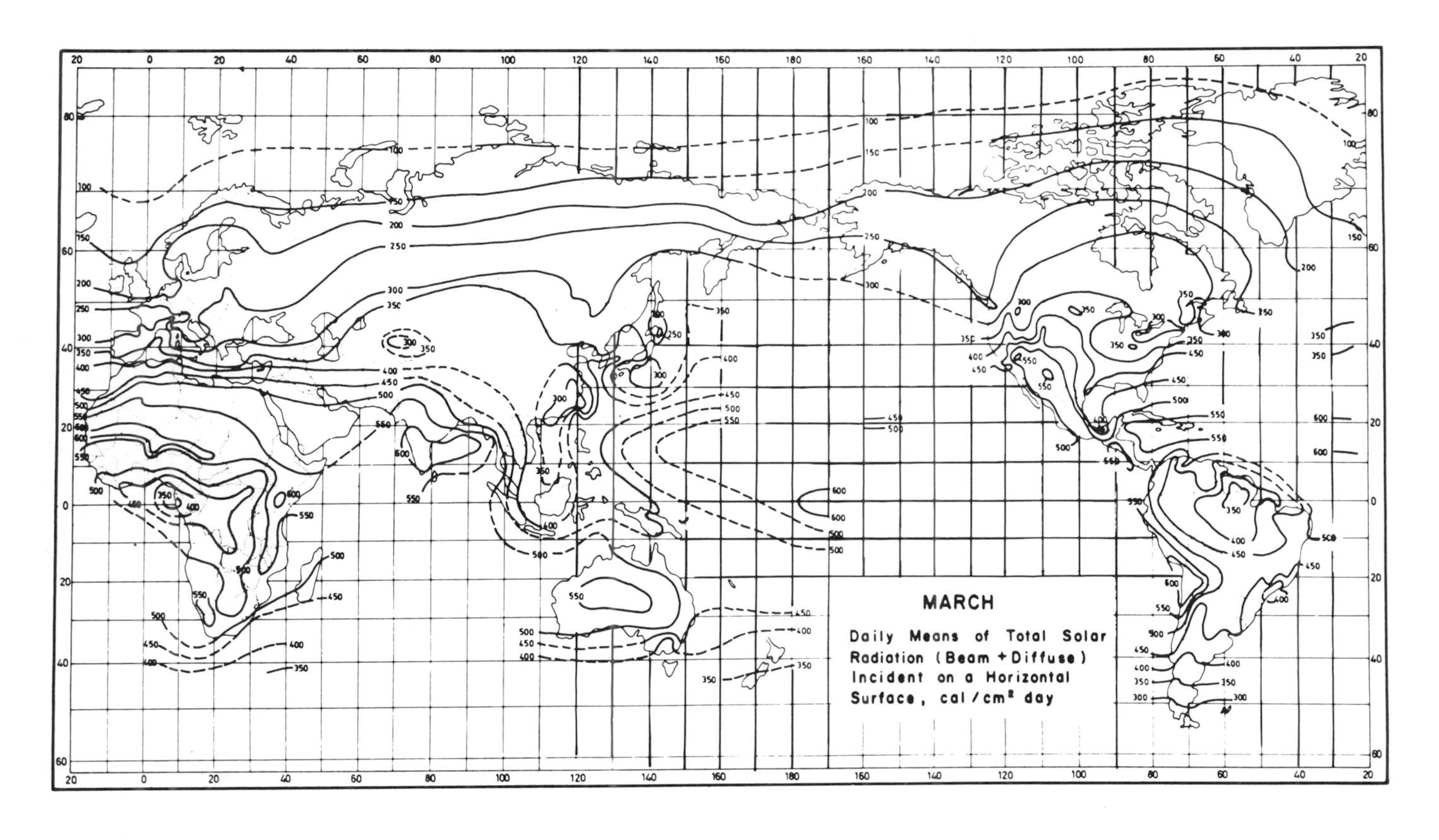

FIGURE 3.8 *Global values of the average daily insolation falling on a horizontal surface for March and June. To convert to MKS units, use* 1 *cal/cm*^{2}*-day* = 0.01163 *kwhr/m*^{2}*-day*. [Reprinted from Ref. 2 by permission of John Wiley & Sons, Inc.]

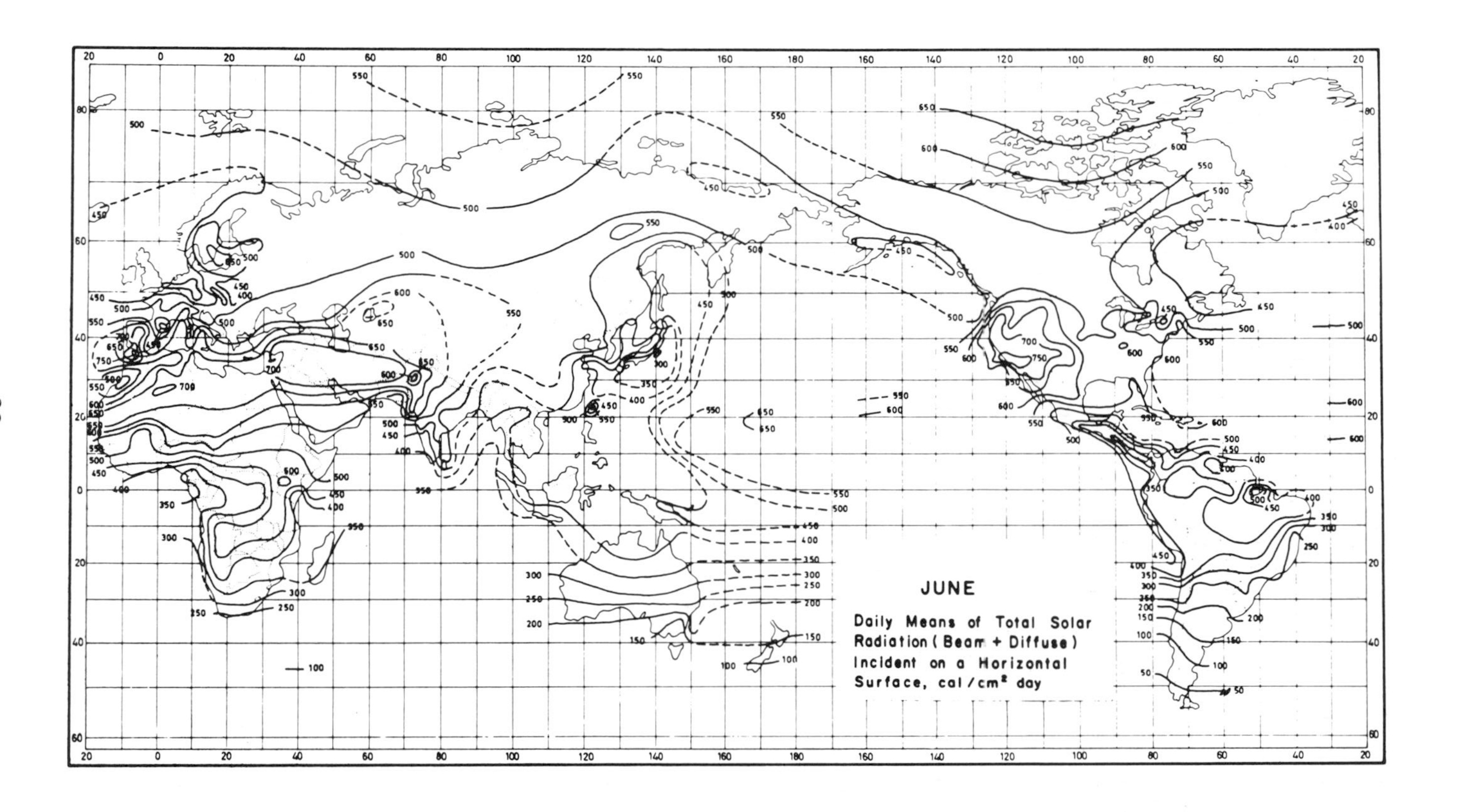
JUNE
Daily Means of Total Solar Radiation (Beam + Diffuse) Incident on a Horizontal Surface, cal/cm² day

Clearly, the results obtained for the diffuse flux depend on the validity of the model used in obtaining Equations 3.15 and 3.16 as well as on the choices for τ, $\tilde{\omega}_0$, and R. The environmental parameters are difficult to measure experimentally and of course vary markedly from day to day. Some results obtained from the theory are shown in Figure 3.7. Note that for the vertical surface the total flux is substantially larger when the ground is a strong reflector because the surface receives much of its energy from reflection. Even for a horizontal surface, the intercepted flux increases with the reflectivity of the terrain, even though such a surface receives no directly reflected radiation. It does, however, receive some extra reflected radiation that is rescattered downward by the atmosphere. Insolation levels are often observed to be higher in regions where the underlying terrain is highly reflective. The diffuse flux constitutes a smaller fraction of the total flux at noon; this fraction increases steadily as we move toward sunset (or sunrise). Even on a clear day, the diffuse radiation may constitute 20 percent of the total flux.

When cloud cover is present, the direct and diffuse fluxes are extremely difficult to compute. Multiple reflections between ground and patches of clouds sometimes intensify the terrestrial insolation to levels higher than those for cloudless days.

The amount of daily and seasonal insolation available to a surface at a given colatitude is of great significance in solar energy technology. Unfortunately, very little reliable worldwide data are available. Typical curves for daily global insolation levels are shown in Figure 3.8; however, these are only statistical estimates.

Measurement of Terrestrial Insolation

Although the preceding analysis gives some insight into the factors that determine terrestrial insolation, it is desirable to have some experimental data with which to compare the theory. Such data are also important for determining the efficiencies of solar energy systems. Data taken over many seasons provide statistical information as to the feasibility of certain solar energy systems in specific geographical areas.

Most instruments for measuring solar energy fall into one of two categories: *photoelectric* and *bolometric* devices. The first group includes devices with receiving elements or sensors whose electrical characteristics change in the presence of solar radiation. For example, *photovoltaics*, such as silicon and selenium cells, generate a voltage when solar energy is incident. Their short-circuit current is used to measure the intensity of the incident radiation. *Photoconductive* detectors, such as cadmium sulphide or cadmium selenide, change resistance in response to electromagnetic radiation. When connected across a battery, the current in the circuit becomes a measure of the intensity level. There are also vacuum tube devices known as phototubes, which have specially coated elements that emit electrons when light is incident on them. The change in conductivity of the tube produced by these photoelectrons is used to determine the intensity of the radiation.

Although solid state photoelectric devices are durable, compact, inexpensive to fabricate, and are minimally affected by ambient conditions, they nevertheless have some major shortcomings. First, they do not always produce a signal that is linear in the insolation level; many tend to saturate or level off at high intensity levels. Second, and more important, photoelectric devices do not have a spectrally flat response over the solar spectrum. This means that equal amounts of solar energy arriving in different spectral regions produce different signals. In particular, if a device is sensitive only to the visible spectrum, it will not detect the presence of or the change in infrared solar energy. Typical response curves for selenium and silicon photovoltaics are shown in Figure 3.9.

The second category includes the bolometric devices. These instruments generally absorb the radiation in a black absorber and use the heat generated to produce a change in the state of the receiver.

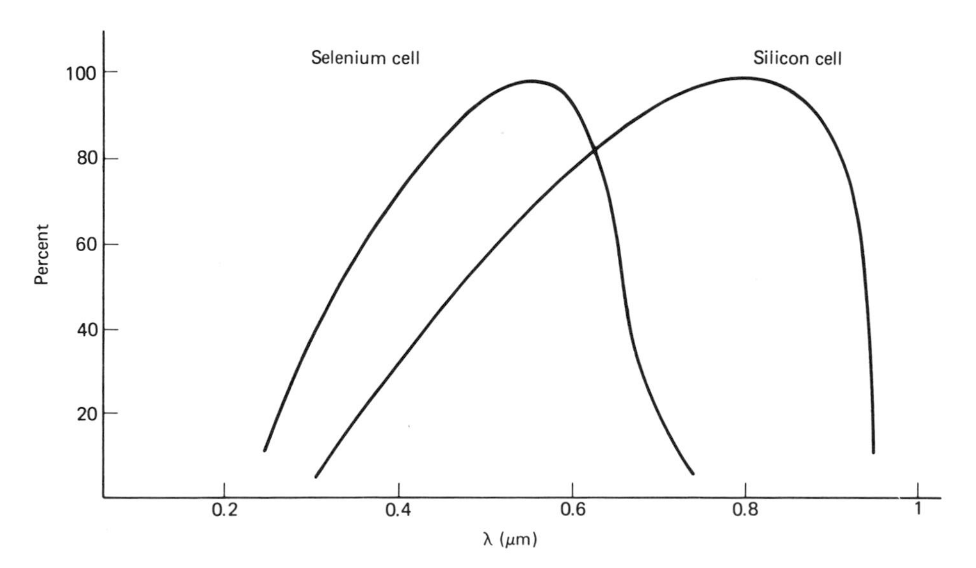

FIGURE 3.9 *A comparison of the relative responsivities of selenium and silicon cells.*

This change is measured and related to the insolation level. A typical bolometric instrument used in the United States is the *black and white pyranometer.* The receiver consists of two adjacent flat surfaces, one black and one white (or silvered). Each has a thermal sensor (usually a thermocouple) affixed. The sensors generate a voltage signal that is directly proportional to the temperature differences between the black and white surfaces. When solar energy is incident, the black surface absorbs the radiation and becomes warmer than the white one which reflects the radiation and remains near the ambient temperature. The greater the insolation level is, the greater the temperature difference and the larger the voltage signal.

The steady-state temperature difference between the surfaces depends on the degree of solar heating of the black surface and also on the heat loss rate to the environment. Because the heat loss is not a simple function of the temperature, there is no apriori reason to expect a linear relationship between radiant intensity and the voltage signal. Pyranometers that are generally calibrated in W/m^2 per millivolt assume such linearity. The output of a pyranometer depends on both the temperature and wind conditions of the surroundings. Consequently, correction charts are usually required when a pyranometer is used under environmental conditions other than those for which the unit was originally calibrated. To protect it from the elements, the receiver is generally covered with a glass dome transparent to solar radiation. A typical black and white pyranometer is shown in Figure 3.10.

Bolometric instruments that operate on the principle of *Angstrom or electrical compensation* avoid many of the difficulties encountered with black and white pyranometers. In electrical compensators a black surface is exposed to sunlight and its temperature noted. The same (or a similar) surface is shielded from the sun and is heated electrically until the same temperature is reached. The electrical power supplied (which is then equal to the absorbed solar power) is measured.

A portable field instrument using the electrical compensation principle is the *autobalancing radiometer* (Figure 3.11). The receiver is composed of two identical black surfaces. Each is a rectangular wafer consisting of a substrate with a black electrically conductive coating such as graphite. The coating acts not only as a black absorber but also as an electrical joule heater. Each wafer has a thermal sensor affixed to detect any temperature difference between them. Furthermore, each wafer is covered with a glass dome and the interior space partially evacuated. The dome is transmissive to solar radiation but opaque to thermal radiation. One wafer is chosen as the heater wafer and its dome is silvered; the other becomes the solar wafer and will

FIGURE 3.10 *A black and white pyranometer. (Courtesy of Eppley Laboratories.)*

receive solar radiation. When the receiver is exposed to solar radiation, the exposed solar wafer becomes warmer than the heater wafer that is shielded from sunlight by its silvered dome. The thermal sensors detect this temperature difference. A dc voltage is applied across the ohmic coating of the heater wafer until the joule heating brings the wafers to the same temperature. In the balanced state both wafers are being heated equally—one wafer by solar energy and the other by joule heating. The incident solar flux is determined from the equation

$$\text{flux} = \frac{V \times I}{\bar{a} \times A} \tag{3.18}$$

where V and I represent the voltage and current supplied to the heater wafer at balance, A is the area of each wafer, and $\bar{a}$ is a correction factor representing the product of the absorptivity of the coating and the transmissivity of the glass dome over the solar wafer. If the coating were perfectly black and the glass perfectly transparent to solar radiation, we would have $\bar{a} = 1$. In operation, an electronic control unit continually adjusts the voltage to keep the system in balance. The unit also performs the necessary computations as required by Equation 3.18.

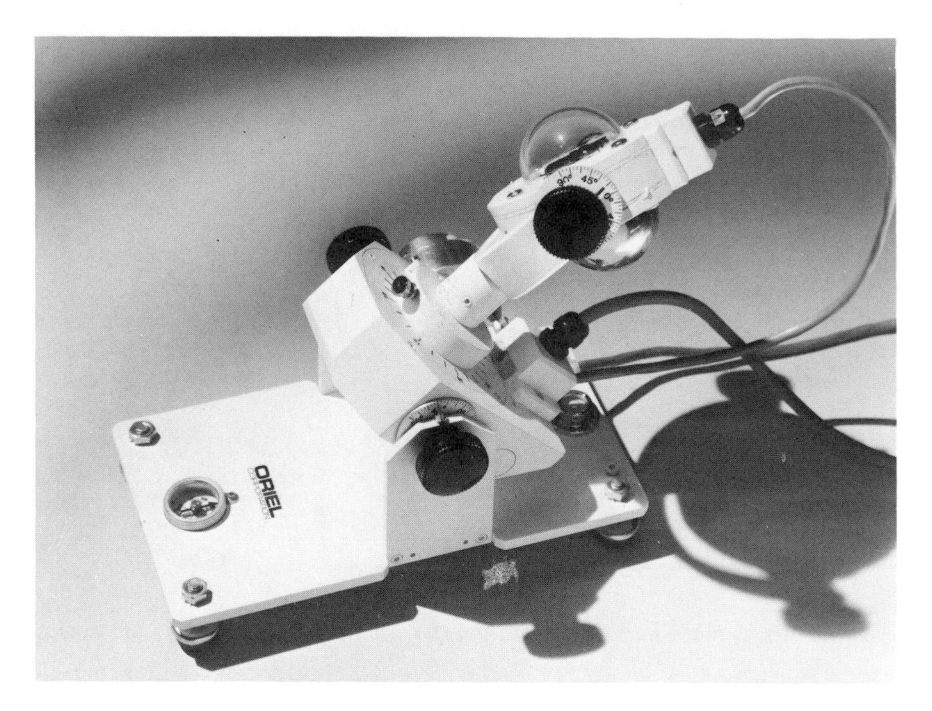

FIGURE 3.11 *An Oriel autobalancing radiometer using electrical compensation. The heater wafer is under the silvered dome. (Courtesy of the Oriel Corporation.)*

Unlike a pyranometer that measures temperature, the autobalancing radiometer actually records electrical wattage. The output signal is, in principle, linear in the insolation level. Because both wafers are exposed to the same environment, the data are minimally affected by changes in ambient conditions. As with all bolometric detectors, it is important that the wafers be as thin as possible in order to reduce thermal inertia (or heat capacity). This allows the unit to respond to rapid changes in the insolation level.

In Figure 3.12 an insolation recording taken with an autobalancing radiometer is compared to one obtained from a selenium cell. The curve corresponding to the output of the cell has been adjusted to coincide with that of the radiometer at solar noon. Note that this selenium cell is sensitive only to visible light and therefore does not sense the rapid decrease in the solar infrared component in the late afternoon.

Regardless of the type of instrument used to measure solar radiation, a receiving element will generally measure the total (direct plus diffuse) insolation incident on it. To measure the direct or diffuse

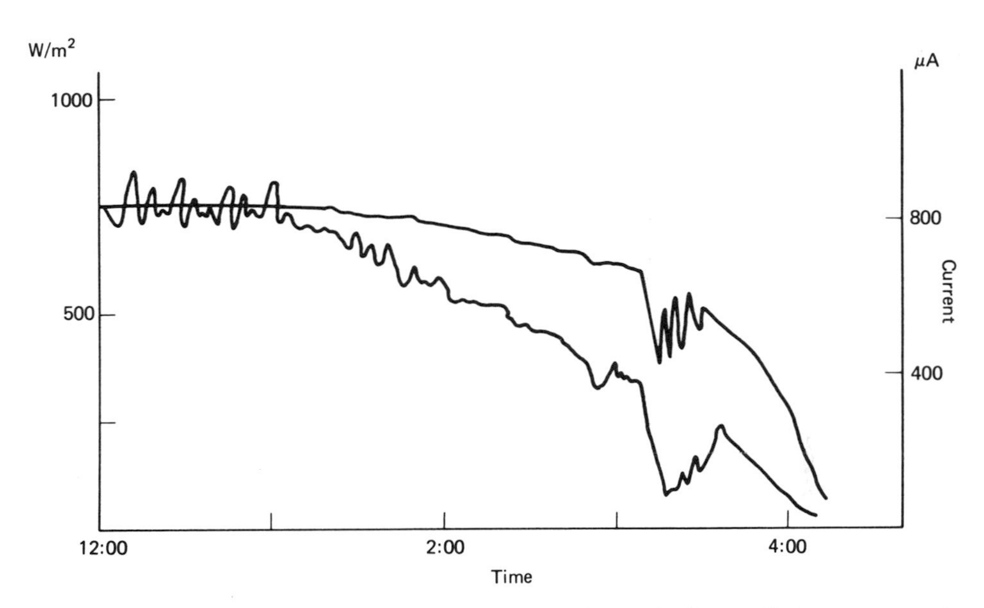

FIGURE 3.12 *An insolation recording taken with a selenium cell (upper curve) compared to one taken with an autobalancing radiometer. The decrease in insolation between* 3:00 *PM and* 4:00 *PM is due to clouds.*

component separately, one must place some form of discriminating element in front of the receiver. It is possible, for example, to measure the diffuse flux alone by placing an opaque disc a distance above the receiver to shield the element from the direct rays. To measure direct flux, one can use a collimating tube. The tube allows rays coming to the solar disc to reach the receiver while obscuring most of the diffuse flux. The tube should, in theory, be a section of a cone whose angle is 0.53 degrees corresponding to the divergence of the sun's rays (Figure 3.13). The longer the tube is, the less diffuse flux reaches the receiving element. In practice, a cylindrical tube whose radius is slightly larger than that of the receiver is sufficient. A radiometer that uses a collimator to measure direct flux is called a *pyrheliometer.* Collimating devices generally require some form of tracking to keep them aligned with the sun's rays.

Thermal Atmospheric Flux

As the final topic of the chapter we consider atmospheric thermal radiation incident on a surface. Although this radiation cannot be called solar energy, it does play an important role in the steady-state operation of solar heating panels.

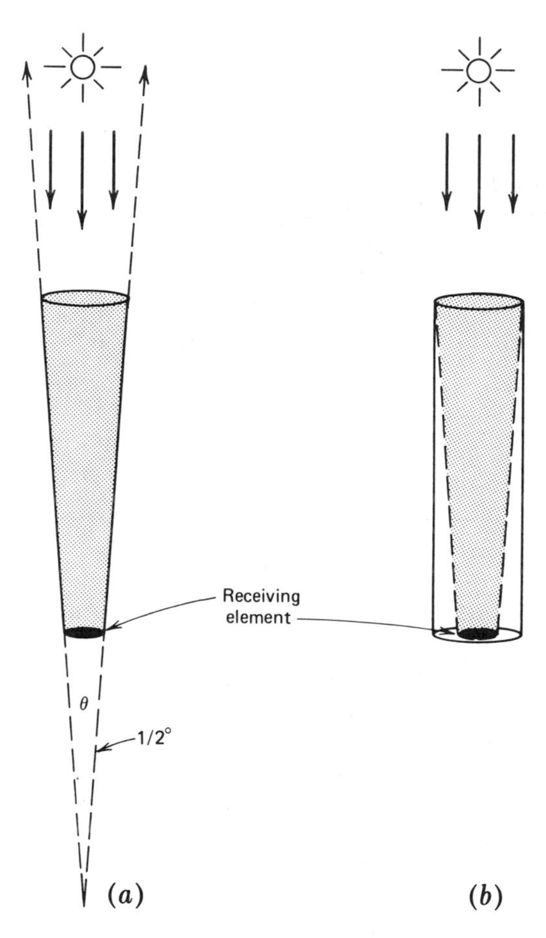

FIGURE 3.13 *(a) A conical collimator used to obscure the diffuse component of solar radiation. (b) An equivalent cylindrical collimator.*

In Chapter 1 it was shown that any body that is a good absorber of electromagnetic radiation is also a good emitter of thermal radiation. The amount and spectral distribution of the emitted radiation depends on the temperature and on the absorptivity (or emissivity) of the body. The atmosphere and the underlying terrain absorb solar radiation and emit thermal radiation. These emitting sources are generally at a temperature of ~300 K. A black-body spectrum at this temperature contains most of its energy in the range $2\,\mu m\ \lambda \leq 20\,\mu m$. Only a negligible fraction of the solar energy arrives in this wavelength interval. It is customary to call the solar energy "short wavelength" radiation ($0.3\,\mu m\ \lambda \leq 2\,\mu m$) in contrast to the thermal energy, which is termed "long wavelength" radiation.

Consider a differential layer of the atmosphere of optical thickness dt_λ, situated at an optical depth t_λ (Figure 3.14). Imagine a ray emitted at an angle θ whose cosine is μ. This ray was emitted by an element of slant thickness dt_λ/μ. Consequently, the intensity of the ray can be represented by

$$dI_\lambda^{(i)(\text{thermal})} = \frac{B_\lambda(T)}{\pi}\, dt_\lambda/\mu \qquad (\mu = \cos\theta) \tag{3.19}$$

where $B_\lambda(T)$ is the Planck function given in Chapter 1 and T is the temperature of the layer. This ray will travel toward the ground and be attenuated by the atmospheric layers below. The intensity of the ray as it arrives at ground is

$$dI_\lambda^{(\text{thermal})} = \{\exp[-(\tau_\lambda - t_\lambda)/\mu]\}\, dI_\lambda^{(i)(\text{thermal})} = \frac{B_\lambda(T)}{\pi}\{\exp[-(\tau_\lambda - t_\lambda)/\mu]\}\frac{dt_\lambda}{\mu} \tag{3.20}$$

where $(\tau_\lambda - t_\lambda)$ is the optical thickness of the air between the emitted layer and the ground. The total intensity is found by adding the contributions from all the layers of the atmosphere. Integrating

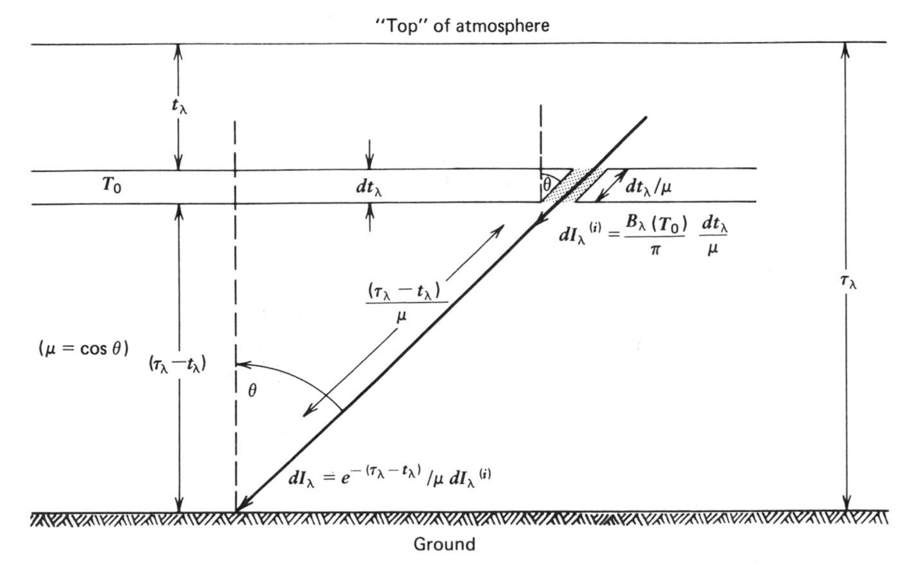

FIGURE 3.14 *A spectral ray of thermal radiation emitted by a layer of the atmosphere at temperature T_0 and its subsequent attenuation by layers below.*

Equation 3.20, we find

$$I_\lambda^{(\text{thermal})} = \frac{1}{\pi}\int_0^{\tau_\lambda} B_\lambda(T)\exp[-(\tau_\lambda - t_\lambda)/\mu]\,dt_\lambda/\mu \qquad (0 \le \mu \le 1) \tag{3.21}$$

Since T is a function of the altitude, the integral can be performed only after the temperature profile is given. If we assume that the atmosphere is isothermal, that is, $T = T_0$, then Equation 3.21 can be integrated as

$$I_\lambda^{(\text{thermal})} = \frac{1}{\pi}B_\lambda(T_0)(1 - e^{-\tau_\lambda/\mu}) \qquad (0 \le \mu \le 1) \tag{3.22}$$

Furthermore, if the optical thickness of the atmosphere can be approximated by a constant equal to some average value over the spectral range $2\,\mu\text{m} \le \lambda \le 20\,\mu\text{m}$, that is, $\bar{\tau}_\lambda = \tau$, then we may write Equation 3.22 as

$$I_\lambda^{(\text{thermal})} = \frac{1}{\pi}B_\lambda(T_0)(1 - e^{-\tau/\mu}) \tag{3.23}$$

The downward thermal flux falling on a horizontal surface is obtained using Equation 3-14 and integrating over a downward hemisphere, giving

$$\begin{aligned} F_{\downarrow\lambda}^{(\text{thermal})} &= 2\pi\int_0^1 I_\lambda \mu\,d\mu \\ &= 2B_\lambda(T_0)\int_0^1 (1 - e^{-\tau/\mu})\mu\,d\mu \\ &= 2B_\lambda(T_0)[\tfrac{1}{2} - E_3(\tau)] \end{aligned} \tag{3.24}$$

where $E_3(\tau)$ is called the *exponential integral* (or Gold function) of the *third order*. To find the total thermal radiation for the entire spectrum, we integrate over all wavelengths and find for a gray isothermal atmosphere

$$\boxed{F_{\downarrow}^{(\text{thermal})} = [1 - 2E_3(\tau)]\sigma T_0^4} \tag{3.25}$$

where we have used the fact that $\int_0^\infty B_\lambda(T_0)\,d\lambda = \sigma T_0^4$.

As the atmosphere becomes opaque to thermal radiation, that is, $\tau \to \infty$, the exponential integral tends to zero and the thermal flux is given by σT_0^4. In this limit the atmosphere is a perfect absorber; consequently, it is also a perfect emitter and emits like a black body. As the atmosphere becomes transparent, $\tau \to 0$, the exponential integral tends to $E_3(\tau) \underset{\tau\to 0}{\to} 0.5$. The resulting flux vanishes because a nonabsorbing body cannot radiate thermal energy regardless of its temperature.

It is convenient to express Equation 3.25 as

$$\boxed{F_{\downarrow}^{(\text{thermal})} = \sigma T_{\text{sky}}^4} \tag{3.26}$$

where the *sky temperature* is

$$T_{\text{sky}} = [1 - 2E_3(\tau)]^{1/4} T_0 \tag{3.27}$$

The sky temperature is the temperature to which a black body would have to be raised in order to emit the same flux as that of the atmosphere. It can be seen that the sky temperature is smaller or

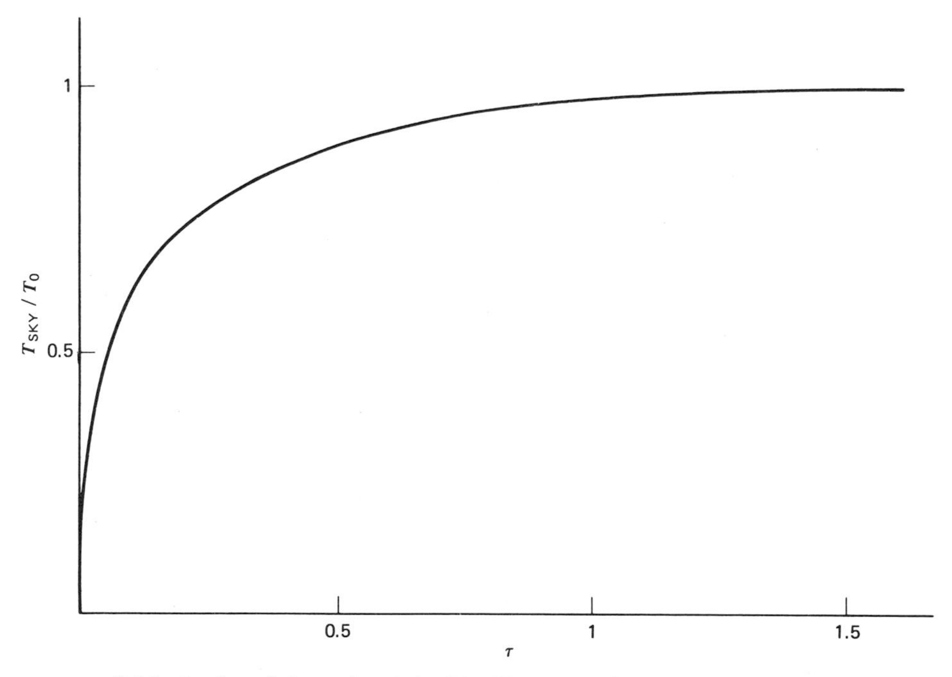

FIGURE 3.15 *A plot of the ratio of the T_{sky}/T_0 versus thermal optical thickness for an isothermal, plane-stratified, gray atmosphere. The curve is obtained from Equation 3.27.*

equal to the kinetic air temperature T_0. The more transparent the atmosphere is to thermal radiation, the lower the sky temperature. The sky temperature approaches the kinetic air temperature when the atmosphere becomes opaque. We have plotted T_{sky}/T_0 versus τ in Figure 3.15 using Equation 3.27. Under typical conditions, T_{sky} is generally less than 10 K below the ambient air temperature.

The steady-state temperature of a horizontal surface is determined, partly by the sky temperature. On a clear dry night when the sky temperature is low, the steady-state temperature of an exterior surface is radiatively cooled to a temperature below that of the ambient air. This radiative cooling is common at night in desertlike regions where clear and dry skies prevail. In solar energy technology the sky temperature plays an important role in the steady-state behavior of solar heating panels. We stress that terrestrial thermal radiation is not truly isotropic. In fact, for a vertical surface much of the intercepted thermal radiation comes from the surrounding terrain. Thus the sky temperature depends on both the terrain and the orientation of the surface as well as on the atmosphere itself.

Having completed our discussion of terrestrial insolation, we turn our attention to the way solar radiation can be harnessed by man.

PROBLEMS

3-1. (a) Show, using Equation 3.5, that the density profile of an atmosphere with a temperature profile $T = T_0 - \Gamma Z$ is given by

$$\rho = \frac{\rho_0}{(1-\beta z)} e^{[\ln(1-\beta z)/\beta H]}$$

where

$$\beta = \frac{\Gamma}{T_0} \qquad \text{and} \qquad H = \frac{RT_0}{\bar{M}g}$$

(b) Show that for an isothermal atmosphere ($\beta \to 0$), the preceding expression reduces to Equation 3.6.

(c) Using the expression of (a), find the densities at altitudes of 4, 8, and 10 km, provided $T_0 = 300$ K, $\rho_0 = 1.29$ kg/m^3, and $\Gamma = 6$ K/km. Compare these results with those obtained using Equation 3.6.

3-2. (a) Show that the density of an exponential atmosphere at an altitude $z = nH$ is $(0.37)^n$ of the sea level value.
(b) Show that if the entire mass of such an atmosphere were redistributed with a uniform density equal to its sea level value, it would extend only as high as its scale height H.
(c) Show that the fraction of the entire mass of an exponential atmosphere contained in the region between sea level and an altitude $z = nH$ is $(1 - e^{-n})$.

3-3. (a) A gray atmosphere has an optical thickness $\tau = 0.2$ for solar radiation. Find the direct solar flux reaching a horizontal collector at ground situated at a colatitude of $L' = 55°$ at two-hour intervals about solar noon at the equinoxes and the solstices.
(b) Plot the results from (a) and estimate the average daily flux at the equinoxes and the solstices.

3-4. Repeat Problem 3-3 for a collector with a southerly 45° tilt.

3-5. An atmosphere consists of constituents that absorb heavily in four infrared bands. Each band is 0.05 μm wide and is centered about the wavelengths 0.9, 1.1, 1.3, and 1.5 μm. Within the bands the optical thickness is 2; elsewhere the optical thickness is 0.1.
(a) Plot τ_λ versus λ.
(b) Plot the spectral transparency T_λ versus λ.
(c) Plot the direct spectral flux reaching ground at one air mass.
(d) Estimate the total direct flux at 1 air mass.

3-6. A gray atmosphere has a total optical thickness of $\tau = 0.3$ and an albedo for single scattering $\tilde{\omega}_0 = 0.55$. The underlying terrain has a reflectivity of $R = 0.2$.
(a) Find the direct flux on a horizontal surface for zenith angles of $Z = 0°$ and $Z = 70°$.
(b) Estimate the total flux for these angles.
(c) Estimate the diffuse flux for these zenith angles and compare the diffuse flux fractions.

3-7. A black and white pyranometer consists of two surfaces, one black ($a = \epsilon = 1$) and one white ($a = \epsilon = 0$). With the sky and air temperatures approximately equal, the system is exposed to a solar flux F.
(a) If convection and back losses are ignored and if the white surface remains at the air temperature T_a, show that the

temperature difference between the surfaces is

$$\Delta T = \left(\frac{F}{\sigma} + T_a^4\right)^{1/4} - T_a$$

(b) Plot ΔT versus F for $T_a = 273$, 283, and 293 K.

3-8. Assume that a photovoltaic of area 10 cm^2 has a constant spectral responsivity of 200 ma/W from 0.4 μm to 0.7 μm and zero otherwise.

(a) Find the response (i.e., the current) produced by a uniform spectral flux of 1000 W/m^2-μm for 0.5 $\mu\text{m} \leq \lambda \leq 0.8\ \mu\text{m}$ and zero otherwise.

(b) Estimate the response to the spectral flux of the solar constant.

3-9. (a) Show that the radiative transfer between a surface at a temperature, T, and the sky is approximately given by

$$F \simeq 4\sigma\epsilon T_{sky}^3(T - T_{sky})$$

provided

$$|(T - T_{sky})/T_{sky}| \ll 1$$

where ϵ is the thermal emissivity of the surface.

(b) Assume that a cooled horizontal surface gains heat by convection from the surrounding air at a rate

$$F = h(T_a - T)$$

where T is the temperature of the surface, T_a is the air temperature, and h is the heat transfer coefficient. Show that the radiative cooling at night for such a surface produces a temperature difference given approximately by

$$\Delta T = T_a - T = T_a - \frac{hT_a + 4\sigma\epsilon T_{sky}^4}{h + 4\sigma\epsilon T_{sky}^3}$$

provided $(T_a - T_{sky})/T_{sky} \ll 1$.

(c) If $h \simeq 2$ W/m^2-K, $\epsilon = 0.5$, $T_{sky} = 273$ K, and $T_a = 283$ K, find the temperature difference between the surface and the surrounding air.

3-10. Using Equation 3.27, find the sky temperature for a gray isothermal atmosphere whose (thermal) optical thickness is τ = 2 and whose air temperature is 10°C. [*Note.* $E_3(2) = 0.03$.]

REFERENCES

1. Chandrasekhar, S., *Radiative Transfer*, Dover, New York (1960), Chapter I.
2. Duffie, J. A. and W. A. Beckman, *Solar Engineering of Thermal Processes*, Wiley-Interscience, New York (1980), Chapter 2.
3. Fleagle, R. G. and J. A. Businger, *An Introduction to Atmospheric Physics*, Academic, New York (1963), Chapters 1, 2, and 4.
4. Goody, R. M. and J. C. Walker, *Atmospheres*, Prentice-Hall, Englewood Cliffs, N.J. (1972), Chapters 1–3.
5. Humphreys, W. J., *Physics of the Air*, Dover, New York (1964), Part I—Chapters I–VI, Part IV—Chapter VII.
6. Kondratyev, K. Y., *Radiation in the Atmosphere*, Academic, New York (1969), Chapters 1–5.
7. Lunde, P. J., *Solar Thermal Engineering*, Wiley, New York (1980), Chapter 3.
8. Meinel, A. B. and M. P. Meinel, *Applied Solar Energy*, Addison-Wesley, Reading, Mass. 1976, Chapters 2 and 3.

CHAPTER 4

Elements of Heat Transfer

Solar heating panels generate heat when solar radiation is absorbed by a blackened absorber plate. In steady-state operation this heat must be equal to the thermal energy leaving the panel. The energy leaves either as useful heat extracted by a transfer fluid or alternatively by thermal leakage to the surroundings. An efficient panel is one that enhances heat transfer to the fluid while minimizing heat transfer to the cooler surroundings. In this chapter we will survey the modes by which heat is transferred from one region to another with particular emphasis on thermal losses from flat plate collectors.

It is a law of nature (the second law of thermodynamics) that heat always tends to flow from hotter to colder regions. There are three modes by which this transfer can occur, namely, *conduction*, *convection*, and *radiation*. Conduction is the transfer of heat through matter in which energy but not mass is transferred. The highly agitated atoms of warmer regions transfer some of their energy to their less agitated neighbors in cooler regions through atomic interactions. However, the atoms themselves are essentially fixed and cannot migrate through the material. The conduction process can be described as the diffusion of thermal energy in matter without the flow of mass.

In contrast, convection is the transfer of heat through matter produced by the transport of mass. Consequently, convection occurs only in fluids, that is, in liquids and gases. When one region of a fluid is made hotter than another, pressure and density gradients result. These gradients generate convection cycles that carry warmer fluids to cooler regions and vice versa. When the source of heating is removed, the cycles continue to mix the fluid until a uniform temperature is achieved. Solids cannot convect heat; however, some liquids both conduct and convect heat well. For example, water and air are poor thermal conductors, whereas mercury, which is a metallic liquid, is both a good conductor and a convector of heat.

Radiative heat transfer is unique in that it does not require any matter to transfer heat from warmer to colder regions. In fact, a vacuum is the most efficient medium for radiative transfer. As shown in Chapter 1, radiative transfer is produced by traveling electromagnetic waves. Any medium that does not transmit these waves will not permit the transfer of heat by radiation. For example, opaque media or reflective surfaces represent barriers to the transmission of thermal radiation.

All three modes of heat transfer play an important role in the operation of solar heating panels. For example, the absorber plate of a flat plate collector loses thermal energy from its front face primarily by convection and radiation and from its back side primarily by conduction.

Conduction of Heat

Consider two surfaces of arbitrary size and shape maintained at different temperatures, T_2 and T_1. Suppose the region between them is filled with a uniform insulating material. Heat will flow from the warmer to the cooler surface through the insulation in a rather complicated pattern (Figure 4.1). The heat flow at any point within the insulator can be represented by the *heat flux* vector $\boldsymbol{J}$ whose direction indicates the direction of heat flow and whose magnitude gives the heat energy per unit time crossing unit area. Like radiative flux, the units of $\boldsymbol{J}$ are typically watts/m^2, cal/sec-cm^2, or Btu/hr-ft^2. The vector $\boldsymbol{J}$ is determined by the law of conduction[1]

$$\boldsymbol{J} = -K\frac{dT}{ds}\hat{\boldsymbol{n}} \qquad \text{(law of conduction)} \tag{4.1}$$

[1]The expression $dT/ds(\hat{\boldsymbol{n}})$ is called the *temperature gradient* and is written formally as ∇T. Heat is said to flow along the negative temperature gradient.

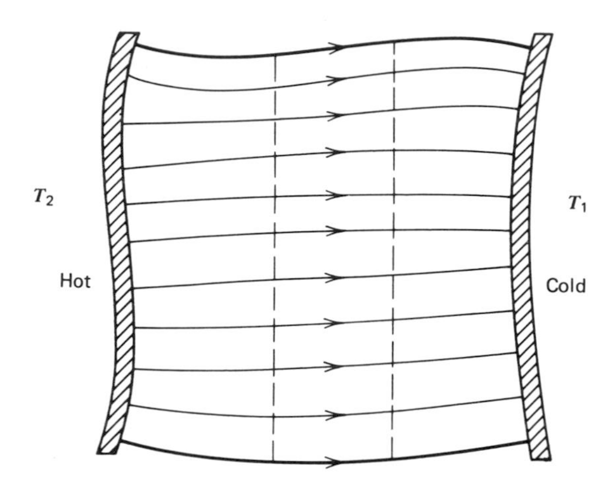

FIGURE 4.1 *The steady-state heat flow pattern through a conductive medium. Heat flows from the hotter to the colder surface. For an arbitrary shape the flow pattern is complex; however, the heat crossing the first surface is equal to that crossing the second (dashed lines).*

where $\hat{n}$ is a unit vector in the direction in which the temperature increases most rapidly, dT/ds is the rate of change of temperature with distance along that direction, and K is the *thermal conductivity* of the insulator. Values of K for various materials are listed in Table 4.1. The negative sign in Equation 4.1 suggests that heat flows in the direction in which the temperature decreases most rapidly with distance. Furthermore, the magnitude of the flow is proportional to the rate of decrease.

Consider two surfaces within the insulator as shown in Figure 4.1. At the first surface, heat is entering the enclosed region while at the second it is leaving. If more heat enters than leaves this region, temperatures at points within the interior will rise and vice versa. When the insulator is first inserted between the hot and cold surfaces, the temperatures at points within the insulator do indeed change with time. However, eventually a steady state is reached in which the temperatures become constant in time. Hence, *in the steady state, the net heat flux crossing any set of surfaces which completely enclose a region within an insulator must be zero.*

The objective in conduction problems, as shown in Figure 4.1, is to compute the rate at which heat is being transferred from the hotter to the colder surface. The more complex the shape of the insulator is, the more difficult the solution. We will consider two geometries for which the solution is straightforward, namely, the planar and cylindrical cases.

TABLE 4.1 *The Thermal Conductivities of Various Materials*

	K (W/m-°C)
Metals	
Aluminum	205
Copper	385
Lead	347
Mercury	8.37
Silver	406
Steel	50.2
Construction Materials (Typical values)	
Brick	0.147
Concrete	0.8
Cork	0.04
Felt	0.04
Glass	0.8
Rock wool	0.04
Wood	0.14
Gases	
Air	0.0239
Helium	0.142
Hydrogen	0.142

Planar Geometry—Conduction Through Slabs

Consider a slab of material of thickness d whose opposite faces at $x = 0$ and $x = d$ are maintained at T_2 and T_1 ($T_2 > T_1$), respectively. In this case the heat flows along parallel lines from $x = 0$ to $x = d$ (Figure 4.2). The conduction Equation (Equation 4.1) can be applied to this case to give the magnitude of the flux as

$$J = \frac{\dot{Q}}{A} = K\frac{dT}{dx} \tag{4.2}$$

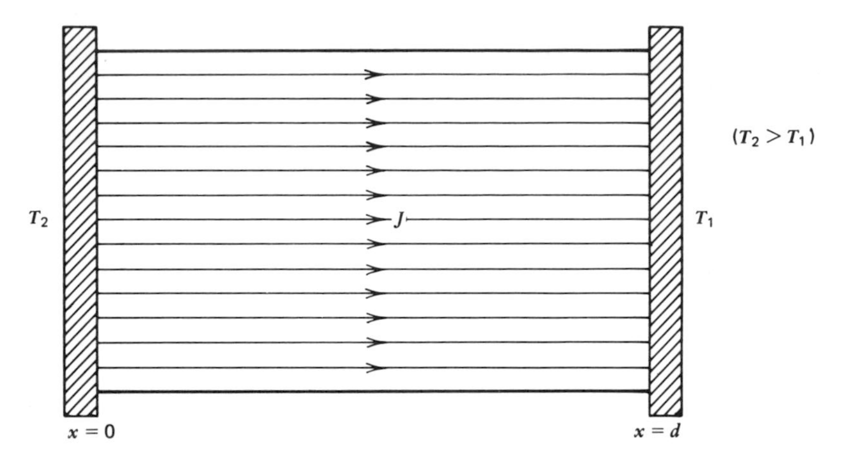

FIGURE 4.2 *The steady-state heat flow pattern through a homogeneous conductive medium. The opposite faces of the slab are at temperatures T_2 and T_1; the heat flow pattern consists of parallel lines.*

In the steady state the heat flow rate, $\dot{Q}$, crossing any plane within and parallel to the slab is constant. Furthermore, the areas of these planes are also constant so that Equation 4.2 can be integrated to give

$$\boxed{J = \frac{\dot{Q}}{A} = \frac{K}{d}(T_2 - T_1)} \qquad \text{(conduction through a slab)} \qquad (4.3)$$

EXAMPLE

A concrete wall ($K = 0.84$ W/m-°C) is 3 m × 4 m × 25 cm thick. The temperatures of the interior and exterior surfaces are observed to be at 11 and 7°C, respectively. Find the rate at which heat flows through the wall.

Using Equation 4.3, we have

$$\dot{Q} = \frac{(3 \times 4)(0.84)}{(0.25)}(11 - 7) = 161 \text{ W} = 39 \text{ cal/sec} = 557 \text{ Btu/hr}$$

It is often convenient to characterize the insulating quality of a slab by its U-value. This value is defined as

$$\boxed{U = \frac{K}{d}} \qquad (4.4)$$

Using the U-value, we can write Equations 4.3 as

$$\boxed{J = \frac{\dot{Q}}{A} = U(T_2 - T_1)} \qquad (4.5)$$

The lower the U-value is, the better, the insulation. It is left as an exercise to show that when a series of different slabs are used face to face, the overall U-value, $\bar{U}$, is given by

$$\boxed{\frac{1}{\bar{U}} = \frac{1}{U_1} + \frac{1}{U_2} + \cdots + \frac{1}{U_n}} \qquad (4.6)$$

where U_1, U_2, and U_n refer to the individual slabs.

EXAMPLE

An absorber plate in a flat plate collector is operating at a temperature of $T = 80°C$ and rests on 15 cm of fiberglass insulation ($K_f = 0.05$ W/m-°C), which is itself supported by a plywood ($K_w =$ 0.08 W/m-°C) sheet of thickness 1.25 cm. If the temperature of the exterior face of the plywood is observed to be at 25°C, find the heat flux rate conducted through the back face of the collector.

The U-values of the fiberglass and the wood are, respectively,

$$U_{\text{fib}} = \frac{0.05}{0.15} = 0.333 \text{ W/m}^2\text{-°C}$$

$$U_{\text{wood}} = \frac{0.08}{0.0125} = 6.4 \text{ W/m}^2\text{-°C}$$

Using Equations 4.5 and 4.6, we find

$$\frac{1}{\bar{U}} = \frac{1}{0.333} + \frac{1}{6.4}$$

or

$$\bar{U} = 0.317 \text{ W/m}^2\text{-°C}$$

and

$$\frac{\dot{Q}}{A} = (0.317)(80 - 25) = 17 \text{ W/m}^2$$

Heat Conduction Between Concentric Cylinders

Consider a pair of long concentric cylindrical surfaces of radii r_{in} and r_{out} whose temperatures are, respectively, T_{in} and T_{out}. The space between the cylinders is filled with a material whose thermal conductivity is K (Figure 4.3). If the inner surface is at a higher temperature than the outer one, heat will flow outward in a radial pattern. We can apply Equation 4.1 to this case as

$$\frac{\dot{Q}}{A} = -K\frac{dT}{dr} \tag{4.7}$$

where dr is a differential element of displacement in the radial direction. The surface area across which the heat flows increases with r according to $A = 2\pi rL$, where L is the length of the system. In the steady state the heat flow rate $\dot{Q}$ crossing these areas is constant and independent of r. Hence we can write Equation 4.7 as

$$\frac{\dot{Q}}{2\pi rL} = -K\frac{dT}{dr}$$

which upon integration gives

$$\boxed{\dot{Q} = \frac{2LK(T_{in} - T_{out})}{\ln \frac{r_{out}}{r_{in}}}} \quad \text{(conduction through cylindrical insulation)} \tag{4.8}$$

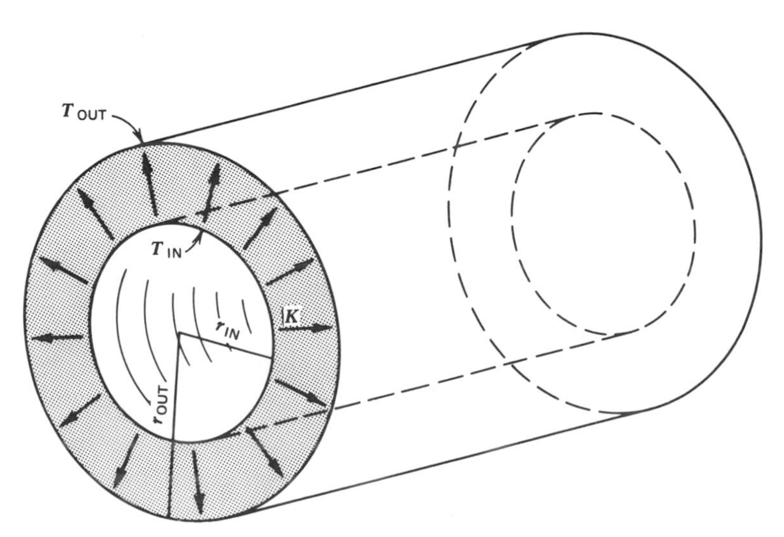

FIGURE 4.3 *The steady-state heat flow pattern through a homogeneous conductive medium between two concentric cylinders.*

This can be written as

$$\boxed{\frac{\dot{Q}}{L} = \dot{Q}_L = U_L(T_{\text{in}} - T_{\text{out}})} \tag{4.9}$$

where

$$\boxed{U_L = \frac{2\pi K}{\ln(r_{\text{out}}/r_{\text{in}})}}$$

is called the U-value *per unit length.* Again, it is left as an exercise to show that when N layers of concentric insulation are used around a pipe, the overall value $\bar{U}_L$ is given by

$$\boxed{\frac{1}{\bar{U}_L} = \frac{1}{U_{L_1}} + \frac{1}{U_{L_2}} + \cdots + \frac{1}{U_{L_N}}} \tag{4.10}$$

where U_{L_1}, U_{L_2}, and U_{L_N} refer to the individual insulators.

EXAMPLE

A plastic ($K = 0.8$ W/m^2-°C) pipe is carrying water at 60°C from a solar heating panel. The inner and outer radii are 0.5″ and 0.75″ respectively. The pipe is wrapped with cylindrical sponge ($K_S =$ 0.04 W/m-°C) insulation to a radius of 1.5″. If the outer surface of the insulation is at 25°C, find $\bar{U}_L$ for the pipe and determine the heat conduction rate per unit length to the surroundings.

The U_L values of the pipe and insulation are, respectively,

$$U_{L_1} = \frac{(2\pi)(0.8)}{\ln(3/2)} = 12.4 \text{ W/m-°C}$$

and

$$U_{L_2} = \frac{(2\pi)(0.04)}{\ln(2)} = 0.36 \text{ W/m-°C}$$

The overall value $\bar{U}_L$ is

$$\frac{1}{\bar{U}_L} = \frac{1}{12.4} + \frac{1}{0.36}$$

or

$$\bar{U}_L = 0.35 \text{ W/m-°C}$$

Thus the heat rate conducted per unit length is

$$\dot{Q}_L = (0.35)(60 - 25) = 12 \text{ W/m}$$

Convection

Convection plays a significant role in the transfer of heat from a solar panel to the surrounding air. It is important to distinguish between *natural* and *forced* convection. Natural convection occurs when the flow patterns are generated by instabilities produced by temperature gradients within the fluid. Forced convection is a mode of heat transfer that results from fluid flow produced by such external agents as pumps or blowers.

Convection is a more complex phenomenon than conduction because of its dependence on the many parameters of the fluid involved in the heat transfer process. Due to its complexity, a rigorous equation analogous to the conduction Equation 4.1 is very difficult to establish. We will attempt to obtain semiquantitative estimates for convective heat losses and establish those factors that affect convection coefficients.

Natural Convection from Heated Surfaces to Open Air

The simplest case of natural convection is that which occurs from a large flat surface to an unbounded fluid such as still air. This process is particularly important in the heat transfer from the warm glazing of a solar heating panel to the cooler surrounding air. The amount of heat transferred from a heated surface to still air depends on the orientation of the surface and on such parameters as the density, humidity, viscosity, specific heat, and thermal conductivity of the air. Thus it is understandable why semiquantitative relationships obtained from experimental data are often the only expressions available to describe natural convection. It is always possible to express the heat flux leaving a surface to the cooler air using the relation

$$J = \frac{\dot{Q}}{A} = h_\infty(T - T_a) \tag{4.11}$$

where T and T_a are the temperatures of the surface and the air, respectively, and h_∞ is called the open air *convection* or *heat transfer coefficient.* The subscript on h_∞ is used here to denote that the heat transfer is to an unbounded fluid. Note that Equation 4.11 does not suggest that the heat flux is strictly proportional to the temperature difference $(T - T_a)$; h_∞ itself depends on T and T_a, as well as on the other factors described. Based on experiments, it has been determined that the convection coefficient from a flat smooth surface to open air is approximately proportional to the fourth root of the temperature difference and can be estimated by (see Ref. 6, p. 90)

$$h_\infty \simeq C(T - T_a)^{1/4} = C\Delta T^{1/4} \tag{4.12}$$

where $C = C^{(\text{hor})} = 2.5\ \text{W/m}^2\text{-°C}^{5/4}$ and $C = C^{(\text{vert})} = 1.77\ \text{W/m}^2\text{-°C}^{5/4}$ are constants for horizontal and vertical surfaces, respectively.

The dependence of h_∞ on ΔT for horizontal and vertical surfaces has been plotted in Figure 4.4. Values of h_∞ for intermediate tilts can be obtained by interpolation. The coefficients $h_\infty^{(\text{vert})}$ and $h_\infty^{(\text{hor})}$ vary from 3 to 5 $\text{W/m}^2\text{-°C}$ and 4.5 to 7 $\text{W/m}^2\text{-°C}$, respectively, over the temperature range $\Delta T = 10$ to 60°C.

With winds, the open air convection coefficients may be many times larger than those given for still air. Even with 10 mph winds, the values of h_∞ may exceed 20 $\text{W/m}^2\text{-°C}$. Shielding collectors from winds will considerably reduce thermal losses and increase operating efficiency.

EXAMPLE

The air temperature in a room is 20°C when the outside air temperature is −10°C. Assuming the air to be still, find the flux lost through a 0.0034-m thick glass window ($K_g = 0.84$ W/m-°C). Also, estimate the temperature of the interior and exterior surfaces of the glass.

Because the glass is thin, we can in a first approximation neglect the temperature drop across the window. Furthermore, because the surface is vertical, symmetry suggests that the glass temperature T_g is equal to the average of the temperatures of the room and the outside air or

$$T_g = \frac{[20 + (-10)]}{2} = 5°\text{C}$$

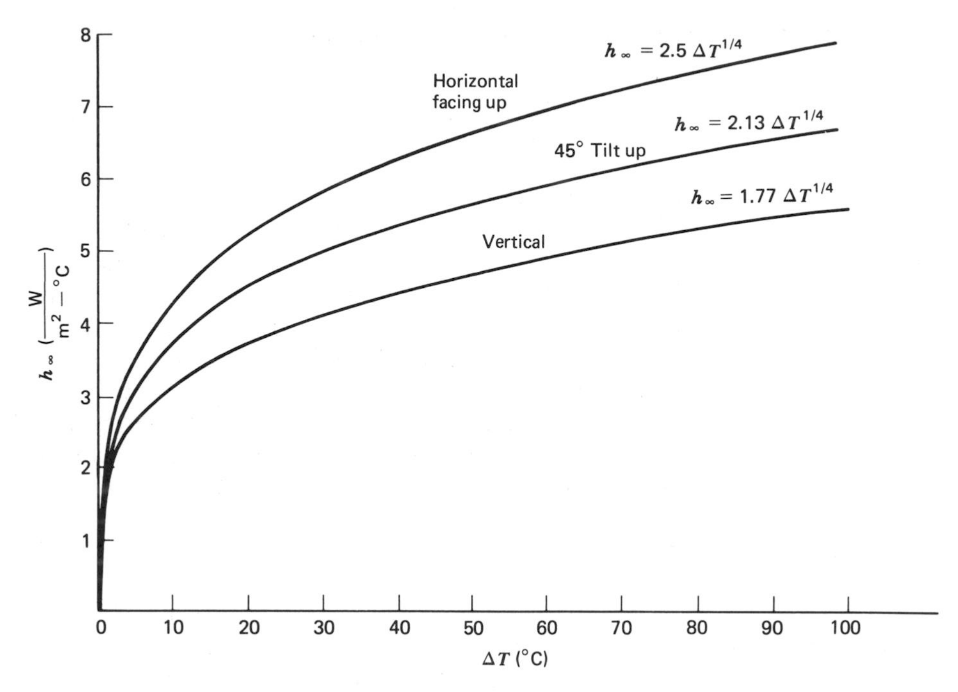

FIGURE 4.4 *The natural convection coefficients for heat transfer from large flat surfaces at various tilts to open still air. The coefficients increase as the temperature difference between the surfaces and the air increases.*

Using Equation 4.12 for a vertical surface, we find the interior and exterior convection coefficients to be equal and given by

$$h_\infty = (1.77)(20 - 5)^{1/4} \simeq 3.48\ \mathrm{W/m^2\text{-}°C}$$

From Equation 4.11 the convected flux is found to be

$$J = (3.48)(20 - 5) \simeq 52\ \mathrm{W/m^2}$$

In the steady state this flux must be equal to the conducted heat through the glass. Using Equation 4.3, we find

$$\Delta T_g = \frac{d}{K} J = \frac{(0.0034)(52)}{0.84} = 0.21°\mathrm{C}$$

Consequently, the inner and outer faces are at temperatures

$$T_g^{(\mathrm{int})} \simeq 5 - \frac{0.21}{2} = 4.9°\mathrm{C}$$

and

$$T_g^{(\text{ext})} \simeq 5 + \frac{0.21}{2} = 5.1°\text{C}$$

Note that the overall U-value for the window in the example can be written

$$\frac{1}{\bar{U}} = \frac{1}{h_\infty} + \frac{d}{K} + \frac{1}{h_\infty} = \frac{2}{h_\infty} + \frac{d}{K} = \frac{2}{3.48} + \frac{0.0034}{0.84}$$

The conduction term d/K is negligible compared with the convection terms so that

$$\bar{U} \simeq \frac{h_\infty}{2} = \frac{3.48}{2} = 1.74\ \text{W/m}^2\text{-°C}$$

The flux transfer is

$$J \simeq \bar{U}(T_{\text{ext}} - T_{\text{int}}) \simeq (1.74)[20 - (10)] = 52\ \text{W/m}^2$$

which agrees with the preceding result. The window's ability to reduce heat losses is not due to the insulating value of glass. Even if the glass were a perfect thermal conductor, the heat transfer would be minimally affected. The glass is effective because it limits convective loss by keeping the cold outside air from reaching the interior of the room.

EXAMPLE

Estimate the heat flux lost by the room in the previous example if winds increase the external convection coefficient to $h_\infty^{(\text{ext})} = 20\ \text{W/m}^2\text{-°C}$. Assume that $h_\infty^{(\text{int})}$ remains at $3.48\ \text{W/m}^2\text{-°C}$.

In the steady state the interior and exterior fluxes are equal and so we write

$$J^{(\text{int})} = J^{(\text{ext})}$$
$$3.48(20 - T_g) = 20[T_g - (-10)]$$

Solving for the temperature of the window, we find

$$T_g = -5.6°\text{C} \qquad \text{and} \qquad J = 3.48[20 - (-5.6)] = 89\ \text{W/m}^2$$

Thus as a result of the winds, the glass temperature, in the example, drops to −5.6°C and the heat flux increases to 89 W/m^2. Because the temperature of the glass is below the ice point, it is now possible for water vapor in the room to condense and freeze on the window.

Natural Convection Across Air Gaps Between Parallel Planes

When two plane surfaces separated by an air gap of thickness d are at different temperatures, heat will be transferred from the warmer to the cooler surface by convection across the gap. This situation occurs in solar heating panels where heat is transferred from the hot absorber plate to the cooler glazing and also when it is transferred from one glazing to another in multiglazing units.

The problem of convection between parallel planes is more complicated than that of heat transfer from a single plane to open air because a new variable has been introduced, namely, the width of the air gap, d. We can still express the convective heat transfer using an equation analogous to Equation 4.11

$$\boxed{J = h_d(T_2 - T_1)} \tag{4.13}$$

where T_2 and T_1 refer to the temperatures of the surfaces and h_d is the heat transfer coefficient for the air gap. This coefficient depends on factors such as the density (ρ), viscosity (η), thermal conductivity (K), volumetric coefficient of thermal expansion (β), and specific heat at constant pressure (C_p) of the air. All these parameters depend on the temperature of the air within the gap. The coefficient h_d also depends on the tilt, Δ, of the planes and on the width, d, of the air gap.

We can estimate the value of h_d by noting that when d is large, the convection involves transfer of heat from the warmer surface to bulk air and then from bulk air to the cooler surface. If the surfaces are vertical, the bulk air temperature would be equal to the average temperature of the two surfaces. The transfer coefficient would be

$$\frac{1}{h_d} \simeq \frac{1}{h_\infty} + \frac{1}{h_\infty}$$

or

$$h_d \simeq \tfrac{1}{2} h_\infty \qquad \text{(large } d\text{)} \tag{4.14}$$

Using this approximation, we observe that h_d should be in the range 2 to 4 W/m^2-°C and 3 to 5 W/m^2-°C for vertical and horizontal surfaces, respectively. A more detailed analysis follows.

To establish the dependence of h_d on the temperature and width of the gap, we introduce the following dimensionless quantities:

$$\begin{aligned} Nu &= h_d d/K && \text{(Nusselt number)} \\ Pr &= C_p\eta/\rho k && \text{(Prandtl number)} \\ Gr &= g\beta(\Delta T)d^3\rho^2/\eta^2 && \text{(Grashof number)} \end{aligned} \tag{4.15}$$

where g is the acceleration of gravity. These quantities are all pure numbers and have the same values whether expressed in cgs, mks, or English units. However, all these numbers are functions of the temperature of the fluid. If an empirical formula could be found to connect these numbers, an expression relating h_d (in Nu) to all the other parameters would result. In essence, we are looking for a relationship of the form

$$Nu = \Phi(Pr, Gr, \Delta) \tag{4.16}$$

where Δ is the tilt of the system. For gases in general and air in particular, the Prandtl number is approximately constant and independent of temperature so that it may be ignored in Equation 4.16. A simple power law for Equation 4.16 is

$$Nu = CGr^m \tag{4.17}$$

where C and m are parameters that depend on the tilt of the system and on the general range of the Grashof number. For most problems relating to heat transfer in solar flat plate collectors, the constants are in the range $0.02 < C < 0.2$ and $0.25 < m < 0.4$; the range of the Grashof number is $10^4 \leq Gr \leq 10^7$.

Using Equation 4.15, we can solve Equation 4.17 for h_d giving

$$h_d = CK\alpha^m \frac{\Delta T^m}{d^{(1-3m)}} \tag{4.18}$$

where

$$\alpha(T) = g\beta\rho^2/\eta^2$$

The air in the gap is assumed to be at atmospheric pressure. It is convenient to write Equation 4.18 as

$$\boxed{h_d = C'F(\bar{T})\frac{\Delta T^m}{d^{(1-3m)}}} \tag{4.19}$$

where C' and m depend on tilt and $F(\bar{T})$ is a function that depends on the mean Celsius temperature in the gap, $\bar{T} = (T_1 + T_2)/2$. The function $F(\bar{T})$ can be written (see Ref. 1, p. 80)

$$F(\bar{T}) = 1 - (0.0018)(\bar{T} - 10)$$

The width of the gap d in Equation 4.19 is expressed in centimeters. Duffie and Beckman (Ref. 1) following Tabor's results give the following constants for Equation 4.19.

Case I	(horizontal)	$C' = 1.613$	$m = 0.281$	$10^4 < Gr < 10^7$
Case II	(45° tilt)	$C' = 1.14$	$m = 0.310$	$10^4 < Gr < 10^7$
Case III	(vertical)	$C' = 0.82$	$m = 0.327$	$10^5 < Gr < 10^7$
	(vertical)	$C' = 0.57$	$m = 0.381$	$10^4 < Gr < 10^5$

Using Equation 4.19, we have plotted h_d versus ΔT for surfaces at various tilts and spacings and compared the results with h_∞ (Figure 4.5). Note that h_d is approximately half of the open air value and that it increases somewhat as d decreases.

EXAMPLE

Find the convective heat flux from an absorber at 80°C to a glazing at 20°C in a horizontal heating panel. The air gap is 10 cm wide.

Using the constants of Case I, we find from Equation (4.19)

$$h_d = [1.613F(\bar{T})]\frac{\Delta T^{0.281}}{d^{0.157}}$$

Since $\bar{T} = (80 + 20)/2 = 50$°C and $\Delta T = 80 - 20 = 60$°C, we have

$$h_d = (1.613)(0.928)\frac{(60)^{0.281}}{(10)^{0.157}} = 3.29 \text{ W/m}^2\text{-°C}$$

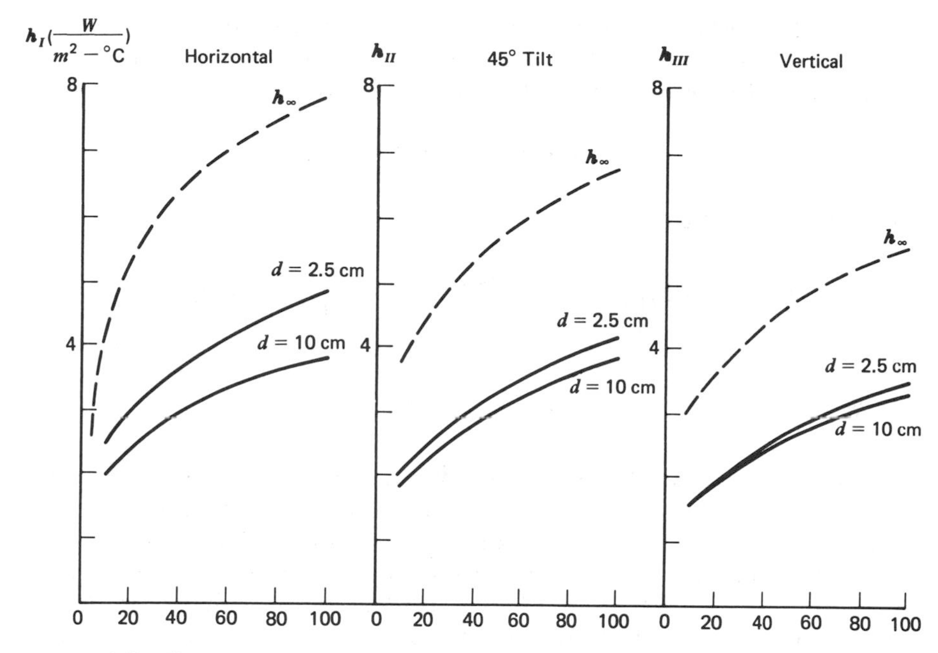

FIGURE 4.5 *The natural convection coefficients for heat transfer across an air gap of thickness d between two large flat surfaces. For the horizontal and 45° cases, it is assumed here that the lower surface is hotter than the upper one so that the heat flow is upward. Note that the heat transfer increases as the gap width is decreased. The dashed curves represent the open air coefficients and are shown for comparison.*

Finally, using Equation 4.13, we obtain

$$J = (3.29)(60) = 198 \text{ W/m}^2$$

The difficulty with convection problems in general is that the heat transfer equations are nonlinear in the temperatures. We illustrate this with the following example.

EXAMPLE

A single-glazed horizontal flat plate collector has its glazing 10 cm above the absorber. Assuming that radiative transfer can be ignored, find the convected flux from the absorber and determine the glazing temperature T_g when the plate and air temperatures are $T_p = 80°C$ and $T_a = 20°C$, respectively. Assume that there are no winds present.

In the steady state the flux transferred from the plate to the glazing must equal that from the glazing to the air. Applying Equations 4.11 and 4.13 to this case, we find

$$J = h_d(80 - T_g) = h_\infty(T_g - 20)$$

where according to Equations 4.19 and 4.12

$$h_d = (1.163)\left[1 - 0.0018\left(\frac{80 + T_g}{2} - 10\right)\right]\frac{(80 - T_g)^{0.281}}{10^{0.157}}$$

and

$$h_\infty = 2.5(T_g - 20)^{1/4}$$

Because these equations are nonlinear, they must be solved by numerical methods. We guess at a value of T_g and determine h_d and h_∞. Then we check to see if this value balances the heat transfer relation. Successive trials show the correct glazing temperature to be $T_g = 40.9°C$. This leads to values of $h_d = 2.86\ W/m^2\text{-}°C$ and $h_\infty = 5.35\ W/m^2\text{-}°C$. Substituting these values into either Equation 4.11 or 4.13, we obtain $J = 112\ W/m^2$.

If the glazing in the preceding example were removed, the flux transferred by natural convection to the open air would be

$$\begin{aligned} J &= h_\infty(T_p - T_a) \\ &= [2.5(T_p - T_a)^{1/4}](T_p - T_a) \\ &= 6.96(60) = 417\ W/m^2 \end{aligned}$$

Note that the convective heat transfer produced by the glazing is reduced and that the thermal conductivity of the glazing does not enter into the analysis. This shows that the glazing is effective because it keeps the cool outside air away from the hot absorber plate rather than because of its insulating value.

It must be stressed that the results in a convection problem depend on the forms assumed for the convection coefficients. A variety of expressions have been suggested to describe convective heat transfer. Those presented here, namely Equations 4.12 and 4.19, adequately describe the convection process for our purposes.

For a complete picture of the way in which solar panels lose heat to the surroundings, it is necessary to consider the transfer of heat by thermal radiation.

Radiative Heat Transfer from Heated Surfaces to the Sky

Whenever a surface is heated to some temperature T (in kelvins), it will generally emit radiation to the environment. Furthermore, it will also absorb some thermal radiation from the surroundings. The net radiative flux from the surface can be expressed as the difference between the flux emitted and the flux absorbed,

$$J = J_{\text{emit}} - J_{\text{abs}}$$

As presented in Chapter 1, the radiative flux emitted by a surface can be expressed as

$$J_{\text{emit}} = \epsilon \sigma T^4$$

where T is in kelvins and ϵ is the average thermal emissivity of the surface over the thermal range of interest. Since T is typically ~ 300 K, the thermal emissivity is represented by its average over the spectral region $2\ \mu\text{m} < \lambda < 20 \mu\text{m}$ and must not be confused with the average for the solar spectrum ($0.3\ \mu\text{m} < \lambda < 2\ \mu\text{m}$). The absorbed flux is

$$J_{\text{abs}} = aJ_{\text{inc}}$$

where a is the thermal absorptivity of the surface and J_{inc} is the thermal flux falling on the surface. Using Kirchhoff's relation, we set $\epsilon = a$, and find the net radiative flux from the surface to be

$$J = \epsilon[\sigma T^4 - J_{\text{inc}}] \tag{4.20}$$

The simplest application of Equation 4.20 is that for a heated surface exposed to the open sky. As shown in Chapter 3, the thermal flux incident on a surface from the sky depends on both the temperature as well as on the opacity of the atmosphere to thermal radiation. From Equation 3.26, the thermal flux from the sky can be expressed as $J_{\text{inc}} = \sigma T^4_{\text{sky}}$ where the sky temperature may be somewhat lower than the actual air temperature. Using this result in Equation 4.20, we find that the net transfer of radiation from a heated surface to the sky is

$$\boxed{J = \epsilon\sigma[T^4 - T^4_{\text{sky}}]} \tag{4.21}$$

A similar result is obtained when a small heated object at a temperature T is placed in a large enclosure. The incident flux on the object becomes $J_{inc} = \sigma T_{enc}^4$ so that the net flux transfer is

$$\boxed{J = \epsilon\sigma[T^4 - T_{enc}^4]} \tag{4.22}$$

EXAMPLE

A glazing whose emissivity is $\epsilon = 0.85$ is observed to be at T = 30°C (303 K). Compare the convective and radiative losses to the surroundings when $T_{sky} \simeq T_a = 10°C$ (283 K) and when winds produce a convective heat transfer coefficient of $h_\infty = 20$ W/m²-°C.

The convected flux is

$$J_1 = h_\infty(T - T_a) = 20(30 - 10) = 200 \text{ W/m}^2$$

The radiated flux is

$$J_2 = \epsilon\sigma(T^4 - T_{sky}^4) = (0.85)(5.67 \times 10^{-8})(303^4 - 283^4)$$
$$= 97 \text{ W/m}^2$$

Radiative Transfer Between Parallel Planes

The radiative transfer between two large, parallel, closely-spaced planes is somewhat more complicated than the preceding situation, because the flux incident on each surface may include multiple reflections. Consider two planes in which the one on the left is at a higher Kelvin temperature, T_2, than the one on the right, which is at T_1 (Figure 4.6). The air between the surfaces is assumed to be totally transparent to thermal radiation. In between the plates the steady-state thermal radiative flux can be characterized as right-or left-going, J_R or J_L. The net flux transferred from the hotter surface (on the left) is, from Equation 4.20,

$$J = \epsilon_2(\sigma T_2^4 - J_L) \tag{4.23}$$

We will assume that the surfaces can absorb or reflect but cannot transmit thermal radiation. Most glazings are virtually opaque to thermal radiation. The left-going flux incident on surface 2 consists of the directly emitted flux from surface 1 plus the flux emitted from

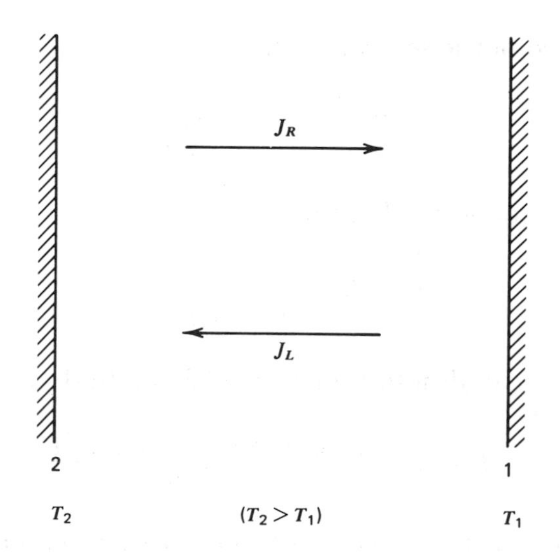

FIGURE 4.6 *The right- and left-going radiative fluxes between two large parallel planes. Since* $T_2 > T_1$, *the net flux must be toward the right so that* $J_R > J_L$.

and then successively reflected by each surface. Thus J_L can be written as

$$J_L = J_L' + J_L''$$

where

$$\begin{aligned} J_L' &= \epsilon_1\sigma T_1^4 + r_1r_2\epsilon_1\sigma T_1^4 + \cdots + (r_1r_2)^n\epsilon_1\sigma T_1^4 + \cdots \\ &= \epsilon_1\sigma T_1^4 \sum_{n=0}^{\infty} (r_1r_2)^n \end{aligned}$$

and

$$\begin{aligned} J_L'' &= \epsilon_2\sigma T_2^4 r_1 + \epsilon_2\sigma T_2^4 r_1(r_1r_2) + \cdots + \epsilon_2\sigma T_2^4 r_1(r_1r_2)^n + \cdots + \\ &= \epsilon_2 r_1\sigma T_2^4 \sum_{n=0}^{\infty} (r_1r_2)^n \end{aligned}$$

The coefficients r_1 and r_2 are the reflectivities of the surfaces. The total left-going flux is

$$\begin{aligned} J_L &= J_L' + J_L'' \\ &= (\epsilon_1\sigma T_1^4 + \epsilon_2 r_1\sigma T_2^4) \sum_{n=0}^{\infty} (r_1r_2)^n \end{aligned}$$

The infinite geometric series is summed as

$$\sum_{n=0}^{\infty} (r_1 r_2)^n = \frac{1}{1 - r_1 r_2}$$

so that the left-going flux becomes

$$J_L = \frac{(\epsilon_1 \sigma T_1^4 + \epsilon_2 r_1 \sigma T_2^4)}{(1 - r_1 r_2)}$$

Substituting this result into Equation 4.23, we find

$$J = \epsilon_2 \sigma T_2^4 - \frac{\epsilon_2(\epsilon_1 \sigma T_1^4 + \epsilon_2 r_1 \sigma T_2^4)}{1 - r_1 r_2}$$

Because the surfaces are assumed to be opaque to thermal radiation, we may write $r_1 = 1 - \epsilon_1$ and $r_2 = 1 - \epsilon_2$. After some simplification we find

$$J = \frac{\epsilon_1 \epsilon_2 \sigma (T_2^4 - T_1^4)}{\epsilon_1 + \epsilon_2 - \epsilon_1 \epsilon_2}$$

or

$$\boxed{J = \frac{\sigma(T_2^4 - T_1^4)}{\left(\frac{1}{\epsilon_1} + \frac{1}{\epsilon_2} - 1\right)} = \epsilon_{1-2}\sigma(T_2^4 - T_1^4)} \qquad (4.24)$$

where

$$\frac{1}{\epsilon_{1-2}} = \frac{1}{\epsilon_1} + \frac{1}{\epsilon_2} - 1$$

This important result shows that if one of the two surfaces is a perfect reflector of thermal radiation (i.e., $\epsilon_1 = 0$ or $\epsilon_2 = 0$), then no radiative transfer occurs between the surfaces. If both are perfect absorbers, the transfer rate is maximum and equal to

$$J = \sigma(T_2^4 - T_1^4)$$

EXAMPLE

Estimate the convective and radiative heat transfer from the absorber plate ($T_p = 80°C = 353$ K) to the glazing ($T_g = 20°C =$

293 K) in a horizontal flat plate collector. The glazing has $\epsilon_g = 0.8$ and is 10 cm above the plate, which has $\epsilon_p = 0.5$.

From a previous example we found that the convection coefficient for this case is $h_d = 3.29$ W/m²-°C. The convected flux is, from Equation 4.13,

$$J = h_d(T_p - T_g) = (3.29)(80 - 20) = 198 \text{ W/m}^2.$$

The radiative flux is, from Equation 4.24,

$$J = \frac{\sigma(T_p^4 - T_g^4)}{\left(\frac{1}{\epsilon_p} + \frac{1}{\epsilon_g} - 1\right)}$$

$$= \frac{(5.67 \times 10^{-8})(353^4 - 293^4)}{\left(\frac{1}{0.8} + \frac{1}{0.5} - 1\right)} = 206 \text{ W/m}^2$$

Radiative transfer Equations 4.21 and 4.24 can be put in a form similar to those for convection, Equations 4.5 and 4.11. Noting that

$$T_2^4 - T_1^4 = (T_2^2 + T_1^2)(T_2 + T_1)(T_2 - T_1)$$

we may write Equations 4.21 and 4.24 as

$$J = h_\infty^{(r)}(T - T_{sky}) \tag{4.25a}$$

and

$$J = h_d^{(r)}(T_2 - T_1)$$

where the radiation coefficients are

$$\boxed{h_\infty^{(r)} = \epsilon\sigma(T^2 + T_{sky}^2)(T + T_{sky})}$$

and

$$\boxed{h_d^{(r)} = \frac{\sigma(T_2^2 + T_1^2)(T_2 + T_1)}{\left(\frac{1}{\epsilon_2} + \frac{1}{\epsilon_1} - 1\right)}} \tag{4.25b}$$

If the temperature differences are not very large, the radiation coefficients can be approximated by

$$\boxed{h_\infty^{(r)}(\bar{T}) \simeq 4\epsilon\sigma\bar{T}^3} \tag{4.26a}$$

where

$$\bar{T} = \frac{T + T_{sky}}{2}$$

and by

$$\boxed{h_d^{(r)}(T) \simeq \frac{4\epsilon\sigma\bar{T}^3}{\left(\frac{1}{\epsilon_1} + \frac{1}{\epsilon_2} - 1\right)} = 4\epsilon_{1-2}\sigma\bar{T}^3} \tag{4.26b}$$

where

$$\bar{T} = \frac{T_1 + T_2}{2}$$

EXAMPLE

Find the approximate radiation coefficient for heat transfer from a black surface at 323 K to the sky when $T_{sky} = 283$ K. Also, find the approximate coefficient for radiative transfer between two black surfaces at 353 and 323 K, respectively.

From Equations 4.26a and 4.26b we have

$$h_\infty^{(r)} = 4(5.67 \times 10^{-8})\left(\frac{323 + 283}{2}\right)^3 = 6.3 \text{ W/m}^2\text{-°C}$$

and

$$h_d^{(r)} = 4(5.67 \times 10^{-8})\left(\frac{353 + 323}{2}\right)^3 = 8.76 \text{ W/m}^2\text{-°C}$$

Linearized Equations of Heat Transfer

The heat transfer equations for convection and radiation involve nonlinear functions of temperature, which makes the solution of heat transfer problems difficult. Consider the heat transfer from the absorber of a single-glazed flat plate to the surroundings. In the steady state the total transfer from the plate to the glazing must be equal to the transfer from the glazing to the surroundings. We may therefore write, using Equations 4.11, 4.13, and 4.25,

$$h_d(T_p - T_g) + h_d^{(r)}(T_p - T_g) = J \qquad \text{(from plate to glazing)}$$

and

$$h_\infty(T_g - T_a) + h_\infty^{(r)}(T_g - T_{sky}) = J \qquad \text{(from glazing to surroundings)}$$

(4.27)

where T_p, T_a, and T_{sky} are given, h_d and h_∞ are convection coefficients, and $h_d^{(r)}$ and $h_\infty^{(r)}$ are radiation coefficients. Because all four coefficients depend on temperature, the equations represent a pair of simultaneous *nonlinear* equations for the unknowns T_g and J. They can be solved by iteration, that is, by trying successive values of T_g until one is found that makes J the same in both equations. The equations can also be solved graphically by plotting J versus T_g for each equation and obtaining the point of intersection. In many applications it is sufficient to approximate Equation 4.27 by assuming that the coefficients are slowly varying functions of temperature. Consequently, coefficients may be replaced by constants, U's, provided that the temperature range is not too broad. If we also make the approximation that $T_{sky} \simeq T_a$, Equation 4.27 becomes

$$U_d^{(c)}(T_p - T_g) + U_d^{(r)}(T_p - T_g) = J$$

and (4.28)

$$U_\infty^{(c)}(T_g - T_a) + U_\infty^{(r)}(T_g - T_a) = J$$

where the heat transfer coefficients are now constants. The superscript identifies the coefficient as representing either convection or radiation, whereas the subscript indicates whether the transfer is between planes or to open air. Because the equations are linear, it is simple to eliminate T_g and to show that the heat transfer is given by

$$J = \bar{U}_c(T_p - T_a) \tag{4.29}$$

The overall coefficient for the collector is

$$\frac{1}{\bar{U}_c} = \frac{1}{\bar{U}_d} + \frac{1}{\bar{U}_\infty}$$

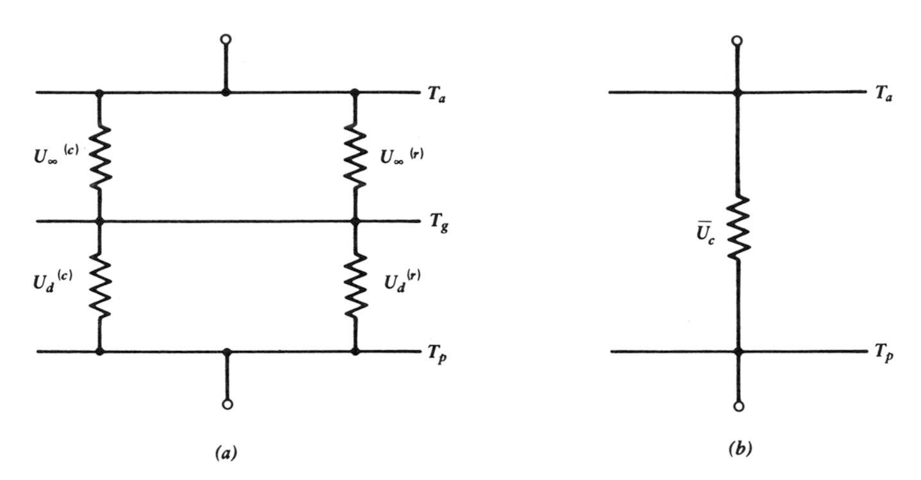

FIGURE 4.7 *Illustrating the resistor equivalent network for heat transfer from a single-glazed panel. Each resistor is labeled by its conductance rather than its resistance. (a) The four resistors representing radiative and convective transfer from the plate to the glazing and from the glazing to open air. (b) The single resistor representing heat losses from the panel.*

where

$$\bar{U}_d = U_d^{(c)} + U_d^{(r)}$$

and

$$\bar{U}_\infty = U_\infty^{(c)} + U_\infty^{(r)}$$

The procedure for finding the overall U-values from the individual U-values is similar to that for obtaining the net resistance of a linear network of resistors (Figure 4.7) and can be extended to any number of surfaces.

EXAMPLE

A flat plate solar heating panel contains two glazings. If $T_p = 80°C$ and $T_{sky} = T_a = 10°C$, find the overall transfer coefficient for the system, the temperature of each glazing, and the flux loss from the absorber. The coefficients for heat transfer from the plate to the inner glazings are $U_{d,1}^{(c)} = 2.5$ W/m^2-°C and $U_{d,1}^{(r)} = 5$ W/m^2-°C. Those for heat transfer from one glazing to the other are $U_{d,2}^{(c)} = 2.3$ W/m^2-°C and $U_{d,2}^{(r)} = 6$ W/m^2-°C. The coefficients for heat transfer from the outer glazing are $U_\infty^{(c)} = 8\ \mathrm{W/m^2\text{-}^\circ C}$ and $U_\infty^{(r)} = 7\ \mathrm{W/m^2\text{-}^\circ C}$.

The overall U-value is obtained from

$$\bar{U}_{d,1} = 2.5 + 5 = 7.5\ \mathrm{W/m^2\text{-}^\circ C}$$

$$\bar{U}_{d,2} = 2.3 + 6 = 8.3\ \mathrm{W/m^2\text{-}^\circ C}$$

and

$$\bar{U}_\infty = 7 + 8 = 15\ \mathrm{W/m^2\text{-}^\circ C}$$

as

$$\frac{1}{\bar{U}_c} = \frac{1}{7.5} + \frac{1}{8.3} + \frac{1}{15}$$

or

$$\bar{U}_c = 3.12\ \mathrm{W/m^2\text{-}^\circ C}$$

Thus

$$J = 3.12(80 - 10) = 218\ \mathrm{W/m^2}$$

The temperature of the inner glazing is found using

$$J = \bar{U}_{d,1}(T_p - T_{g,1})$$

which gives

$$T_{g,1} = T_p - \frac{J}{\bar{U}_{d,1}} = 80 - \frac{218}{7.5} = 51^\circ\mathrm{C}$$

The temperature of the outer glazing can be found using

$$J = \bar{U}_\infty(T_{g,2} - T_a)$$

which gives

$$T_{g,2} = \frac{J}{\bar{U}_\infty} + T_a = \frac{218}{15} + 10 = 25^\circ\mathrm{C}$$

Although these results are only approximate, they illustrate the relative roles that convection and conduction play in the heat transfer process of solar heating panels. Once the overall transfer coefficient is known, the heat transfer from the absorber to the surroundings can be computed using Equation 4.29. We will use the linearized forms Equations 4.28 and 4.29 extensively in our analysis of heat losses from solar panels.

Heat Exchangers

The transfer of heat from a hot bath to a fluid flowing through the bath is an important process in solar heating applications. For example, the hot bath could be the absorber plate of a solar heating panel and the fluid flowing past the plate would be the transfer fluid used to extract heat. The hot bath could also be the tank used to store solar heat.

The system to be described here is called a *single current heat exchanger.* To simplify matters, we will assume that the bath temperature, T_B, remains constant. The heat exchanger consists of a pipe that is immersed in the hot bath. The fluid enters the pipe at an inlet temperature $T_{f,i}$ and leaves at an outlet temperature $T_{f,e}$. The rate at which heat is extracted by the fluid is

$$\boxed{\dot{Q} = \dot{m}C(T_{f,e} - T_{f,i})} \tag{4.30}$$

where $\dot{m}$ is the mass flow rate and C is the specific heat of the fluid. If $\dot{m}$ and C are expressed in kg/sec and J/kg-°C, respectively, the extraction rate will be in watts.

To find the outlet temperature and extraction rate for a given inlet temperature and flow rate, we consider a differential element of the pipe (Figure 4.8). The heat transfer rate from the bath to the fluid within the element dl is

$$d\dot{Q} = U_L(T_B - T)dl$$

where U_L is the heat transfer coefficient per unit length of the pipe and T is the steady-state temperature of the element. Using Equation 4.30, we express the increase in temperature of the fluid as it traverses this element as

$$d\dot{Q} = \dot{m}C\,dT = U_L(T_B - T)dl$$

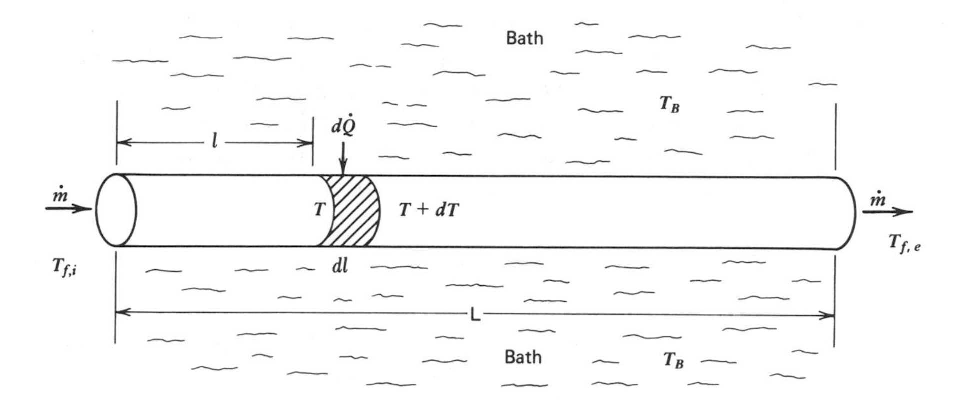

FIGURE 4.8 *Illustrating heat transfer from a hot bath to an element of fluid flowing in a pipe of a single current heat exchanger.*

or

$$\frac{dT}{T - T_B} = -\frac{U_L}{\dot{m}C} dl$$

If the bath temperature is the same everywhere along the pipe and if C and U_L can be approximated as also being constant along the pipe, the preceding relationship can be integrated to give

$$\ln \frac{T_B - T_{f,e}}{T_B - T_{f,i}} = -\frac{U_L L}{\dot{m}C}$$

or

$$\boxed{T_{f,e} = T_B - (T_B - T_{f,i})e^{-H'/\dot{m}C}} \tag{4.31}$$

where $H' = U_L L$ is a characteristic constant of the exchanger. Substituting this result into Equation 4.30, we find

$$\boxed{\dot{Q} = \dot{m}C(T_B - T_{f,i})(1 - e^{-H'/\dot{m}C})} \tag{4.32}$$

The relations given in Equations 4.31 and 4.32 are valid provided that U_L is constant along the pipe. In practice, U_L depends in detail on the diameter of the pipe and on the temperature, velocity, and type of fluid used in the pipe. For a fluid velocity below a critical value the flow will be *laminar*, which means that the fluid will slide in concentric cylindrical layers within the pipe. Heat transfer is not particularly efficient and U_L is small. When the fluid velocity exceeds the critical value, the fluid flow becomes *turbulent* and begins to churn, thereby increasing the value of U_L. The critical velocity is determined by the Reynolds number, which is

$$Re = \frac{vD\rho}{\eta} \tag{4.33}$$

where v, ρ, and η are the velocity, density, and viscosity of the fluid, respectively, and D is the diameter of the pipe. In long smooth pipes, laminar flow results when $Re \ll 2000$ and turbulent flow begins when $Re \gg 2000$. For a given fluid and pipe, Equation 4.33 can be solved for the critical velocity. While turbulence does increase heat transfer, it also increases the pipe's resistance to the flow of fluid. Details on the relationship between U_L and the fluid parameters can be found in texts on heat transfer. For simplicity we will assign a constant value to U_L for the exchanger.

EXAMPLE

Water ($C = 4186$ J/kg-°C) enters a long convoluted pipe at a temperature of 10°C. The walls of the pipe are maintained at 50°C by a heat bath. The pipe is 3 m long and carries water at a rate of 10 g/sec. The flow is laminar and the average heat transfer coefficient is $U_L = 6$ W/m-°C. Find the temperature of the water when it exits and determine the heat extraction rate.

Setting $H' = U_L L = (6)(3) = 18$ W/°C and $\dot{m} = 0.01$ kg/sec in Equation 4.31, we find

$$T_{f,e} = 50 - (50 - 10)\exp[-18/(0.01)(4186)] = 24°\text{C}$$

Using Equation 4.30, we have

$$\dot{Q} = (0.01)(4186)(24 - 10) = 586 \text{ W}$$

Repeating the preceding example for a reduced flow rate of 5 g/sec, we find $T_{f,e} = 33°$ and $\dot{Q} = 482$ W. Thus by reducing the rate at which fluid flows in the single current heat exchanger, one increases the output temperature; however, the heat collection rate is decreased. This phenomenon, as we will see see in Chapter 6, plays an important role in

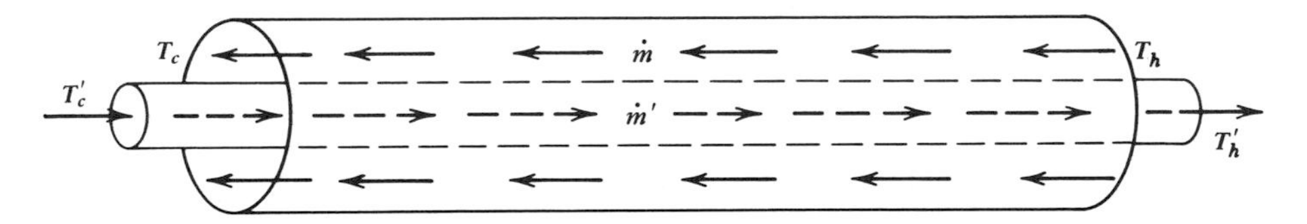

FIGURE 4.9 *Illustrating a section of a countercurrent heat exchanger. Heat is transferred from a hotter current to a cooler countercurrent through the wall of the inner pipe.*

the operation of solar heating panels. As we lower the flow rate of the transfer fluid through a heating panel, we increase the output temperature but decrease the heat collection rate.

It is often desirable to transfer heat from a hot fluid to a cooler one by allowing the two fluids to flow past each other. This is usually done by allowing the fluids to flow in opposite directions in the separate regions of concentric pipes (Figure 4.9). This device is known as a *countercurrent heat exchanger*. It is left as an exercise to show that the heat transfer rate for a pipe of length L is given by

$$\boxed{\dot{Q} = H'\left\{\frac{(T_h - T_h') - (T_c - T_c')}{\ln[(T_h - T_h')/(T_c - T_c')]}\right\} = H'\Lambda} \qquad (4.34)$$

where $H' = U_L L$. Here U_L is the effective heat transfer coefficient per unit length, and T_h, T_h', T_c, and T_c' refer respectively to the hot and cold temperatures of the fluids entering and leaving the exchanger. The quantity in brackets Λ is called the *log mean* temperature difference in the exchanger.

EXAMPLE

Heated fluid ($C = 4500$ J/kg-°C) leaves a solar panel and transfers its heat through a countercurrent heat exchanger to a stream of water. The fluid is cooled from 60 to 20°C while the water is heated from 10 to 40°C. If $U_L = 10$ W/m-°C and $L = 3$ m, find the log mean temperature difference in the exchanger, the heat transfer rate, and the flow rates of each fluid in the exchanger.

The log mean temperature difference is

$$\Lambda = \frac{(60 - 40) - (20 - 10)}{\ln\dfrac{(60 - 40)}{(20 - 10)}} = 14.4°\text{C}$$

so that

$$\dot{Q} = (10)(3)(14.4) = 432 \text{ W}$$

The mass flow rates follow from Equation 4.30 as

$$\dot{m} = \frac{432}{(4500)(60-20)} = 0.0024 \text{ kg/sec}$$

$$= 144 \text{ g/min} \qquad \text{(collector fluid)}$$

$$\dot{m}' = \frac{432}{(4186)(40-10)} = 0.00344 \text{ kg/sec} = 206 \text{ g/min} \qquad \text{(water)}$$

The elements of heat transfer just presented provide us with the means of understanding the thermal behavior of solar heating panels. In order to explain fully the operating principles of a solar collector, we need to understand the optical principles behind solar heating. All the incident radiation is not actually absorbed by the plate—some is absorbed by the glazing and some is reflected back to the surroundings. In the next chapter we will deal with the optics of solar panels as they apply to both flat plates and concentrators.

PROBLEMS

4-1. (a) Find the heat flux conducted through a concrete wall ($K_c = 0.84$ W/m-°C) 25 cm thick when the temperatures of the interior and exterior surfaces are observed to be at $T = 10$ and $T = 5$°C, respectively.
(b) What is the U-value for the concrete?

4-2. Fiberglass ($K_f = 0.05$ W/m-°C) 18 cm thick is added to the inside of the concrete wall in Problem 4-1.
(a) Find the U-value of the fiberglass insulation.
(b) Determine the overall value, $\bar{U}$, for the concrete-fiberglass combination.
(c) When the fiberglass is added, the interior and exterior surface temperatures become 15 and 0°C, respectively. Find the flux transfer through the wall.
(d) Find the temperature at the fiberglass-concrete interface.

4-3. A sealed container made of a certain material has a surface area of $0.5\ m^2$ and a thickness of 0.03 m. It contains 300 g of crushed ice at 0°C when it is immersed in boiling water at 100°C. It is found that it takes 30 min for all the ice to melt. Find the thermal conductivity of the material. (The latent heat of melting for ice is 333 J/g.)

4-4. A 2 m section of a plastic pipe ($K_p = 0.7$ W/m-°C) has inner and outer diameters of 1.5 and 2 cm, respectively. It is holding water at 80°C. If the outer surface of the pipe is observed to be at 76°C, find the heat transfer rate from the water to the surroundings.

4-5. The pipe described in Problem 4-4 is wrapped with fiberglass ($K_f = 0.05$ W/m-°C) to an outside diameter of 5 cm.

(a) Find the overall value of $\bar{U}_L$ for the pipe and insulation.
(b) The heat flow rate is now observed to be only 8 W. What is the temperature of the outer surface of the fiberglass?

4-6. The surface temperature of an outside wall is 10°C when the air temperature is −10°C.

(a) If the outside air is still, find the open air convection coefficient h_∞. [Use Equation 4.12].
(b) Find the heat flux from the wall to the surrounding air assuming there is no radiative transfer.
(c) If the wall is made of concrete ($K_c = 0.8$ W/m-°C) 9 cm thick, find the temperature of the interior surface.

4-7. An exterior surface of a wall is at 15°C when the outside air is at −5°C.

(a) Find the heat transfer coefficient h_∞ for convection to still air and determine the convective flux transfer. [Use Equation 4.12]
(b) When winds are blowing, the convection coefficient increases to 10 W/m^2-°C. Find the convective flux transfer if the wall remains at 15°C.
(c) Find the equivalent still air temperature that would result in the same heat transfer.

4-8. A window has interior and exterior convection coefficients of 4 and 10 W/m^2-°C, respectively.

(a) Find the overall convection coefficient for the window. (Neglect conductive and radiative transfer.)
(b) Find the convective flux transfer if the inside and outside air temperatures are 20 and 0°C, respectively.
(c) Find the temperature of the window.
(d) A storm window is placed outside the window; the con-

vection coefficient between the two windows is 2 W/m^2-°C. Assuming that the interior and exterior convection coefficients remain the same, find the overall transfer coefficient of the system and the heat flux.
(e) Find the temperatures of the inner and outer windows.

4-9. A horizontal surface whose thermal emissivity is $\epsilon = 1$ is placed outdoors on a clear night when the air and sky temperatures are $T_a = 2$°C and $T_{sky} = -4$°C, respectively. If no winds are blowing, find the steady-state temperature of the surface. (Neglect heat transfer from the underside of the surface.) [Use Equation 4.12]

4-10. A double-glazed window is used in a home to minimize heat losses. The space between the windows is evacuated. It is observed that the inner and outer panes are at 15 and 0°C, respectively. If the emissivities of the surfaces of the panes are each $\epsilon = 0.5$, find the heat flux through the window. (Assume that the windows are opaque to thermal radiation.)

4-11. A solar panel is double glazed. The coefficients for heat transfer between the absorber and the inner glazing can be approximated as $U_1^{(c)} = 3$ W/m^2-°C and $U_1^{(r)} = 5$ W/m^2-°C, whereas the coefficients across the gap of the double glazing are $U_2^{(c)} =$ 2 W/m^2-°C and $U_2^{(r)} = 4$ W/m^2-°C. The coefficients for heat transfer from the outer glazing to the surroundings are approximated as $U_\infty^{(c)} = 8$ W/m^2-°C and $U_\infty^{(r)} = 4.5$ W/m^2-°C.
(a) Draw a resistor network that represents the system and find $\bar{U}_c$.
(b) If the absorber and ambient temperatures are $T_p = 70$°C and $T_a = T_{sky} = 10$°C, respectively, find the heat flux from the absorber and the temperature of each glazing. Determine how much flux is transferred by convection and by radiation in each region.
(c) Repeat (b) if the air between the glazings is removed, that is, $U_2^{(c)} = 0$.

4-12. A large solar storage tank contains toxic fluid at 70°C. A coiled copper pipe is inserted in the tank and used to warm a stream of water from 10 to 50°C. The tube is 5 m long and has a heat transfer coefficient per unit length of $U_L = 12$ W/m-°C.
(a) What flow rate is being used to warm the water to the desired value?
(b) What is the heat extraction rate from the tank?
(c) If the flow rate is doubled, what will the exit temperature of

the water and the heat extraction rate be? (Assume that the tank temperature remains constant.)

4-13. Derive Equation 4.34 for the countercurrent heat exchanger.

4-14. Heat is to be transferred from an ethyl glycol solution to a countercurrent of water. The water current is to be raised from 10 to 40°C at a flow rate of 5 gal/hr. The ethyl glycol solution ($C = 4000$ J/kg-°C) is heated by a solar heating panel to a temperature of 60°C. If the heat exchanger has $U_L L = 50$ W/°C, find the flow rate of the ethyl glycol solution, its temperature when it returns to the solar panel, and the heat transfer rate.

REFERENCES

1. Duffie, J. A. and W. A. Beckman, *Solar Energy Thermal Processes*, Wiley-Interscience, New York (1974), Chapter 4.
2. Holman, J. P., *Heat Transfer*, McGraw-Hill, New York (1976), Chapters 2, 3, 5–8, 10.
3. Kreith, F., *Principles of Heat Transfer*, 3rd ed., Intext, New York (1973), Chapters 6–7, 11.
4. Lunde, P. J., *Solar Thermal Engineering*, Wiley, New York (1980), Chapters 2 and 5.
5. Meinel, A. B. and M. P. Meinel, *Applied Solar Energy*, Addison-Wesley, Reading, Mass (1976), Chapters 10 and 11.
6. Zemansky, M. W., *Heat and Thermodynamics*, 4th ed., McGraw-Hill, New York (1957), Chapter 5.

CHAPTER 5

The Optics of Collectors

The performance of a solar heating panel depends to a great extent on the optical properties of its components. These include glazings and absorbers in the case of flat plates and mirrors and lenses used in concentrators. Optical properties of materials generally depend on the wavelength of the radiation involved. We will be concerned with the optical behavior of systems in two distinct regions of the electromagnetic spectrum. The solar or short wavelength region consists of wavelengths in the interval $0.2\ \mu m \leq \lambda \leq 2\ \mu m$. The thermal or long wavelength region includes those wavelengths in the range $2\ \mu m \leq \lambda \leq 20\ \mu m$. Most solar glazings that are nearly transparent to solar energy are virtually opaque to thermal radiation. On the other hand, certain absorber coatings that are excellent absorbers of solar radiation are actually poor emitters of thermal radiation.

A solar heating panel develops heat only when solar energy is actually absorbed by the blackened absorber plate. Because some radiant energy is reflected and absorbed by the glazing system, and because some is reflected by the plate itself, the actual heat produced by the panel is somewhat less than the intercepted radiation. Although glazings reduce the energy reaching the absorber, they are necessary because they reduce the heat losses to the environment substantially. The decrease in optical performance is more than offset by the increase in thermal performance, especially in cold environments.

In a focusing collector, a lens, a mirror, or a combination is used to concentrate the sun's rays on to a small absorbing surface. The smaller the area is, the smaller the thermal losses from the surface. Consequently, concentrators can produce higher temperatures with lower heat losses than can be achieved with flat plates. In this chapter we consider the optical elements of solar heating panels, beginning with an analysis of the optical properties of glazings and absorbers in flat plates.

Reflection and Refraction at Dielectric Interfaces

When electromagnetic radiation passes from one medium to another, some reflection takes place at the interface. For example, if a beam of light travels from air into glass or from glass into air, some light will be reflected back into the incident medium at the surface of the glass. Thus even if a solar glazing did not absorb any solar radiation, some energy would be lost by reflection.

It is convenient to define the overall transmittance and reflectance of a sheet of glazing by

$$\boxed{T = \frac{F_{\text{trans}}}{F_{\text{inc}}} \qquad \text{and} \qquad R = \frac{F_{\text{refl}}}{F_{\text{inc}}}} \tag{5.1}$$

respectively, where F_{inc} is the flux incident on the glazing and F_{trans} and F_{refl} are the transmitted and reflected fluxes, respectively. To simplify the analysis, we will assume that the fluxes in Equation 5.1 are all beamlike. Although the solar flux contains some diffuse radiation, the results to be developed here are nevertheless sufficiently accurate for our purposes. The transmittance of a glazing is less than unity because of both reflection at the surface and bulk extinction in the interior of the glazing.

The glazing material is characterized by two parameters, the bulk attenuation or *extinction* coefficient, k_λ, and the *index of refraction*, n_λ. The latter is defined by

$$n_\lambda = \frac{c}{v} \tag{5.2}$$

where c is the speed of light in a vacuum and v is the speed of light within the glazing. We will assume for all practical purposes that air behaves as a vacuum. Because light travels more slowly in glass than in air, n_λ is necessarily greater than unity. All materials exhibit some degree of *dispersion,* which means that n_λ varies with wavelength to some extent. Equivalently, different spectral components travel with different speeds within the glazing. For ordinary crown glass the index of refraction is large for ultraviolet wavelengths but falls to approximately 1.5 in the visible and near infrared regions of the solar spectrum (Figure 5.1). To simplify the analysis, we will use spectrally averaged values and drop the subscripts on both k and n.

The angle between the incident ray and the normal to the interface is called the angle of incidence, θ_i (Figure 5.2). The plane determined by the incident ray and the normal is called the *plane of incidence* (POI). The angle between the reflected rays and the normal is called the angle of reflection θ_{rl}. According to the law of reflection, we have

$$\boxed{\theta_i = \theta_{rl}} \quad \text{(law of reflection)} \tag{5.3}$$

The rays that enter the glazing are called refracted rays. The angle

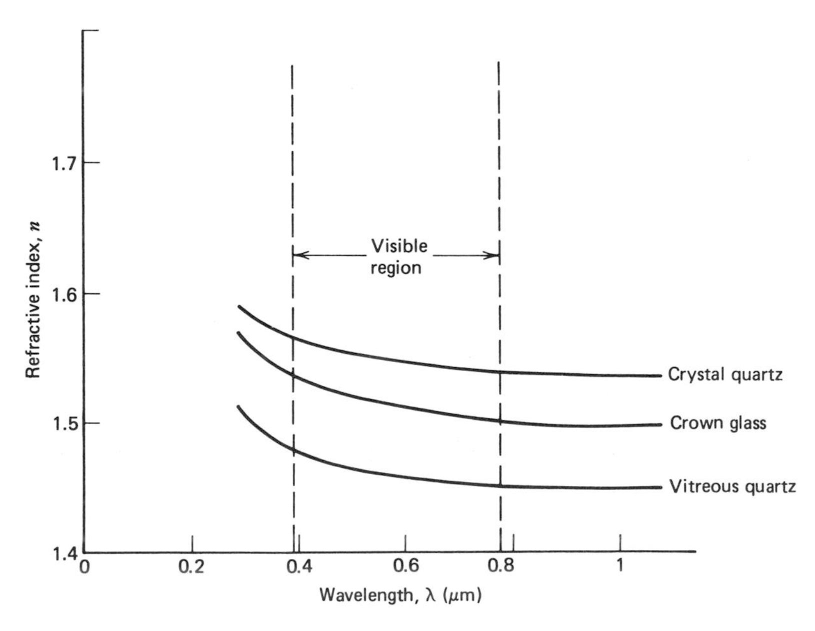

FIGURE 5.1 *The variation of the index of refraction with wavelength for different glasses.*

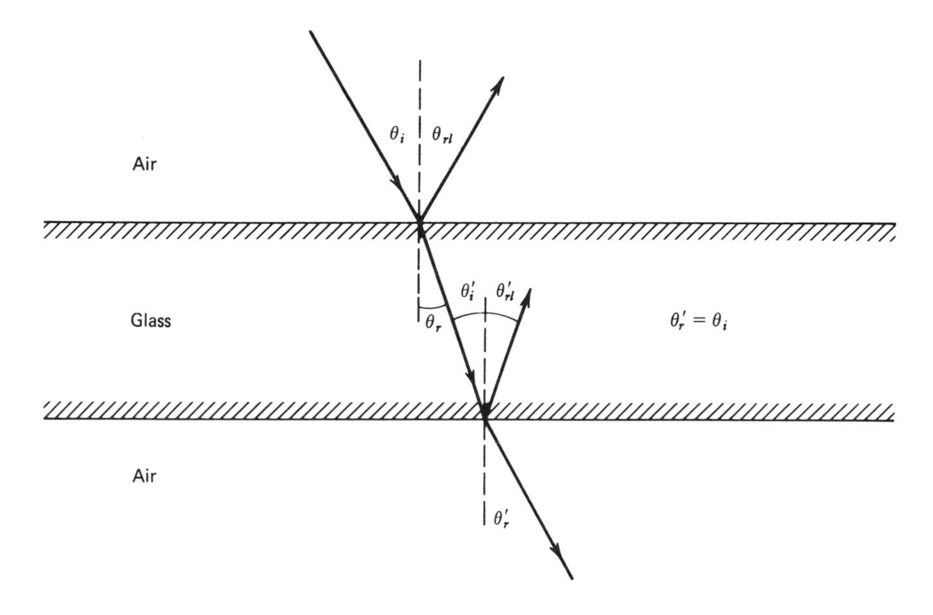

FIGURE 5.2 *The reflection and refraction of rays at the upper and lower surfaces of a sheet of glass. Note that the incident and emergent beams are parallel.*

between the refracted rays and the normal is called the angle of refraction (Figure 5.2). The direction of propagation of the refracted rays within the glazing is given by the law of refraction (also called *Snell's law*)

$$\boxed{n_{\mathrm{i}} \sin \theta_{\mathrm{i}} = n_{\mathrm{r}} \sin \theta_{\mathrm{r}}} \qquad \text{(law of refraction)} \tag{5.4}$$

where i and r refer to the incident and refracting media, respectively. For light traveling from air to glass, we set $n_{\mathrm{i}} = 1$ and $n_{\mathrm{r}} = n_{\mathrm{g}}$. When rays enter glass from air, they bend toward the normal. If they enter air from glass, the bend away from the normal. When a beam of light passes through a sheet of glazing, it is refracted at the upper face and then refracted again at the lower face so that the emergent beam is parallel to the incident beam (Figure 5.2).

The laws of reflection and refraction give the directions of reflected and refracted rays, respectively. The amount of flux reflected and transmitted across the glass–air interface depends on the index of refraction of the glazing, n_{g}, the angle of incidence, θ_{i}, and the state of polarization of the incident beam. Because electromagnetic radiation is a transverse wave, it can be described in terms of its polarization state. The wave vibration can be resolved into two components, one parallel and one perpendicular to the POI. The coefficients for reflection at the interface of a dielectric material for the two polarization states are given by the *Fresnel equations*:

$$r_{\parallel} = \left[\frac{n_r^2 \cos\theta_i - n_i\sqrt{n_r^2 - n_i^2 \sin^2\theta_i}}{n_r^2 \cos\theta_i + n_i\sqrt{n_r^2 - n_i^2 \sin^2\theta_i}}\right]^2 \tag{5.5a}$$

and

$$r_{\perp} = \left[\frac{n_i \cos\theta_i - \sqrt{n_r^2 - n_i^2 \sin^2\theta_i}}{n_i \cos\theta_i + \sqrt{n_r^2 - n_i^2 \sin^2\theta_i}}\right]^2 \tag{5.5b}$$

The subscripts $_{\parallel}$ and $_{\perp}$ refer to polarization states parallel and perpendicular to the POI, respectively. These coefficients refer to surface reflection only. As we will presently see, the actual reflectance of a sheet of glazing is determined by computing the total flux resulting from successive multiple reflections from the upper and lower surfaces. At normal incidence ($\theta_i = 0$) there is no longer any distinction between the two polarization states and it can be shown that both Equations 5.5a and 5.5b reduce to

$$r_{\parallel} = r_{\perp} = \left(\frac{n_r - n_i}{n_r + n_i}\right)^2 \quad \text{(normal incidence)} \tag{5.6}$$

If we assume that direct solar radiation is approximately unpolarized (or randomly polarized) so that 50 percent of the energy arrives in each polarization state, the average reflection coefficient at the interface can be taken to be

$$\bar{r} = \tfrac{1}{2}(r_{\perp} + r_{\parallel}) \tag{5.7}$$

Figure 5.3 is a plot of Equations 5.5 and 5.7 for glass ($n_g = 1.5$), from which we see that $\bar{r}$ is smallest and approximately 0.04 at normal incidence. It increases steadily and approaches unity (total reflection) at glancing angles. Note that at $\theta_i = 56°$, $r_{\parallel}$ vanishes so that the reflected solar beam is polarized perpendicular to the POI. This angle is called the *polarization* or *Brewster* angle for the glass. For solar glazings in general, the average surface reflection coefficient $\bar{r}$ is usually below 10 percent for angles of incidence less than 60°. It increases dramatically as this angle increases from 60 to 90°.

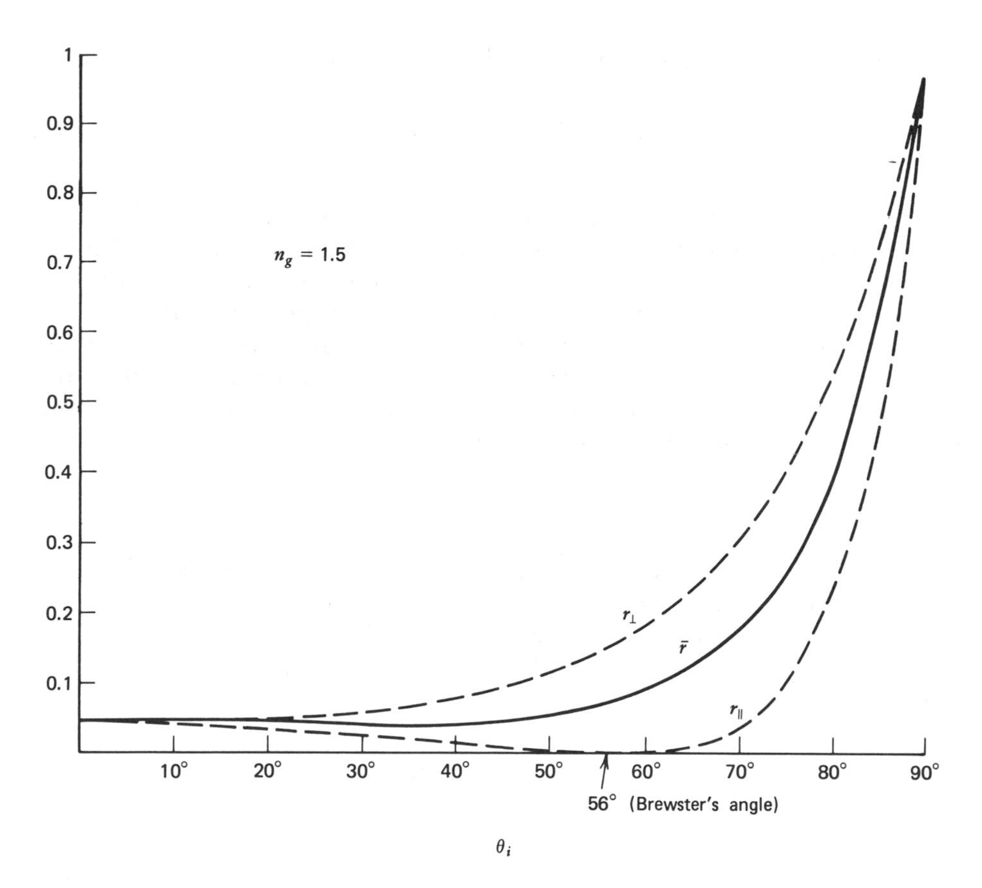

FIGURE 5.3 *The variation of the surface reflectivity with the angle of incidence for an air–glass interface. The index of refraction of the glass is $n_g = 1.5$. Note that at the Brewster angle for this glass ($\theta_B = 56°$) no light polarized parallel to the plane of incidence is reflected.*

The interface transmission coefficients can be written $t_\perp = 1 - r_\perp$ and $t_\parallel = 1 - r_\parallel$. These parameters determine the flux fractions entering the refracting medium.

Bulk Extinction

The radiation that enters the glazing travels obliquely through the sheet, making an angle of θ_r with the normal. It is attenuated primarily by absorption within the glazing. If the sheet thickness is s, then the flux fraction reaching the lower interface is given by

$$\boxed{\alpha = \exp(-ks/\cos\theta_r) = \exp(-nks/\sqrt{n^2 - \sin^2\theta_i})} \tag{5.8}$$

where k and n are the extinction coefficient and index of refraction of the glazing, respectively. In obtaining Equation 5.8, we used Snell's law, Equation 5.4. The quantity α is the bulk transmissivity of the glazing for a single transit of the beam. The bulk extinction of the beam is often produced by impurities in the glazing material. It is smallest at normal incidence and increases as θ_i approaches 90°. The effect is similar to attenuation in the atmosphere. A typical extinction coefficient for a solar glazing is $k \simeq 0.1\ \text{cm}^{-1}$.

EXAMPLE

Find the bulk transmissivity of a glazing 0.3 cm thick whose extinction coefficient is $k = 0.09\ \text{cm}^{-1}$ for rays incident at 60°. The index of refraction of the glazing is $n = 1.53$.

Using Equation 5.4, we find the angle of refraction within the glazing to be

$$\sin\theta_r = \frac{\sin 60°}{1.53} = 0.566$$

or

$$\theta_r = 34°$$

Substituting into Equation 5.8, we find

$$\alpha = \exp[-(0.09)(0.3)/\cos 34°] = 0.97$$

Consequently, 97 percent of the flux passing through the upper surface reaches the lower surface after a single transit.

It is interesting to compute the net flux transmitted when the beam emerges after a single transit. The incident beam is partially transmitted across the upper interface, attenuated as it crosses the sheet, and then partially transmitted across the lower interface as it finally emerges after the single transit. Careful inspection of Equations 5.4 and 5.5 shows that the interface reflection and transmission coefficients at the upper and lower faces of the glazings are equal.

Thus after a single transit through the sheet, the fraction transmitted is given by

$$\left.\frac{F_{\text{trans}}}{F_{\text{inc}}}\right] = T^{(0)}_{\text{single transit}} = t\alpha t = \alpha t^2 = \alpha(1-r)^2$$

where t and r are the surface transmission and reflection coefficients, respectively, and α is the bulk transmissivity. The superscript on T indicates that the result is for a single transit only. To simplify the notation, we have omitted the subscripts $_{\parallel}$ and $_{\perp}$ on t. It is understood that there are two distinct values of $T^{(0)}$—one for each polarization state.

Overall Transmittance and Reflectance of a Single Glazing

The single transit result previously obtained does not adequately describe the phenomenon of overall transmittance through a glazing because it ignores the multiple internal reflections at the upper and lower surfaces of the glazing. The emergent beam can be regarded as the sum of an infinite number of beams resulting from multiple internal reflections (Figure 5.4). To simplify the analysis, we will

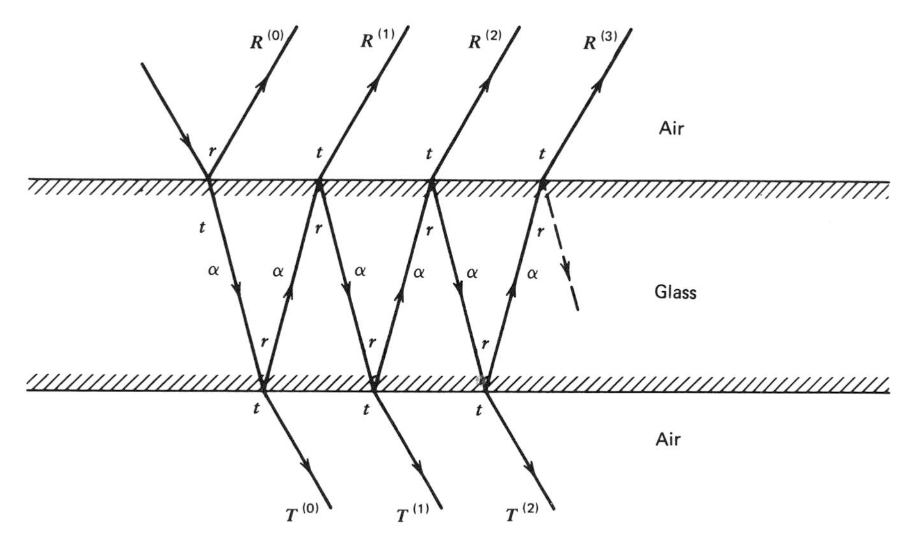

FIGURE 5.4 *Illustrating how the transmitted and reflected radiation from a glazing is affected by multiple reflections at the upper and lower surfaces.*

count in complete circuits after the first transit. Thus the transmission factor for the first transit (zeroth circuit) is $T^{(0)} = t\alpha t$. After one complete circuit, the beam has been decreased by a factor $t\alpha r\alpha r\alpha t = t^2\alpha(\alpha^2 r^2)$. After n circuits we find

$$T^{(n)} = t^2\alpha(\alpha^2 r^2)^n$$

The overall transmittance of the glazing is

$$T = \sum_{n=0}^{\infty} T^{(n)} = t^2\alpha \sum_{n=0}^{\infty} (\alpha^2 r^2)^n$$

The infinite goemetric series is evaluated as

$$\boxed{T = \frac{\alpha t^2}{1-\alpha^2 r^2} = \frac{\alpha(1-r)^2}{1-\alpha^2 r^2}} \tag{5.9}$$

A similar computation shows the overall reflectance to be

$$\boxed{R = r\left[1 + \frac{\alpha^2 t^2}{1-\alpha^2 r^2}\right] = r\left[1 + \frac{\alpha^2(1-r)^2}{1-\alpha^2 r^2}\right]} \tag{5.10}$$

To obtain the polarization averaged transmittance for solar radiation, we evaluate Equations 5.9 and 5.10 for $T_\parallel$, $T_\perp$, $R_\parallel$, and $R_\perp$ separately using $r_\parallel$ and $r_\perp$ as given by Equation 5.5. We find the average transmittance and reflectance for the glazing to be

$$\boxed{\bar{T} = \tfrac{1}{2}(T_\parallel + T_\perp)} \quad \text{and} \quad \boxed{\bar{R} = \tfrac{1}{2}(R_\parallel + R_\perp)} \tag{5.11}$$

It is straightforward to show that if bulk extinction is absent (i.e., $\alpha = 1$), then $\bar{R} + \bar{T} = 1$. We have plotted T and R as a function of θ_i for a typical glazing in Figure 5.5.

If the bulk transmissivity of a single glazing is near unity, we may set $\alpha = 1$ in the denominator of Equation 5.9 so that Equation 5.11 can be written, after some simplification,

$$\boxed{\bar{T} \simeq \alpha\left\{\frac{1}{2}\left[\left(\frac{1-r_\parallel}{1+r_\parallel}\right) + \left(\frac{1-r_\perp}{1+r_\perp}\right)\right]\right\}} \quad (\alpha \sim 1) \tag{5.12}$$

From Equation 5.12, it follows that the overall transmittance of a

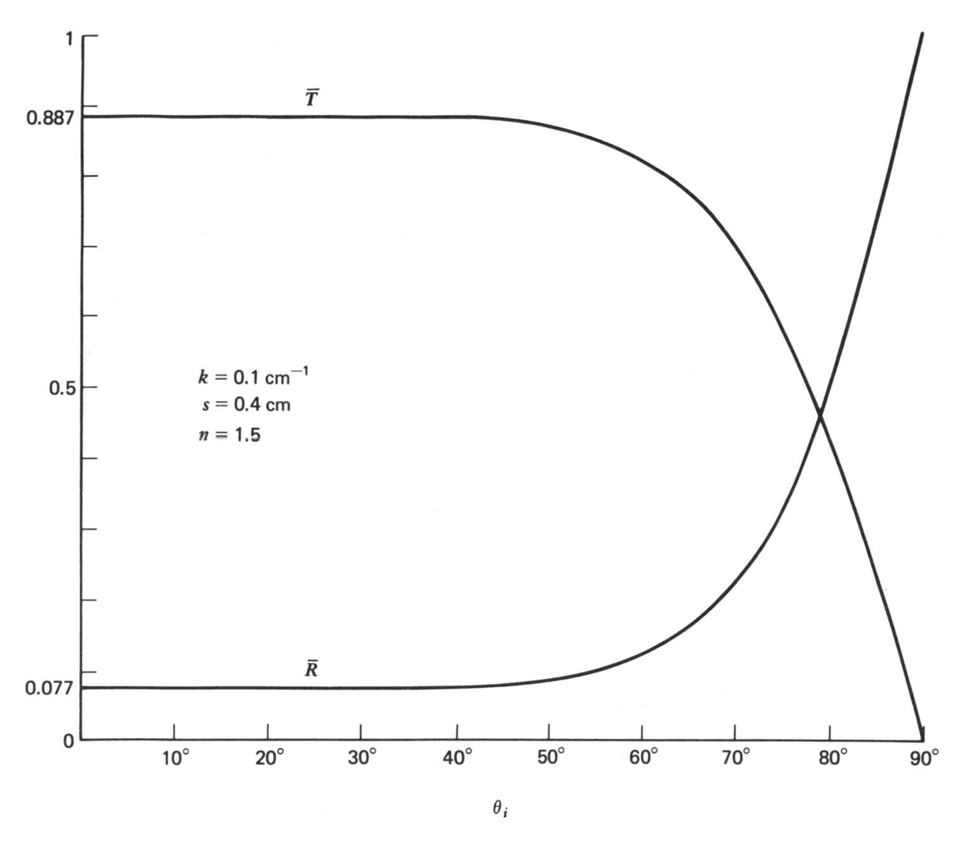

FIGURE 5.5 *The variation of the overall transmittance and reflectance of a single glazing with the angle of incidence. Note that since the glazing is absorbing energy,* $R + T \neq 1$.

single glazing is approximately a product of two factors—extinction and reflection.

Multiple Glazings

In solar heating panels it is sometimes necessary to use more than one glazing in order to reduce thermal losses. For a double glazing, the overall transmittance is obtained by adding the effects of the multiple reflections between the individual glazings. Using mathematical procedures similar to those for a single glazing, we can obtain the transmittance and reflectance of a double glazing composed of two identical glazings in a straightforward manner. Making the replacements $r \rightarrow R$, $t \rightarrow T$, and $\alpha \rightarrow 1$ in Equations 5.9 and 5.10, we find

$$\boxed{T_2 = \frac{T^2}{1-R^2}} \quad \text{and} \quad \boxed{R_2 = R\left(1 + \frac{T^2}{1-R^2}\right)} \tag{5.13}$$

when T, R and T_2, R_2 refer to the single and double glazings, respectively.

Applying Equation 5.13 to each polarization state, we find the overall polarization-averaged transmittance and reflectance of a double glazing to be

$$\boxed{\bar{T}_2 = \tfrac{1}{2}(T_{2,\parallel} + T_{2,\perp})} \quad \text{and} \quad \boxed{\bar{R}_2 = \tfrac{1}{2}(R_{2,\parallel} + R_{2,\perp})} \tag{5.14}$$

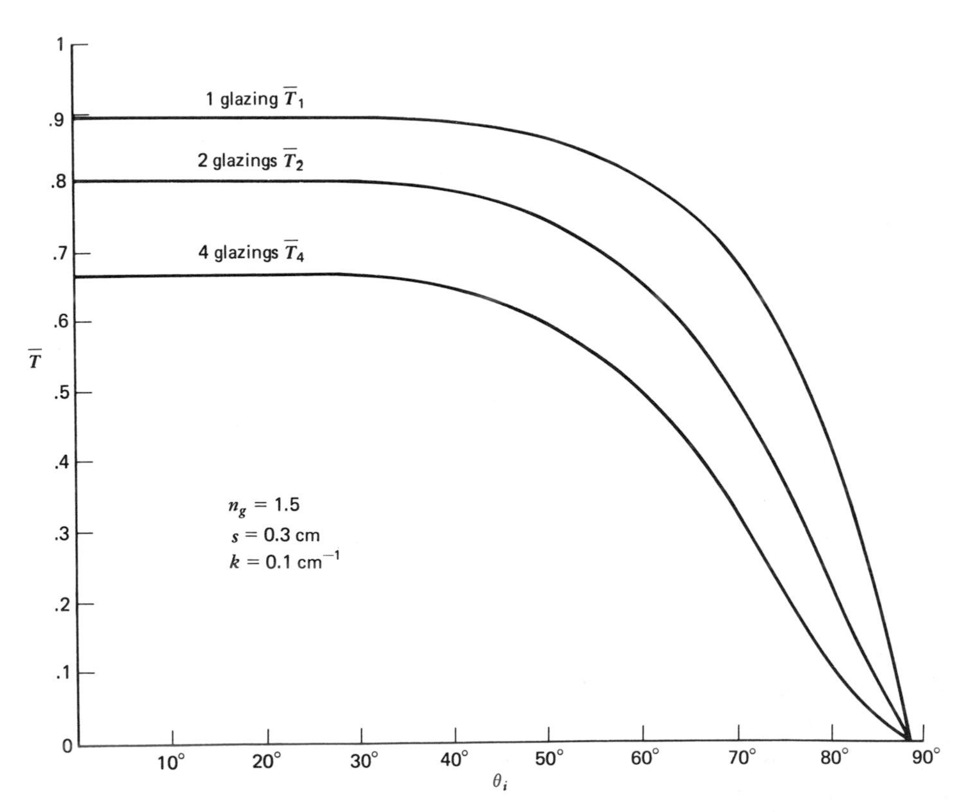

FIGURE 5.6 *Plots of the overall transmittance versus angle of incidence for single, double, and quadruple glazings. Each glazing has* $n_g = 1.5$, $s = 0.3$ *cm, and* $k = 0.1$ *cm*$^{-1}$.

Equation 5.13 can be extended to a quadruple galzing by making the replacements $R_2 \rightarrow R_4$, $T_2 \rightarrow T_4$, and $R \rightarrow R_2$, $T \rightarrow T_2$. The transmittances of single, double, and quadruple glazings are plotted in Figure 5.6. Note that at or near normal incidence, where the single glazing reflectance R is small, the transmittance in Equation 5.13 is

$$T_2 \simeq T^2$$

This can be generalized for n glazings as

$$\boxed{T_n \simeq T^n} \qquad \text{(small } R\text{)} \tag{5.15}$$

If the single-glazing transmittance is $T = 0.9$, then $T^2 \simeq 0.81$ and $T^4 \simeq 0.66$; consequently, more than two glazings can substantially reduce the overall transmittance.

Overall Optical Efficiency of a Glazing-Absorber System

As already shown, a glazing system can be characterized by an overall transmittance and reflectance, which we now label as T_g and R_g, respectively. A typical solar flat plate panel has an absorber plate placed under the glazing. The net flux absorbed by the plate is

$$F_{abs} = A_p F'_{inc}$$

where A_p is the absorptance of the plate and F'_{inc} is the flux incident on the plate. This flux includes multiple reflections (Figure 5.7) between the glazing and the plate and can be written

$$\begin{aligned} F'_{inc} &= T_g \sum_{n=0}^{\infty} [(1 - A_p)R_g]^n F_{inc} \\ &= \{T_g/[1 - (1 - A_p)R_g]\} F_{inc} \end{aligned}$$

where F_{inc} is the incident flux on the panel. The optical efficiency of the panel is therefore

$$\boxed{\eta_{opt} = \frac{F_{abs}}{F_{inc}} = \frac{A_p T_g}{1 - (1 - A_p)R_g}} \tag{5.16}$$

If the absorptance of the plate is near unity (which is almost always

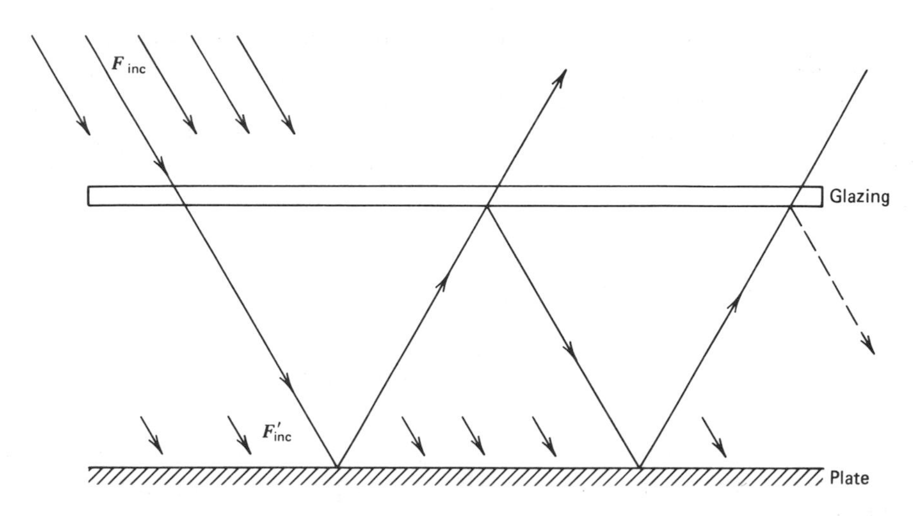

FIGURE 5.7 *The flux incident on the absorber plate as a result of multiple reflections between the plate and the glazing.*

the case), the optical efficiency is approximately

$$\boxed{\eta_{opt} \simeq A_p T_g \qquad (A_p \sim 1)} \tag{5.17}$$

Hence we have the following important result. *The optical efficiency of a flat plate is approximately equal to the product of the absorptance of the plate and the transmittance of the glazing.* If the glazing is totally transparent and the plate is a perfect absorber of solar radiation, the optical efficiency is 100 percent.

EXAMPLE

Find the optical efficiency of a single-glazed panel at normal incidence given the following facts: $R_\parallel = R_\perp = 0.05$, $T_\parallel = T_\perp = 0.89$, and $A_p = 0.95$. Repeat the computation for a double-glazed system.

Using Equation 5.17, we find for the single-glazed system

$$\eta_{opt} = (0.95)(0.89) = 0.85 = 85\%$$

For the double-glazed system, we have from Equation 5.13,

$$T_g = T_2 = (0.89)^2/(1 - 0.05^2) = 0.79$$

Using Equation 5.17, we find the optical efficiency to be

$$\eta_{opt} = (0.95)(0.79) = 0.75 = 75\%$$

Antireflective Coatings for Glazings

If a glazing is sufficiently thin, its bulk transmissivity will be near unity and its transmittance will be limited only by the fact that reflection takes place at the upper and lower surfaces of the glazing. Reflection losses can be a problem especially when multiple glazings are used. It is possible to reduce reflections from the glazing by coating its surfaces with a thin film (Figure 5.8). We will consider only the case of reflection at normal incidence.

The reflection coefficient for light incident normally on a surface is given by Equation 5.6, which for a glazing-air interface is

$$r = \left(\frac{n_g - 1}{n_g + 1}\right)^2$$

For glass ($n_g \simeq 1.5$), r is approximately 4 percent. When a thin film of thickness s_f is deposited on the glass, reflections will occur at both the air-film and the film-glass interfaces. Suppose light of a given wavelength λ_0 is incident normally on the coated glazing. It is possible to make the waves reflected from the upper and lower surfaces of the film interfere destructively so that no net reflection takes place. To do this, we must satisfy two conditions. First, the two waves should be reflected with equal amplitudes. To assure this, we relate the index of refraction of the film to that of the glass by

$$\boxed{n_f = \sqrt{n_g}} \qquad (5.18)$$

Second, the two reflected waves should be exactly 180° out of phase when they are reflected back into air. This condition will be satisfied if the film thickness is exactly

$$\boxed{s_f = \frac{m\lambda_0}{4n_f}} \qquad m = 1, 3, 5, \ldots \qquad (5.19)$$

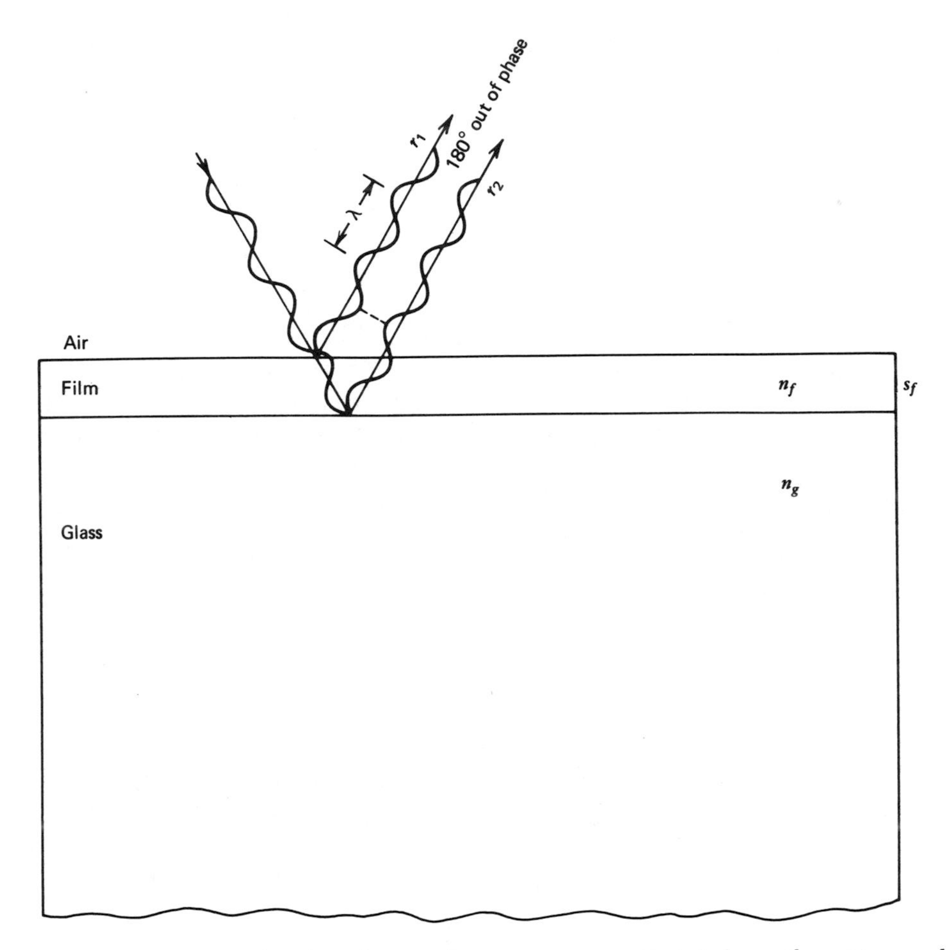

FIGURE 5.8 *Destructive interference between waves reflected from the upper and lower surfaces of an antireflective coating.*

EXAMPLE

Find the index of refraction of a film that will make glass ($n_g = 1.5$) nonreflective. What is the minimum thickness of this film that must be deposited in order to make the glass nonreflective at normal incidence for green light, $\lambda_0 = 0.5\ \mu m$?

According to Equation 5.18, the film index should be

$$n_f = \sqrt{1.5} = 1.22$$

The minimum thickness of the film is obtained by setting $m = 1$

in Equation 5.19. This gives

$$s_f = \frac{1(0.5)}{4(1.22)} = 0.102\ \mu\text{m}$$

If the coating is designed to be antireflective for λ_0, it will in general allow reflections for other wavelengths. It can be shown that at normal incidence, the ratio of the spectral reflectivity of a coated glazing to that of an uncoated glazing is (see Ref. 2, p. 61)

$$\boxed{\gamma_\lambda = \frac{r_\lambda}{r} = \frac{(2p/r)(1+\cos\beta_\lambda)}{1+2p\cos\beta_\lambda + p^2}} \tag{5.20}$$

where

$$p = \left(\frac{n_f - 1}{n_f + 1}\right)^2, \qquad r = \left(\frac{n_g - 1}{n_g + 1}\right)^2, \qquad \text{and} \qquad \beta_\lambda = \frac{\pi\lambda_0}{\lambda}$$

provided Equation 5.18 is satisfied.

The function in Equation 5.20 is zero when $\lambda = \lambda_0$ as expected; it approaches unity both as $\lambda \to \lambda_0/2$ and as $\lambda \to \infty$. When $\gamma_\lambda = 1$, the glazing behaves as if it were uncoated. Equation 5.20 is plotted in Figure 5.9 for a film ($n_f = 1.22$) on glass ($n_g = 1.5$); the film's thickness is 0.102 μm. This film produces zero reflection for $\lambda_0 = 0.5$ m.

The spectrally averaged factor γ for the solar spectrum is obtained using

$$\gamma = \int_0^\infty \gamma_\lambda b_\lambda(5760\ \text{K})\, d\lambda \tag{5.21}$$

where $b_\lambda(5760\ \text{K})$ is a normalized black-body curve at 5760 K (i.e., one with unit area under it). Applying the curve in Figure 5.9 to Equation 5.21, we see that $\gamma \simeq 1/3$; consequently, the surface reflectivity is reduced from 0.04 to 0.0133 for solar radiation. This reduction can improve optical efficiency significantly especially when multiple glazings are used.

EXAMPLE

A solar glazing of bulk transmissivity $\alpha = 0.96$ has a film deposited on its upper and lower surfaces to reduce surface reflection from 0.04 to 0.0133. Compute the overall transmit-

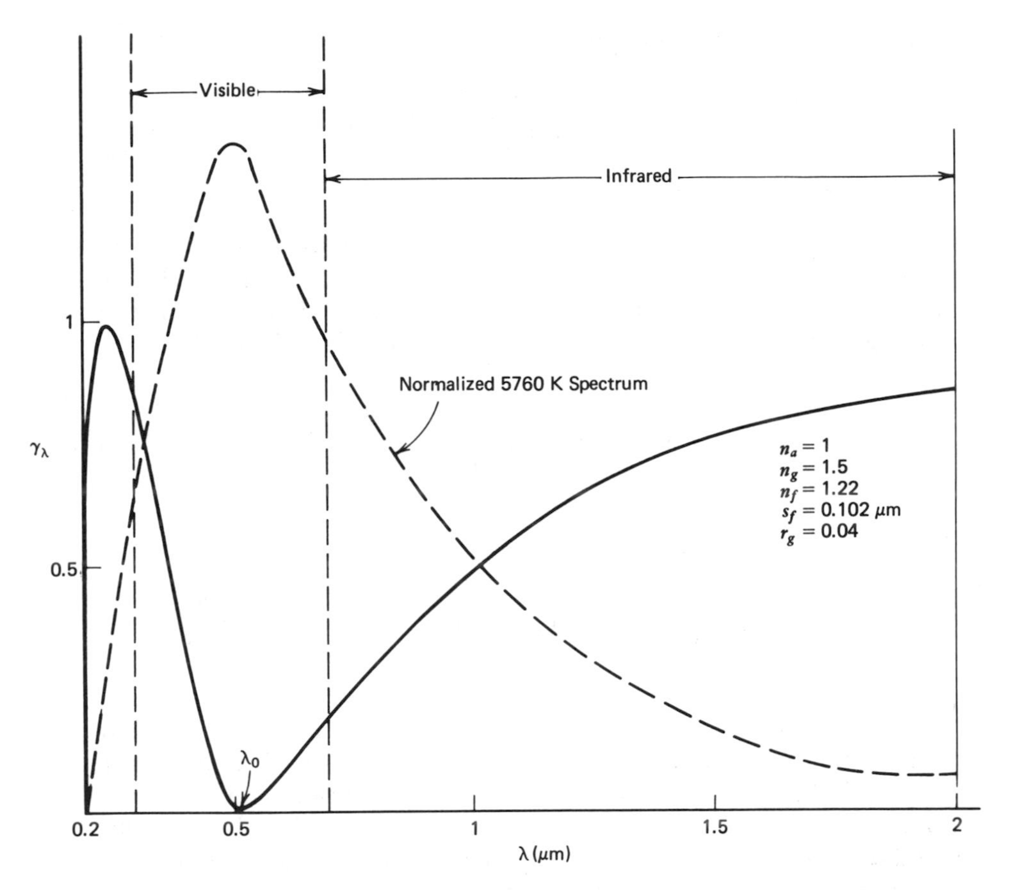

FIGURE 5.9 *A plot of γ_λ versus wavelength for a thin film on glass. Here γ_λ represents the ratio of the reflectivity (at normal incidence) of the system to that of uncoated glass. The coating has been chosen to make the glass anti-reflective for $\lambda_0 = 0.5\ \mu m$. The dashed plot is a 5760 K black-body emission curve normalized so that the area under it is unity.*

tances of two such glazings with and without the coating. Assume normal incidence.

The single-glazing values are obtained from Equations 5.9 and 5.10 as

$$T^{(\text{uncoated})} = 0.887 \qquad \text{and} \qquad R^{(\text{uncoated})} = 0.074$$

and

$$T^{(\text{coated})} = 0.936 \qquad \text{and} \qquad R^{(\text{coated})} = 0.025$$

Using Equation 5.13, we find for the double glazing

$$T_2^{(\text{uncoated})} = 0.77 \qquad \text{and} \qquad T_2^{(\text{coated})} = 0.86$$

From the example it is clear that antireflection films increase the optical efficiency of solar panels, especially of those with more than one glazing. There are, however, a number of problems with coatings. A thin film designed to reduce reflection at normal incidence is not effective at oblique angles. Furthermore, for the film to be antireflective, its index must satisfy Equation 5.18. As seen in the case of glass ($n_g = 1.5$), the film's index should be $n_f = 1.22$. Because a material with this index is difficult to fabricate, a substance such as magnesium fluoride ($n_f = 1.38$) is often substituted. A stack of films of different materials can also be used to reduce reflections. Unfortunately, it is costly to deposit high quality films uniformly on large surfaces. This is true, in particular, when the surfaces are flexible and thermally expand at a rate different from that of the film. Also films tend to deteriorate when exposed to the elements. Recent developments have shown that reflections can also be reduced by etching the surfaces of the glazing.

The Greenhouse Effect

Most solar glazings are highly transparent to solar radiation but are essentially opaque to thermal radiation. This characteristic is particularly useful in a solar heating panel. When a glazing is placed above the absorber plate, the glazing allows solar energy to pass through and be absorbed by the plate. However, the thermal radiation emitted by the plate is trapped and reemitted by the glazing. Part of this energy finds its way back to the plate, thereby increasing its steady-state temperature. This phenomenon is known as the *greenhouse effect.*

To appreciate this effect, consider an absorber plate suspended in space above the atmosphere where convection effects are absent. Let the front face be blackened and oriented toward the sun. Furthermore, let the back face be silvered so that back losses are absent. The steady-state absorber temperature is obtained by equating the absorbed solar flux to the emitted thermal flux; that is,

$$\sigma T_p^4 = 1352 \text{ W/m}^2$$

We are assuming the emissivity of the surface to be unity for all wavelengths. Solving for the plate temperature, we find

$$T_p = T_0 = \left(\frac{1352}{5.67 \times 10^{-8}}\right)^{1/4} = 393 \text{ K} = 120°\text{C}$$

Next, consider what happens when we place a glazing in front of the absorber (Figure 5.10). Suppose the glazing is transparent to solar radiation but appears opaque and black to thermal radiation. Because all the thermal radiation emitted back to space originates from the glazing, we set, using steady-state conditions,

$$\sigma T_g^{\ 4} = 1352 \text{ W/m}^2$$

or

$$T_g = T_0 = 393 \text{ K} = 120°\text{C}$$

Since the heated glass emits equally in both directions, the thermal flux emitted by the glazing and reaching the absorber must also be 1352 W/m^2. Thus the thermal flux from the glazing plus the solar radiation reaching the absorber is $2 \times 1352 = 2704$ W/m^2. Setting this

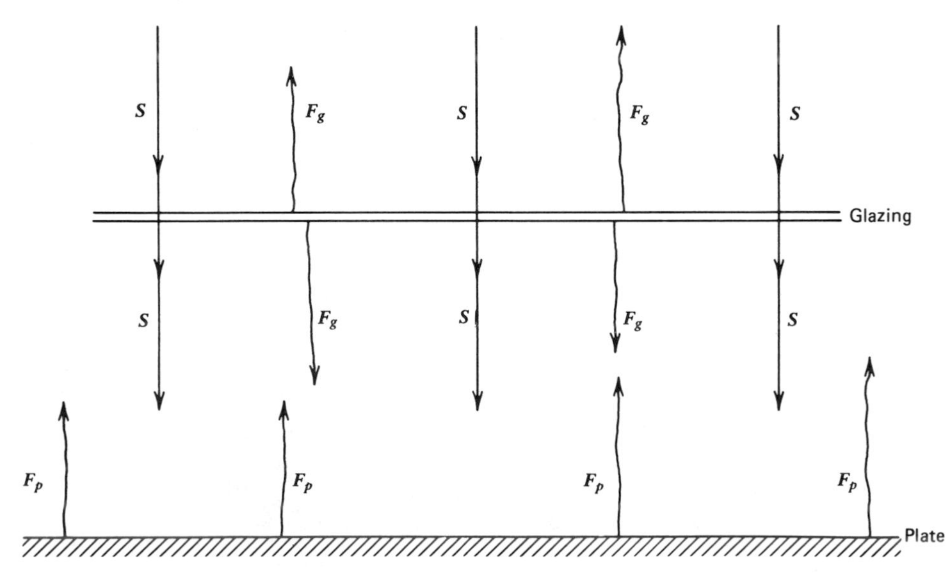

FIGURE 5.10 *An idealization of the greenhouse effect. The thermally opaque glazing traps the thermal flux emitted by the plate (F_p). It then reemits some of this flux (F_g) back to the plate. The solar flux (S) passes through the glass unaffected.*

equal to the flux emitted by the absorber, we find

$$\sigma T_p^4 = 2704$$

or

$$T_p = \left(\frac{2704}{5.67 \times 10^{-8}}\right)^{1/4} = 467 \text{ K} = 194°\text{C}$$

Thus in theory the thermally opaque glazing is capable of raising the temperature of the absorber from 120 to 194°C. In practice, the increase would not be nearly so marked primarily because convection effects would tend to cool the absorber plate.

The greenhouse effect also plays an important role in establishing the earth's climate. The earth's surface and atmosphere act as absorber plate and glazing, respectively. The *atmosphere or greenhouse effect* results in higher surface temperatures than would be possible without an atmosphere.

Selective Absorber Coatings

An effective way to reduce thermal losses from the absorber plate of a solar heating panel is by using selective absorber coatings. An ideal selective coating is one that is a perfect absorber of solar radiation while being a perfect reflector of thermal radiation. Such a coating will make a surface a poor emitter of thermal radiation. The spectral absorptivity or emissivity of an idealized coating is shown in Figure 5.11. The cutoff wavelength, λ_c, should ideally be between the solar and thermal regions of the electromagnetic spectrum, that is, in the range $2\ \mu\text{m} \leq \lambda_c \leq 6\ \mu\text{m}$.

To understand the way in which a selective coating increases the temperature of an absorbing surface, consider what happens when such a surface is suspended in space and exposed to solar radiation. If back losses are absent, the steady-state conditions give

$$\text{solar fluxed absorbed} = \text{thermal flux emitted}$$

$$\int_0^\infty a_\lambda S_\lambda \, d\lambda = \int_0^\infty \epsilon_\lambda B_\lambda(T_p) \, d\lambda \qquad (5.22)$$

where S_λ is the spectral flux of the solar constant, S, and where $B_\lambda(T_p)$ is the Planck function at the plate temperature. By Kirchhoff's law we have $\epsilon_\lambda \equiv a_\lambda$. For an idealized selective absorber—one for

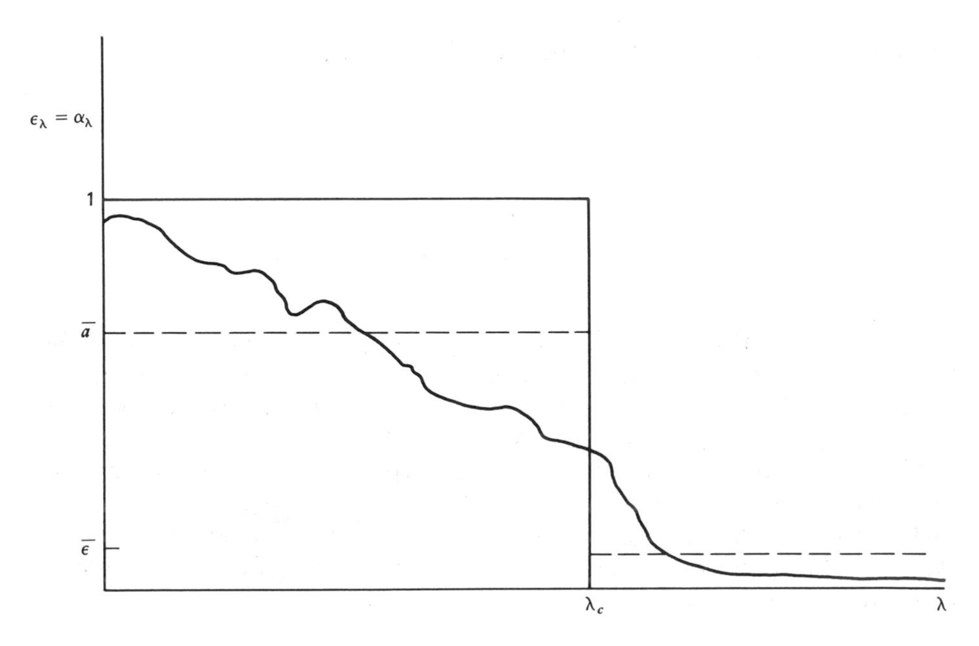

FIGURE 5.11 *The spectral emissivity (or absorptivity) of an ideal selective coating whose cutoff wavelength is λ_c. The irregular curve is typical for a real selective coating. The real coating is characterized by an average absorptivity, $\bar{a}$, (over the solar spectrum) and an average emissivity, $\bar{\epsilon}$, (over the thermal region).*

which $a_\lambda = \epsilon_\lambda = 1$ for $0 \leq \lambda \leq \lambda_c$ and $a_\lambda = \epsilon_\lambda = 0$ otherwise—we find

$$\int_0^{\lambda_c} S_\lambda \, d\lambda = \int_0^{\lambda_c} B_\lambda(T_p) \, d\lambda$$

or

$$[f(\lambda_c T_\odot)]S = [f(\lambda_c T_p)]\sigma T_p^4 \tag{5.23}$$

where $T_\odot = 5760$ K is the sun's spectrum temperature and where $f(x)$ is the function defined in Equation 1.14 and tabulated in Table 1.1. If we assume that λ_c is greater than 2 μm, we may set $f(\lambda_c T_\odot) \simeq 1$ and find

$$S \simeq f(\lambda_c T_p)\sigma T_p^4 \qquad (\lambda_c > 2\ \mu\text{m}) \tag{5.24}$$

Note that if there is no cutoff (i.e., $\lambda_c \to \infty$), then $f \to 1$ and we recover the result for a nonselective coating, or neutral absorber, namely,

$$T_P \to T_0 = \left(\frac{S}{\sigma}\right)^{1/4} = 393 \text{ K} = 120°\text{C}$$

To find T_p, we write Equation 5.24 as

$$f(x)x^4 = x_0^4$$

where $x = \lambda_c T_p$ and $x_0 = \lambda_c T_0 = 393\lambda_c$. Once λ_c is given, Equation 5.25 can be solved numerically for x; T_p can then be found using $T_p = x/\lambda_c$.

EXAMPLE

Estimate the steady-state temperature of an absorber plate suspended in space, exposed to the flux of the solar constant, and coated with an ideal selective absorber whose cutoff is $\lambda_c = 4\ \mu m$. Assume zero back losses.

Setting $x_0 = (4)(393) = 1572\ \mu m$-K in Equation 5.25, we find

$$f(x)x^4 = (1572)^4$$

The solution is $x = 2492\ \mu m$-K. This can be verified using the fact that $f(2492) = 0.158$ (see Table 1.1) and noting that

$$(0.158)(2492)^4 \simeq (1572)^4$$

Thus

$$T_p = \frac{2492\ \mu m\text{-K}}{4\ \mu m} = 623\ K = 350°C$$

Note the remarkable increase in plate temperature (i.e., from 120 to 350°C) when a selective coating is used.

In solar heating panels the increase in plate temperature is not so marked because reduced radiation losses are compensated for by increased convection losses. Also, real coatings have solar absorptivities that are less than unity and thermal emissivities that are small but larger than zero. As we see from Figure 5.11, a real coating does not have a sharp cutoff in its spectral characteristics.

A real selective coating is usually characterized by the ratio of its average absorptivity over the solar spectrum ($0.2\ \mu m < \lambda < 2\ \mu m$), $\bar{a}$, to its average emissivity over the thermal region of the spectrum ($2\ \mu m < \lambda < 20\ \mu m$), $\bar{\epsilon}$. The higher the ratio, $\bar{a}/\bar{\epsilon}$, is, the more effective the coating. The ratio for a neutral absorber is of course unity.

Concentrators

In practice, the steady-state temperatures of simple flat plate collectors rarely, if ever, exceed 150°C, even under ideal conditions. In severe winter conditions, operating temperatures are well below 100°C because thermal losses to the surroundings limit the maximum possible output values. The smaller the overall heat transfer coefficient is to the surroundings, the higher the temperature. Consequently, flat plates with multiple glazings, partially evacuated interiors, and selective absorber coatings tend to reach higher temperatures than simply designed flat plates. Because all three modes of heat transfer from a heated surface are proportional to the area of the surface, a substantial reduction of heat loss will result if this area is small. Thus if the intercepted solar power can be concentrated and directed on to a small absorber area, a very high temperature will be achieved.

Concentrating solar collectors can produce higher temperatures than flat plates and can be operated with lower thermal losses and higher thermal efficiencies. However, concentrators have some major drawbacks. First, as we will see, they can only harness the monodirectional or direct component of solar radiation. Because the diffuse component often constitutes more than 20 percent of the total incident flux, the optical efficiency of a concentrator is generally lower than that of a flat plate. A more severe limitation is that a simple concentrator can function only if it is properly oriented with respect to the direct solar beam. Consequently, some type of tracking apparatus must be used. This adds to the complexity and cost of the system. Therefore concentrators are used primarily where very high temperatures are required; they are not normally found in space heating and hot water supply systems.

Focusing of direct solar radiation can be achieved with either mirrors or lenses. Lenses have certain drawbacks. They are generally costly to fabricate, especially for large collector systems, and can also be heavy and difficult to install.[1] Since radiation passes through a lens, some energy is absorbed, thereby reducing optical efficiency. Mirrors, on the other hand, are less costly and lighter in weight. Since they concentrate radiation by reflection, mirrors absorb minimally. We will

[1]There is a type of *flat* lens known as a *Fresnel lens* that is cheaper to make, lighter in weight, and much thinner than a conventionally curved lens with the same area. A Fresnel lens is made by molding a series of grooves or facets on the surface of a thin sheet of transparent plastic. Such a lens is less efficient than a conventional lens because certain sections of grooves are ineffective in concentrating light especially when the beam radiation is incident obliquely to the lens' surface.

consider the optics of concentrating mirrors. The results developed here for mirrors can be adapted to lenses with minor modifications.

Parabolic and Circular Mirrors

To understand the way in which a curved mirror concentrates direct solar energy, we consider a two-dimensional mirror whose contour is a parabola. The mathematical equation of the parabola is

$$x^2 = 4fy \tag{5.26}$$

It is plotted in Figure 5.12*a*.

The symmetry line (in this case the y axis) is called the *axis* of the parabola. Any ray traveling parallel to the axis is called a *paraxial* ray. The point f on the axis is called the focus of the parabola. Consider any point on the parabola. It can be shown that a paraxial line drawn to that point makes the same angle with the mirror as a line drawn from that point to the focus. Consequently, all paraxial rays incident on a parabolic mirror will be directed toward the focus or focal point.

The concentration ratio (cr) of a focusing collector is the ratio of the flux falling on a surface at the focal point to the flux intercepted. If, in fact, a concentrator were able to direct all the intercepted flux to a point, the concentration ratio would be *infinite* because radiant power would be directed on to a surface of zero area. In practice, there are a number of factors that prevent the concentration ratio from being infinite. First, the curvature of a mirror is never perfect. Second, the incident solar beam is not perfectly monodirectional but diverges slightly. Finally, even if the mirror were perfect and the sun's rays perfectly monodirectional, a phenomenon known as diffraction would cause some inherent blurring. Diffraction occurs because radiation is composed of electromagnetic waves. The ray model used here to represent solar radiation is only an approximation. Thus the law of reflection that applies strictly to rays applies only approximately to waves. If the dimensions of the mirror are large compared with the wavelength of the solar radiation, which is usually the case, blurring due to diffraction will be negligible. In most basic solar concentrator applications, aberrations due to diffraction are not a problem. Solar radiation cannot be concentrated to a point primarily because the sun's rays diverge and also because mirrors cannot be shaped like perfect parabolas. Because the off-axis aberrations of a parabolic mirror are difficult to treat, we will consider a simpler reflector, namely, the circular mirror. Aside from its

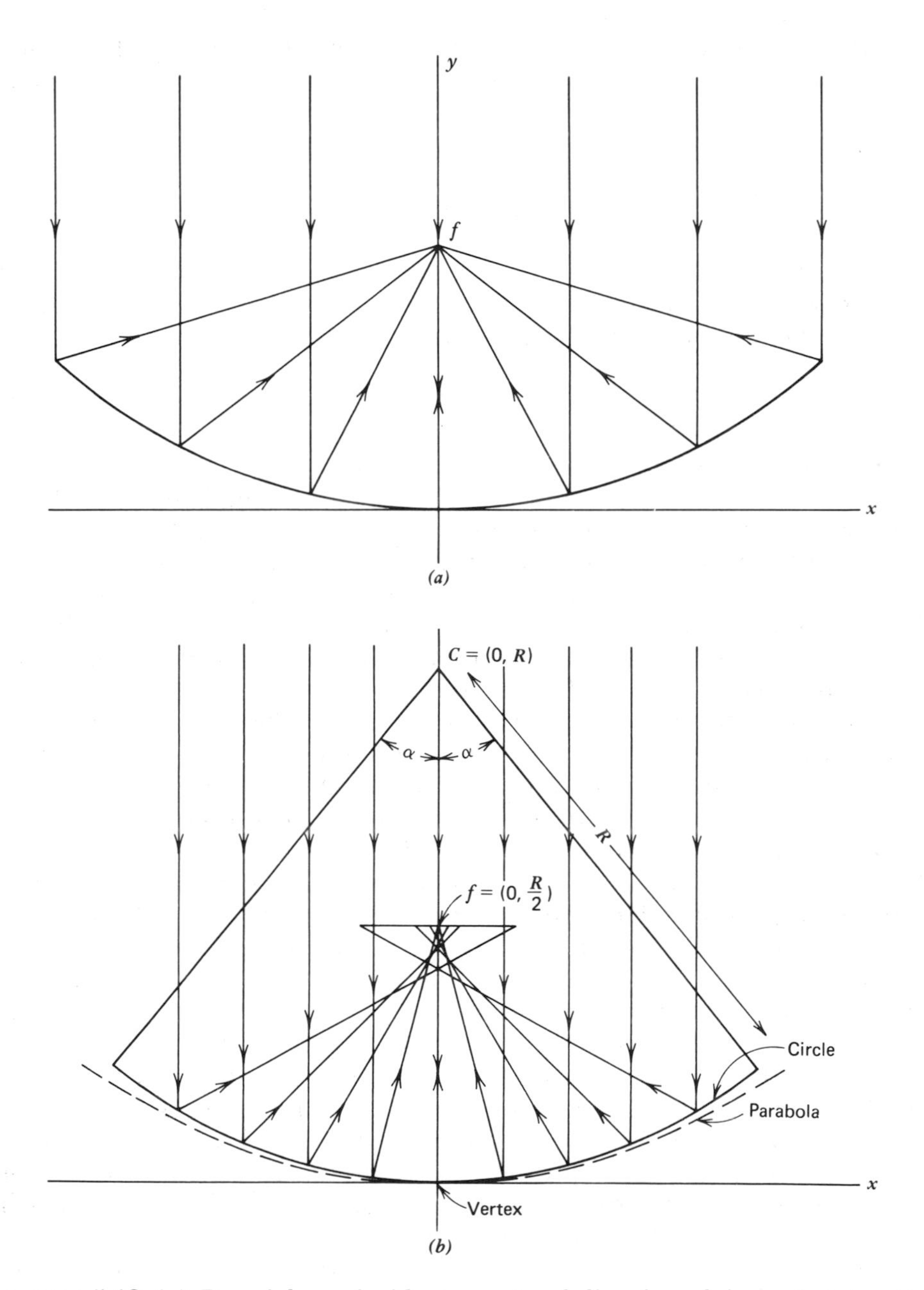

FIGURE 5.12 (*a*) *Paraxial rays incident on a parabolic mirror being reflected to the focal point.* (*b*) *Paraxial rays being reflected from a circular mirror. Rays arriving near the vertex are reflected approximately toward the focal point. Rays reflected from the edges produce considerable aberration. The curved dashed line shows the contour of a parabolic mirror having the same focal point.*

theoretical simplicity, a circular mirror is also easier to fabricate than a parabolic mirror.

A circular mirror is one whose contour conforms to the circular curve (Figure 5.12*b*)

$$\begin{aligned} x^2 + (y - R)^2 &= R^2 \\ x^2 &= 2Ry[1 - (y/2R)] \end{aligned} \tag{5.27}$$

This equation describes a circle of radius R centered at the point $(0, R)$. We can express Equation 5.27 as

$$x^2 = 2Ry\left[\cos^2\frac{\theta}{2}\right]$$

where θ is the angle between the axis of the curve and a line drawn from R to a point (x, y) on the curve. If the mirror is shaped like a small section of the circle, then θ will be small everywhere along the arc and we may set $\cos^2 \theta/2 \simeq 1$ and write

$$x^2 \simeq 2Ry \tag{5.28}$$

Comparing Equations 5.28 and 5.26, we observe that a circular mirror behaves approximately as a parabolic mirror whose focal length is $f = R/2$, provided only a small section of the circle is used. Mirrors made of large sections of a circle exhibit *spherical* aberration (Figure 5.12*b*).

Spherical Mirrors

The two obvious three-dimensional generalizations of the circular mirror are the spherical and cylindrical mirrors. The convex spherical mirror or dish reflector is a small section of a sphere of radius R and focal length $f = R/2$. Paraxial rays converge approximately to the focal point. Convex spherical mirrors can produce sharp images of objects if aberrations due to mirror imperfections and those due to diffraction are ignored.

If an object is placed on the axis of the mirror at a distance p from the vertex of the mirror, the vertex being the intersection point of the axis and the mirror, the image will appear on the axis at a distance q given by the mirror equation

$$\frac{1}{p} + \frac{1}{q} = \frac{1}{f} \tag{5.29}$$

or

$$q = \frac{f}{1 - f/p}$$

The magnification m is defined as the ratio of the size of the image to the size of the object. The magnification is determined from the relation

$$m = \frac{q}{p} = \frac{(f/p)}{(1 - f/p)} \tag{5.30}$$

Suppose now that the mirror is oriented so that the solar disc (the object) is on the axis of the mirror. Because the object distance is very large when compared with the focal distance of the mirror, Equations 5.29 and 5.30 give the position of the solar image and its magnification as

$$q \simeq f$$

and

$$m = \frac{R'_{\odot}}{R_{\odot}} \simeq \frac{f}{p}$$

Thus the image of the solar disc appears at the focal point and its radius is reduced to

$$\boxed{R'_{\odot} \simeq f\frac{R_{\odot}}{p} \simeq (0.0046)f} \tag{5.31}$$

We have taken the ratio of the sun's radius to its distance from the earth to be 0.0046. For a typical mirror of focal length $f = 10$ cm, the radius of the sun's image would be $R'_{\odot} = 0.046$ cm $= 0.46$ mm, provided all aberrations were absent. Actually, Equation 5.31 represents the *minimum* possible size of the solar image; the sun cannot be imaged to a point because its rays are not perfectly parallel. In practice, even minor aberrations result in a blurred image two or three times the value given in Equation 5.31. Since an analysis of blurring is beyond the level of this text, we will use Equation 5.31 to describe the semiquantitative features of concentrators.

When aberrations are absent and the mirror surface is perfectly reflective, all the radiant power intercepted by the mirror will be directed toward the image at the focal point. The concentration ratio is computed using

$$\mathrm{cr} = \frac{\text{flux in image}}{\text{flux intercepted}} = \frac{\text{interception area}}{\text{image area}} = \frac{A}{A'}$$

The interception or *aperture* area is

$$A = \pi s^2 \quad \text{(spherical collector)}$$

where s is the radius of the aperature (Figure 5.13). We may write

$$s = R \sin \alpha = 2f \sin \alpha \tag{5.32}$$

where 2α is termed the *rim angle* subtended by the arc of the mirror. The minimum image area is

$$A' = \pi (R'_{\odot})^2$$

so that the maximum concentration ratio is

$$\mathrm{cr}_{\max} = \frac{\pi s^2}{\pi (R_{\odot})^2} = \left(\frac{2 \sin \alpha}{0.0046}\right)^2 \quad \text{(spherical collector)} \tag{5.33}$$

Thus the maximum possible concentration ratio of a spherical mirror increases as the angle subtended by the mirror increases. However, Equation 5.33 is only valid when α is not very large so that spherical aberration is negligible (see Problem 5-13).

EXAMPLE

A spherical mirror has a radius of $R = 50$ cm and a rim angle of 30°. If aberrations increase the image radius of the solar disc to four times the minimum value, find the concentration ratio of the mirror. Find the average flux and the power falling on a small absorber at the focal point if the direct solar flux is 700 W/m^2.

Setting $\alpha = 15°$ in Equation 5.33, we find

$$\mathrm{cr}_{\max} = \left(\frac{2 \sin 15°}{0.0046}\right)^2 = 12{,}663$$

Since aberrations quadruple the image radius, the concentration

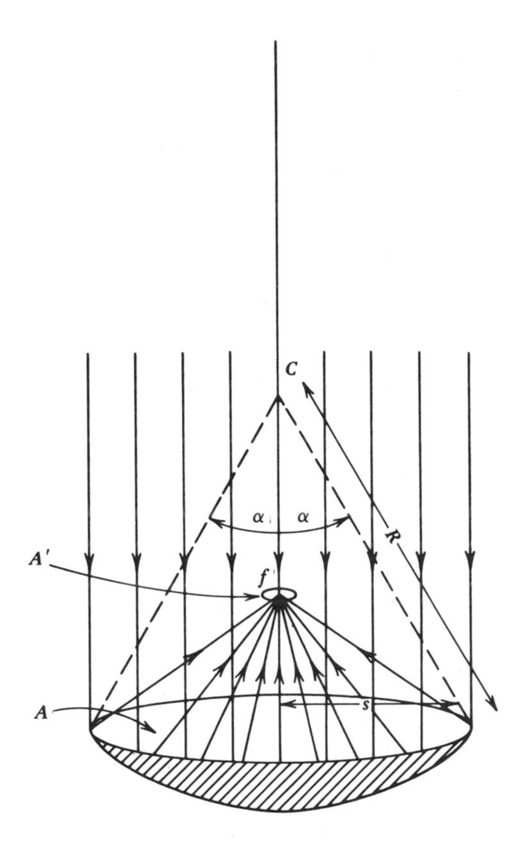

FIGURE 5.13 *A spherical dish mirror of radius R and rim angle 2α. The aperture area is $A = \pi s^2$. An absorber disc of area A′ is situated at the focal point.*

ratio decreases by a factor of 16 so that

$$\text{cr} = \frac{12{,}663}{16} = 791$$

The radiant power arriving at the focal point is equal to the intercepted power, which in turn is given by

$$P = F^{(\text{dir})}\pi s^2 = F^{(\text{dir})}\pi R^2 \sin^2 \alpha$$
$$= (700)(\pi)(0.5)^2 \sin^2 15° = 36.8 \text{ W}$$

The average flux at the focal point is

$$F' = (\text{cr})(F^{(\text{dir})}) = (791)(700) = 5.54 \times 10^5 \text{ W/m}^2$$

Angle of Acceptance

If a spherical mirror is oriented so that the sun is situated on the mirror axis, the solar image will appear approximately at the focal point on the axis. However, if the sun's motion takes it off the mirror axis, the image will also shift off axis. Because the mirror is spherical, the image moves on an arc of radius $f = R/2$ about the center of curvature of the mirror (Figure 5.14). If the sun moves off the axis by an angle ϕ, its image moves off axis a distance

$$\delta = f\phi$$

where ϕ is in radians.

Suppose that the absorber is a disc of radius b centered at the focal point on the axis and facing the mirror. The image of the sun will fall on the absorber if and only if

$$\delta < b$$

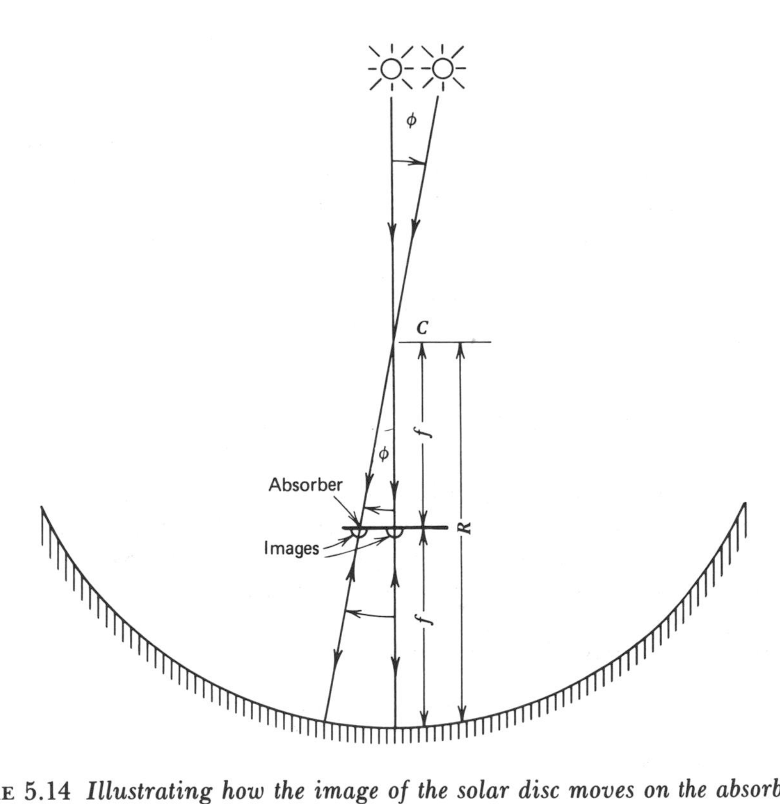

FIGURE 5.14 *Illustrating how the image of the solar disc moves on the absorber disc as the sun moves off axis by an angle ϕ.*

or

$$\phi < \phi_{max} = \frac{b}{f} \tag{5.34}$$

We are assuming that the sun's image is much smaller than the size of the absorber (Figure 5.14). As seen from Equation 5.34, the acceptance angle decreases as the absorber's size decreases. This in turn makes tracking more critical.

EXAMPLE

A spherical mirror of focal length 25 cm has an absorber disc of radius 0.5 cm placed at the focal point on the axis. Find the maximum angle between the sun's rays and the mirror axis that will permit the solar image to fall on the absorber. (Neglect aberrations.)

Using Equation 5.34, we have

$$\phi_{max} \simeq 0.5/25 = 0.02 \text{ rad} = 1.15°$$

Although multielement optical systems can be used to increase the acceptance angle, the basic problem remains. In order to increase this angle, one must increase the absorber area. This in turn reduces the effectiveness of the concentrator because increasing the absorber area increases thermal losses. In fixed arrays it is necessary to make this sacrifice in order that the collectors remain operational for larger fractions of the day.

Cylindrical Trough Concentrators

Optical principles of the circular mirror are applicable to dish concentrators and can also be applied to cylindrical trough concentrators (Figure 5.15). The mathematics is quite similar except that the image of the sun appears on the focal line as a strip of width $W' = 2R'_{\odot}$ and length L, where L is the length of the trough. From Equation 5.31, the minimum width of the image is

$$W' = 2(0.0046)f$$

so that the image area is

$$A' = W'L = 2(0.0046)fL \tag{5.35}$$

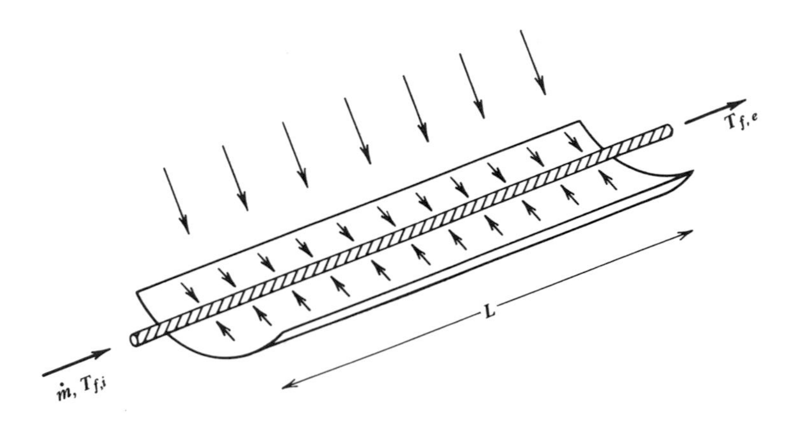

FIGURE 5.15 *A trough collector with an absorber pipe situated along the focal line. Fluid flowing through the pipe is heated by the concentrated flux.*

The interception or *aperture* area is

$$A = WL \quad \text{(cylindrical collector)} \tag{5.36}$$

The width of the trough is $W = 2s$, where s is, according to Equation 5.32,

$$s = 2f \sin \alpha \tag{5.37}$$

and where 2α is the rim angle. Combining Equations 5.35, 5.36, and 5.37, we get the maximum concentration ratio

$$\text{cr}_{\text{max}} = \frac{A}{A'} = \frac{2 \sin \alpha}{0.0046} \quad \text{(cylindrical collector)} \tag{5.38}$$

Comparing Equations 5.33 and 5.38, we observe that the maximum theoretical concentration ratio of a cylindrical trough is equal to the square root of the maximum concentration ratio of a spherical dish with the same rim angle. For practical reasons, the maximum concentration ratio of a dish concentrator is usually limited to ~3600 so that the corresponding limit for a cylindrical trough is ~60. In spite of their lower concentration ratios, cylindrical troughs have distinct advantages. First, they can be implemented in heating panels more easily than dish concentrators because absorber pipes carrying fluids can be placed along their focal lines. Second, cylindrical troughs are more tolerant to tracking error than are dish collectors. Once oriented directly toward the sun, only solar motion perpendicular to the focal

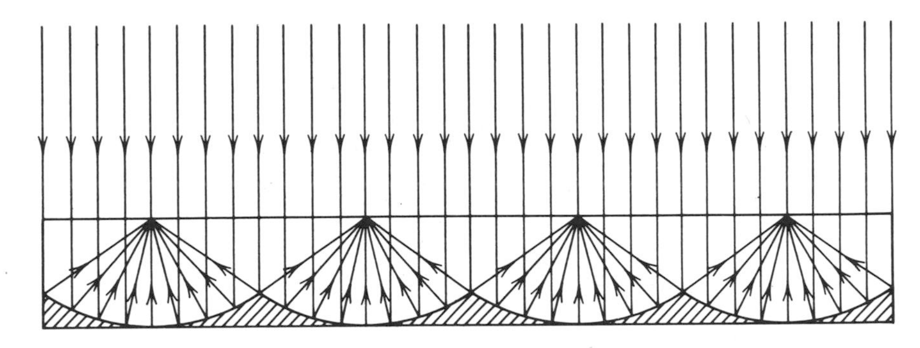

FIGURE 5.16 *Profile of a four-trough collector. A single trough panel with the same aperture area would necessarily be much thicker.*

line limits performance. Although obliquity reduces the amount of intercepted flux, the motion of the sun parallel to the focal line of the trough still preserves the focusing condition. Consider, for example, a cylindrical trough mounted with its axis parallel to the earth's axis of rotation, that is, along the H axis described in Chapter 2. This trough requires only daily tracking in order to follow the hour angle of the sun. It will concentrate the sun's rays regardless of the sun's codeclination and hence needs no seasonal adjustment. Although such adjustments are desirable because they maximize the intercepted flux, they are not necessary to preserve the focusing condition. A spherical dish, however, requires constant seasonal as well as diurnal tracking.

A single trough collector is difficult to incorporate into a low profile heating panel. According to Equation 5.37, the width of the trough is proportional to the focal length for fixed α. In order to decrease the focal length (so that we lower the profile), we must decrease the trough width of the mirror. A set of parallel troughs is usually used in a single heating panel to achieve a low profile (Figure 5.16).

EXAMPLE

A set of four cylindrical troughs are to be used in a heating panel 4 ft wide by 12 in. thick. If the focal line of each trough is to be just below the glazing, find the rim angle, 2α, of each trough and the maximum possible concentration ratio of the system.

Since the trough width is $W = 2s = (4\text{ ft}/4) = 1\text{ ft} = 12\text{ in.}$, we have $s = 6\text{ in.}$

Substituting into Equation 5.37, we obtain

$$\alpha = \sin^{-1}(s/2f) = \sin^{-1}(6''/24'') = 14.5°$$

or

$$2\alpha = 29°$$

From Equation 5.38, we find

$$\text{cr}_{max} = \left(\frac{2}{0.0046}\right) \sin 14.5° = 109$$

There are many different types of focusing collectors, some of which use combinations of mirrors and lenses. Their theory of operation is beyond the scope of this text. A more complete presentation of concentrators is given in Reference 6. Although concentrators are of interest, their major drawback is that they require some form of seasonal or diurnal tracking. The larger the collector array is, the more complex and costly the tracking apparatus. Also, concentrators cannot harness diffuse solar radiation. Their major advantages are that they can produce very high temperatures and that their thermal losses are low. For low temperature applications such as space heating and hot water supply, flat plates are far more common because they are less costly to manufacture and install.

In Chapter 6 we will see how the fundamentals presented in Chapters 4 and 5 apply to solar heating panels. Our emphasis will be on those flat plate heating panels that use liquids as the transfer fluid.

PROBLEMS

5-1. A beam of unpolarized light is incident at an angle of 30° on an opaque glazing (i.e., $\alpha = 0$) whose index is $n = 1.6$. The incident flux is $F_{inc} = 500 \text{ W/m}^2$.
   (a) Find the fluxes $F_{\parallel}$ and $F_{\perp}$ reflected from the surface.
   (b) Find the average reflection coefficient for the surface.

5-2. Direct solar radiation (unpolarized) is incident at an angle of 10° on a plastic glazing ($n_g = 1.55$) of thickness 1.2 cm. The intercepted flux is $F_{inc} = 400 \text{ W/m}^2$. If the extinction coefficient is $k = 0.3 \text{ cm}^{-1}$ and if surface reflection losses are negligible, find the flux leaving the lower surface of the glazing.

5-3. For a glazing whose extinction coefficient is negligible and whose surface reflectivity is small, show that $R \simeq 2r$ and $T \simeq 1 - 2r$. (For high quality, thin glass plates, the overall reflectance is approximately equal to twice the reflectivity of the glass–air surface.)

5-4. A glazing 1.5 cm thick has an index of refraction of $n = 1.5$ and a bulk extinction coefficient of $k = 0.2\ \text{cm}^{-1}$. Find the overall transmittance of the glazing for direct solar radiation incident at angles of 0, 30, and 60°.

5-5. (a) Show that the solar heating rate per unit area produced within a single glazing is

$$F_{\text{heating}} = \tfrac{1}{2}\left[\frac{(1 - r_{\parallel})(1 - \alpha)}{(1 - r_{\parallel}\alpha)} + \frac{(1 - r_{\perp})(1 - \alpha)}{(1 - r_{\perp}\alpha)}\right] F_{\text{inc}}$$

where α is the bulk transmissivity and where F_{inc} is the incident solar flux. Assume F_{inc} to be direct unpolarized radiation.

(b) Find the heating rate per unit area produced in the glazing described in Problem 5-4 for $\theta_i = 0$, 30, and 60° when $F_{\text{inc}} = 300\ \text{W/m}^2$.

(c) Verify that $R + T = 1$ when $\alpha = 1$.

5-6 Using the results of Problem 5-4, find the overall transmittance of a double glazing for direct solar radiation incident at 0, 30, and 60°. Each glazing has $n = 1.5$, $k = 0.2\ \text{cm}^{-1}$, and a thickness $s = 1.5$ cm.

5-7. A glazing of index $n = 1.6$ is coated with a film to make it antireflective at normal incidence.

(a) What should the film index be?

(b) What should its minimum thickness be if it is to produce zero reflection at $\lambda_0 = 600$ nm?

(c) What will the surface reflection coefficients be at 400, 1000, and 2000 nm? [Use Equation 5.20]

5-8. As shown in the text, the greenhouse effect increases the absorber plate temperature because the glazing transmits solar radiation but *absorbs* thermal radiation. Compare this effect with the situation in which the glazing transmits solar radiation but *reflects* thermal radiation at wavelengths longer than $\lambda_c = 4\ \mu\text{m}$.

5-9. Consider a surface coated with a real selective absorber whose absorptivity to emissivity ratio is $\bar{a}/\bar{\epsilon}$. Suppose such a surface were taken above the atmosphere and oriented toward the sun.

(a) Show that the steady-state temperature is given by

$$T_P = \left(\frac{\bar{a}S}{\bar{\epsilon}\sigma}\right)^{1/4}$$

where S is the solar constant. Assume that only radiative losses from the front face occur.

(b) Find the temperature when $\bar{a}/\bar{\epsilon} = 1$.

(c) Find the temperature when $\bar{a}/\bar{\epsilon} = 0.9/0.2$.

5-10. An ideal selective absorber coating has a sharp cutoff at $\lambda_c = 5\ \mu$m. Find the steady-state temperature of a surface using this coating when the incident flux is 1000 W/m^2. Neglect back losses and convection losses.

5-11. A spherical dish mirror has a radius of curvature of 10″ and subtends a rim angle of $2\alpha = 40°$. Direct solar flux of 800 W/m^2 is incident along the axis.

(a) Find the focal point of the mirror.

(b) Neglecting aberrations, find the maximum concentration ratio for the mirror.

(c) Find the power intercepted by the mirror.

(d) If blurring triples the linear dimensions of the solar image at the focal point, find the actual concentration ratio of the mirror and the average solar flux at the focal point.

5-12. Referring to the mirror of Problem 5-11, if a small absorber disc of radius $b = 1$ cm is placed at the focal point and oriented toward the mirror vertex, find the acceptance angle.

5-13. Imagine that a beam of perfectly monodirectional radiation is incident along the axis of a mirror. As discussed in the text, if diffraction is negligible, a perfect parabolic mirror concentrates this radiation to a point at the focus. Suppose instead that a perfect spherical mirror with a rim angle of 2α is used.

(a) Show that the radius, R', of the blur produced by spherical aberration on an absorber disc at the focal point, $f = R/2$, is

$$R' = \frac{R}{2}\left[(2\cos\alpha - 1)\tan 2\alpha - 2\sin\alpha\right]$$

where R is the radius of the curvature of the mirror.

(b) What happens to the blurring as $\alpha \rightarrow 0$? as $\alpha \rightarrow 45°$?

REFERENCES

1. Andrews, C. L., *Optics of the Electromagnetic Spectrum,* Prentice-Hall, Englewood Cliffs, N.J. (1960), Chapter 8.
2. Born, M. and E. Wolf, *Principles of Optics,* Pergamon, New York (1959), Chapter 1, Sections 5 and 6.
3. Brinkworth, B. J., *Solar Energy for Man,* Wiley, New York (1972), Chapter 4.
4. Daniels, F., *Direct Use of the Sun's Energy,* Ballantine, New York (1964), Chapter 4.
5. Duffie, J. A. and W. A. Beckman, *Solar Engineering of Thermal Processes,* Wiley-Interscience (1980), Chapters 5, 8.
6. Meinel, A. B. and M. P. Meinel, *Applied Solar Energy,* Addison-Wesley, Reading, Mass. (1976), Chapters 4–7.
7. Resnick, R. and D. Halliday, *Fundamentals of Physics,* Wiley, New York (1970), Chapter 36.

CHAPTER 6

Solar Heating Panels

We now apply the principles developed in the preceding chapters to the operation of heating panels, with particular emphasis on liquid-type flat plates. Commercial heating panels vary sufficiently from one manufacturer to another so that no single theory can be applied to all systems. It is therefore impossible to develop a rigorous analysis to describe every design detail of a solar collector. In order to illustrate the essential features of flat plates and their principles of operation, we will use a comparatively simple model. Although the results depend somewhat on the choice of model, the analysis provides an understanding of the operation of a broad class of flat plates.

A simple liquid-type flat plate collector is illustrated in Figure 6.1. It consists of a black absorber plate whose absorptivity for solar radiation is near unity. A selective absorber coating of low thermal emissivity is deposited on the plate. The plate is fitted with tubes or channels so that a transfer liquid can extract the heat produced in the plate when solar energy is absorbed. For air-type collectors, fans or ducts are affixed to the plate and air is circulated either by forced or natural convection. The plate is placed in an airtight insulated container and covered with a glazing. Back and side thermal losses are usually negligible when compared with front

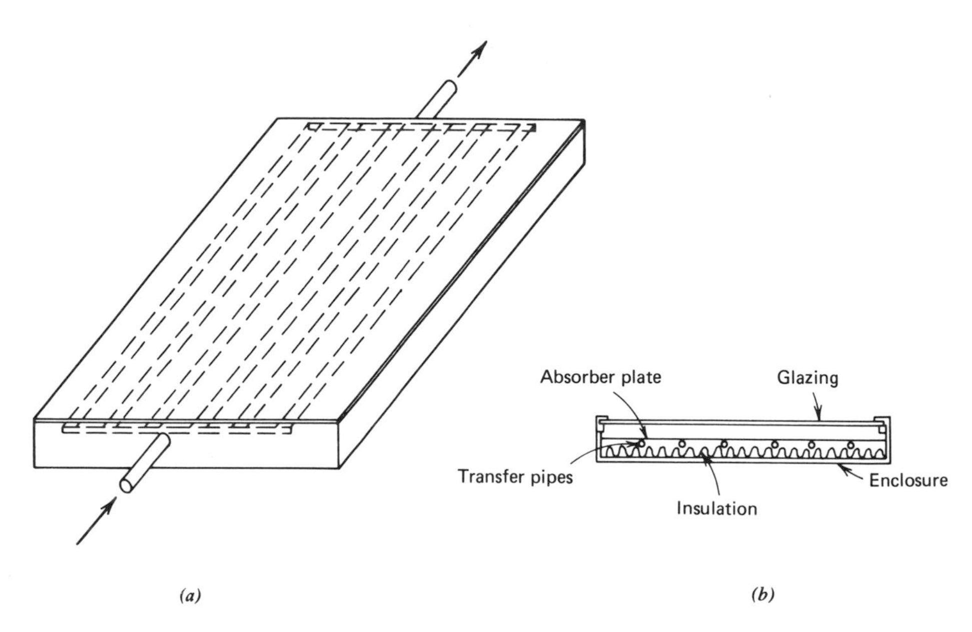

FIGURE 6.1 *(a) A typical flat plate collector showing transfer pipes (dashed lines) under the absorber plate. (b) A cross-sectional view of the same flat plate.*

losses through the glazing. Heating panels can be classified as either active or passive according to whether a pump (or a blower) or natural convection is used to circulate the fluid. We address ourselves to active, liquid-type, flat plates.

The overall efficiency of a solar heating panel is defined as

$$\boxed{\eta = \frac{J_c}{F_{inc}}} \tag{6.1}$$

where

J_c = heating power collected/collector area

and

F_{inc} = solar flux intercepted by the collector

It is convenient to separate the efficiency in Equation 6.1 into two parts—one that characterizes optical performance and a second that describes thermal effectiveness. We write Equation 6.1 as

$$\boxed{\eta = \frac{J_c}{F_{inc}} = \frac{F_{abs}}{F_{inc}} \times \frac{J_c}{F_{abs}} = \eta_{opt} \times \eta_{thermal}} \tag{6.2}$$

where

$$\eta_{opt} = \frac{F_{abs}}{F_{inc}} \qquad \text{and} \qquad \eta_{thermal} = \frac{J_c}{F_{abs}}$$

Here, F_{abs} is the solar flux actually absorbed and is therefore equal to the heating power per unit area produced within the absorber plate.

The optical efficiency was shown (see Equation 5.17) to be given approximately by

$$\boxed{\eta_{opt} \simeq A_p \times T_g} \tag{6.3}$$

where A_p represented the solar absorptance of the plate and T_g the transmittance of the glazing. The optical efficiency can exceed 85 percent even in a relatively unsophisticated panel, provided the angle of incidence of the direct solar radiation is not too large.

The limiting factor in most simple flat plates is the thermal efficiency, $\eta_{thermal}$. Thermal losses to the environment are significant, especially when high output temperatures are required and the panel is operating in a cold and windy environment. In this chapter we will be concerned primarily with thermal performance.

Stagnant Performance of a Solar Heating Panel

When solar energy is incident on a heating panel and heat is not being extracted, the thermal efficiency is zero and all the heat produced by the plate is lost to the surroundings. The absorber plate reaches a uniform temperature $T_{p,s}$ called the static or *stagnation* temperature. For a collector of specific design and tilt, the stagnation temperature of the plate will be a function of both the absorbed solar flux and the ambient conditions, and we may write

$$T_{p,s} = T_{p,s}(F_{abs}, f_{amb}) \tag{6.4}$$

The label f_{amb} represents all relevant ambient parameters such as air temperature, T_a, sky temperature, T_{sky}, and wind velocity. The plate temperature increases as the level of absorbed insolation, F_{abs}, increases and decreases as the ambient conditions become more severe (i.e., cold and windy).

To establish the relation, Equation 6.4, we note that under stagnant conditions—when no useful heat is extracted—the steady state is reached when the absorbed solar flux is equal to the thermal flux loss, that is,

$$\boxed{F_{abs} = J_{loss} \qquad \text{(stagnancy condition)}} \tag{6.5}$$

If the sky temperature is approximately equal to the air temperature, the flux loss can be written

$$\boxed{J_{loss} \simeq \bar{U}_c(T_{p,s} - T_a)} \tag{6.6}$$

where $\bar{U}_c$ is the overall coefficient for heat transfer from the absorber plate to the surroundings. If a linearized form is assumed and $\bar{U}_c$ is taken to be constant, Equations 6.5 and 6.6 can be solved as

$$\boxed{T_{p,s} - T_a \simeq \frac{F_{abs}}{\bar{U}_c}} \tag{6.7}$$

If multiple glazings and selective absorbers are used, $\bar{U}_c$ will be small and $T_{p,s}$ will increase dramatically as the absorbed solar flux increases.

A similar relation to Equation 6.7 can be applied to the glazing temperature of a single-glazed collector. Because the flux transfer from the plate to the glazing and from the glazing to the surroundings must be equal to each other and each must be equal to F_{abs}, we have

$$T_{g,s} - T_a = \frac{F_{abs}}{\bar{U}_{g-a}} \tag{6.8}$$

where $\bar{U}_{g-a}$ is the coefficient for transfer of heat from the glazing to the surrounding air.

The overall heat-transfer or loss coefficient is obtained using

$$\frac{1}{\bar{U}_c} = \frac{1}{\bar{U}_{p-g}} + \frac{1}{\bar{U}_{g-a}} \tag{6.9}$$

where $\bar{U}_{p-g}$ refers to heat transfer from the plate to the glazing. For high quality, single-glazed, solar panels, we usually find that $\bar{U}_{p-g} \ll \bar{U}_{g-a}$. This means that the temperature difference between the glazing and the surrounding air is generally much smaller than the temperature difference between the plate and the glazing.

EXAMPLE

Using Equation 6.7, obtain the relationship between the absorbed flux and the stagnancy temperature of the plate of a collector whose transfer coefficients are $\bar{U}_{p-g} = 10\ \mathrm{W/m^2\text{-}^\circ C}$ and $\bar{U}_{g-a} = 15\ \mathrm{W/m^2\text{-}^\circ C}$. Take $T_a = 10^\circ\mathrm{C}$. Using Equation 6.8, also find the relationship between the glazing temperature and the absorbed flux. Repeat the analysis when a selective coating deposited on the plate reduces the plate-to-glazing coefficient to $\bar{U}_{p-g} = 5\ \mathrm{W/m^2\text{-}^\circ C}$.

In the first case we have

$$\frac{1}{\bar{U}_c} = \frac{1}{10} + \frac{1}{15}$$

or

$$\bar{U}_c = 6\ \mathrm{W/m^2\text{-}^\circ C}$$

From Equation 6.7, we have

$$T_{p,s} = \frac{F_{abs}}{6} + 10 = 0.167 F_{abs} + 10$$

For the glazing we have

$$T_{g,s} = \frac{F_{abs}}{15} + 10 = 0.067 F_{abs} + 10$$

In the second case (i.e., the selective absorber) we have

$$\frac{1}{\bar{U}_c} = \frac{1}{5} + \frac{1}{15}$$

or

$$\bar{U}_c = 3.75\ \mathrm{W/m^2\text{-}^\circ C}$$

so that

$$T_{p,s} = \frac{F_{abs}}{3.75} + 10 = 0.0267 F_{abs} + 10$$

For the glazing we have

$$T_{g,s} = \frac{F_{abs}}{15} + 10 = 0.067F_{abs} + 10$$

Note that although the stagnation temperature of the plate is increased by the selective coating, the stagnation temperature of the glazing remains unaffected. The results of the preceding example are plotted in Figure 6.2.

Curves of the type shown in Figure 6.2 are very useful in analyzing collector performance. Because the stagnation temperature represents an upper limit to the temperature that can be derived from a collector when it is operational, the plots of $(T_{p,s} - T_a)$ versus F_{abs} indicate the maximum temperatures attainable for given insolation levels and ambient conditions. They also provide estimates of

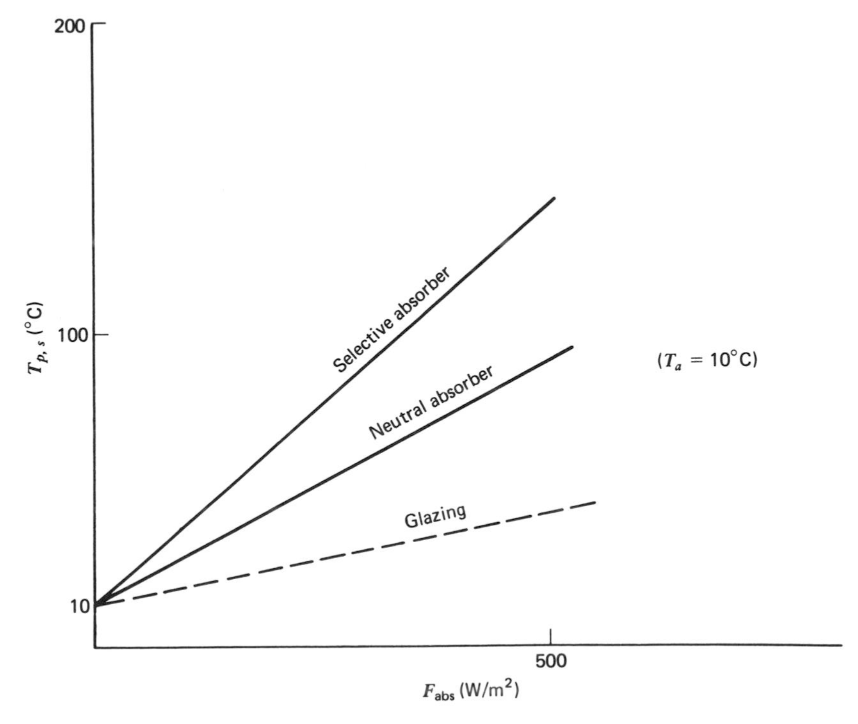

FIGURE 6.2 *Plots of the stagnation temperatures of an absorber plate versus absorbed solar flux. The ambient temperature is $T_a = 10°C$. The selective coating increases the plate temperature; the glazing temperature (dashed lines) is unaffected by the coating.*

thermal losses when the collector is supplying heat and operating at plate temperatures below $T_{p,s}$.

Time Constant of a Collector Under Stagnant Conditions

The time constant of a collector provides an estimate of the time it takes for the absorber temperature to respond to changes in the insolation level. When no heat is being extracted from the collector, the energy balance under nonsteady-state conditions is

$$C_A \frac{dT_p}{dt} = F_{abs} - J_{loss} \simeq F_{abs} - \bar{U}_c(T_p - T_a) \tag{6.10}$$

where C_A is the heat capacity *per unit area* of the plate measured in J/°C-m^2. The term on the left represents the rate at which heat is being stored in the plate.

Let us assume that no solar radiation is incident on the collector and that the plate is in thermal equilibrium with the surroundings; that is, $T_p = T_a$. Next, suppose that the insolation changes suddenly so that the absorbed flux becomes constant and equal to some value, say, F_{abs}. The rise in temperature of the plate as a function of time, t, is given by the solution of Equation 6.10, which is

$$\boxed{(T_p - T_a) = \frac{F_{abs}}{\bar{U}_c}[1 - \exp(-\bar{U}_c t/C_A)]} \tag{6.11}$$

Note that as $t \to \infty$, the new steady state is reached in which

$$(T_p - T_a) \to \frac{F_{abs}}{\bar{U}_c}$$

as expected from Equation 6.7. The time constant for the temperature rise is

$$\boxed{\bar{t} = \frac{C_A}{\bar{U}_c}}$$

After a time equal to $\bar{t}$ the term in the parenthesis on the right-hand side of Equation 6.11 will have reached 0.63 of its final value. After two and three time constants the term in Equation 6.11 will have reached 0.86 and 0.95, respectively, of its final value. The larger the value of $\bar{t}$ is, the more sluggish the collector. It follows that a collector whose heat loss coefficient $\bar{U}_c$ is small and whose heat capacity is large will have a long time constant.

EXAMPLE

Estimate the time constant (under stagnancy conditions) of a collector if its heat loss coefficient is $\bar{U}_C = 6$ W/m^2-°C and if its plate has a heat capacity per unit area

$$C_A = 2000 \text{ J/°C-m}^2$$

Using these values, we have

$$\bar{t} = \frac{C_A}{\bar{U}_C} = \frac{2000}{6} = 333 \text{ sec} = 5.55 \text{ min}$$

When the panel is operational, the time constant is significantly more difficult to determine because heat is being lost to the transfer fluid as well as to the surroundings.

Operational Characteristics of a Flat Plate Collector

A solar heating panel is said to be operational when useful heat is being extracted. The steady-state energy balance equation is

$$\boxed{F_{abs} - J_{loss} = J_C} \tag{6.12}$$

where J_C is the heat flux collected, that is, the amount of useful heat per unit time per unit plate area actually extracted from the panel. If the transfer fluid enters at some temperature $T_{f,i}$ and leaves at a temperature $T_{f,e}$, the collected flux is

$$\boxed{J_C = \dot{\sigma} C_f (T_{f,e} - T_{f,i})} \tag{6.13}$$

where $\dot{\sigma} = \dot{m}/A_P$ is the mass flow rate of the transfer fluid divided by the plate area and C_f is the specific heat of the fluid. In mks units, $\dot{\sigma}$ is in kg/sec-m^2 and C_f is in J/kg-°C.

An analysis of an operational collector is far more complex than that for a stagnant one because more variables are required to describe its performance. Furthermore, when a collector is operational, its plate is no longer at a uniform temperature. The lack of a single well-defined plate temperature makes it difficult to determine the thermal loss from the collector. The temperature profile on the plate depends on its thermal conductivity and on the arrangement of the channels that carry the fluid. Because collector designs differ, no simple analysis can be applied to all flat plates. It is, however, possible to present a general qualitative description for the operation of collectors.

Consider a flat plate collector that is intercepting and absorbing solar energy at a constant rate under fixed ambient conditions. Suppose next that a transfer fluid enters the collector at a fixed temperature $T_{f,i}$. If the mass flow rate per unit area $\dot{\sigma}$ is varied, the output temperature, $T_{f,e}$, and the useful heat flux collected, J_C, will vary. For example, if the flow rate is very small, the fluid extracts very little heat; however, the temperature of the fluid leaving the collector will be high. As a result, the average plate temperature is also high so that thermal losses are large and the collector's efficiency is low. In fact, as $\dot{\sigma} \rightarrow 0$, the plate temperature approaches the stagnation limit; all the heat is lost to the environment and the collector efficiency approaches zero. If, on the other hand, $\dot{\sigma}$ becomes large, the fluid leaves the collector only at a slightly higher temperature than when it enters. The rapidly flowing fluid cools the plate so that thermal losses are small. Therefore although the outlet temperature is not very high, the extraction rate and the thermal efficiency are both increased.

The temperature at which heat is derived is sometimes referred to as its *grade*. When a solar panel operates under a fixed level of insolation and under fixed ambient conditions, the efficiency generally decreases as the grade of heat derived increases. The grade of heat can be adjusted by changing the flow rate of the transfer fluid. Low flow rates produce high grade heat at low efficiency, whereas high flow rates produce low grade heat at high efficiency.

In a typical liquid-type flat plate, an array of channels is affixed to or is part of the absorber plate. Some of the heat developed in the plate is extracted by the transfer liquid. The liquid in turn cools the plate so that its temperature falls below the stagnation limit. However, the plate temperature is not uniform but varies along the plate according to some function $T_p(x, y)$. Points on the absorber plate situated in the vicinity of the channels are somewhat cooler than those in regions between the channels (Figure 6.3). The temperature also increases along the flow, being lowest at the inlet and highest at

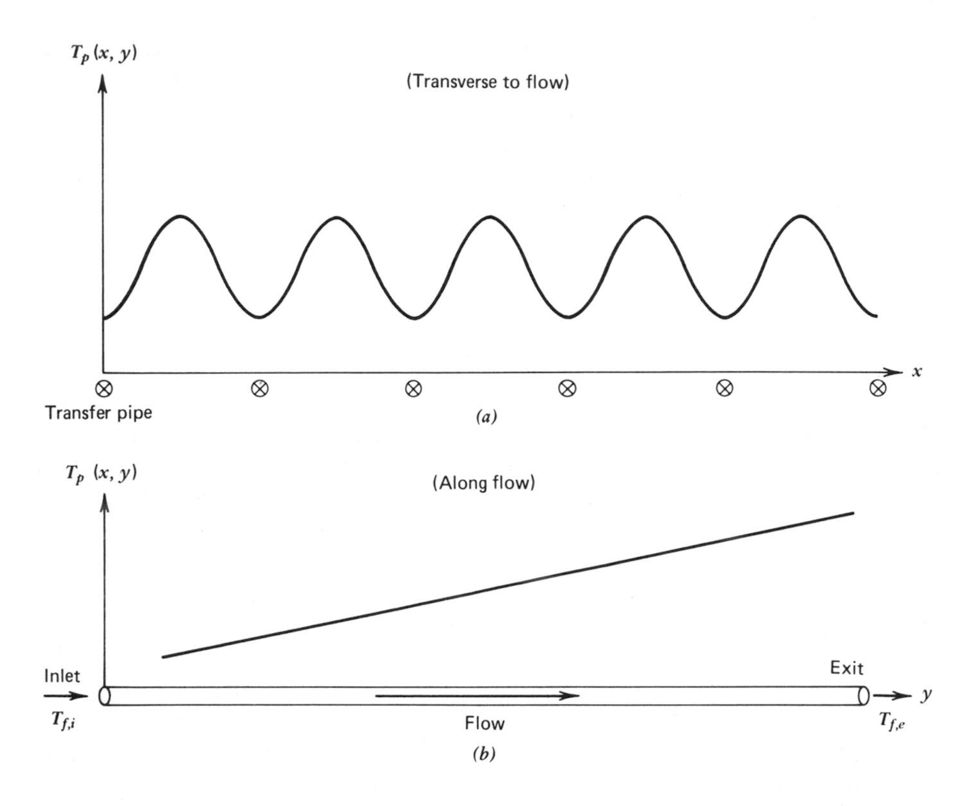

FIGURE 6.3 *(a) The temperature profile on the absorber plate transverse to the flow. (b) The temperature profile along the flow.*

the outlet. The more elaborate the configuration of the channels is, the more complex the pattern $T_p(x, y)$. The steady-state performance of the collector can be described by three equations:

$$\boxed{J_C = F_{abs} - J_{loss}\,(T_p(x, y), f_{amb})} \qquad \text{(energy balance)} \tag{6.14a}$$

$$\boxed{T_{f,e} = T_{f,e}(\dot{\sigma}, T_{f,i}, T_p(x, y))} \qquad \text{(heat exchanger)} \tag{6.14b}$$

$$\boxed{J_C = \dot{\sigma} C_f[T_{f,e} - T_{f,i}]} \qquad \text{(calorimetry)} \tag{6.14c}$$

Equation 6.14a is a statement of conservation of energy and requires that the useful heat collected be equal to the difference between the heat produced in the plate and the heat lost to the surroundings. (Note that the function J_{loss} is difficult to compute because the plate is not at a uniform temperature.) Equation 6.14b is the heat exchanger relation and gives the temperature of the fluid leaving the panel as a function of the flow rate, the fluid temperature at the inlet, and the temperature profile of the plate. This equation can be established only after the collector's design parameters are known. Note that $\dot{\sigma}$ and $T_p(x, y)$ are *not* independent. As we increase the flow rate, the average temperature of the plate decreases accompanied by a decrease in $T_{f,e}$. Equation 6.14c is the calorimetry equation that relates the heat collection rate to the inlet and exit temperatures and to the flow rate of the fluid.

In Equation 6.14 we will assume for the moment that the insolation level, F_{abs}, the ambient parameters, f_{amb}, and the fluid inlet temperature, $T_{f,i}$ are all fixed parameters. The variables are taken to be $\dot{\sigma}$, J_C, $T_{f,e}$ and $T_p(x, y)$. In principle, it should be possible to solve these equations so that each of the last three variables can be expressed as a function of $\dot{\sigma}$. Thus for a fixed insolation level and fluid inlet temperature, and for fixed ambient conditions, the flow rate $\dot{\sigma}$ determines the plate temperature profile, the fluid exit temperature, and the useful heat collection rate.

The solution to Equation 6.14 is difficult to obtain and depends on how the channels are arranged in the plate. To simplify the analysis considerably and to illustrate quantitatively how the variables are related in Equation 6.14, we consider a highly idealized solar flat plate consisting of an absorber plate that is a *perfect* thermal conductor. Such a plate will remain at a uniform temperature regardless of how heat is deposited and extracted. An isothermal plate is particularly simple to treat because the theory of heat transfer can be applied directly. Furthermore, the plate acts as a uniform bath of temperature, T_p, for the transfer fluid. For an isothermal plate, Equation 6.14 can be written

$$\boxed{J_C = F_{abs} - \bar{U}_c(T_p - T_a)} \tag{6.15a}$$

$$\boxed{T_{f,e} = T_{f,i} + (T_p - T_{f,i})[1 - \exp(-H/\dot{\sigma}C_f)]} \tag{6.15b}$$

$$\boxed{J_C = \dot{\sigma}C_f(T_{f,e} - T_{f,i})} \tag{6.15c}$$

In Equation 6.15a it is assumed that $T_a = T_{sky}$ and that the thermal flux loss from the collector can be approximated by a linearized form whose coefficient is $\bar{U}_c$. Equation 6.15b is obtained from the single current heat exchanger equation (Equation 4.31). Note, however, that the absorber plate is not a true heat bath because the temperature T_p decreases as the heat extraction rate increases. The coefficient for the transfer of heat from the plate to the fluid is characterized by the constant H. In this simple model the collector is characterized by two parameters—the heat loss coefficient $\bar{U}_c$ and the heat exchanger coefficient H. An efficient collector is one having a low value of $\bar{U}_c$ and a high value of H. The low $\bar{U}_c$ value results in lower thermal losses, whereas a large value of H effects the efficient exchange of heat to the transfer fluid.

It is straightforward to eliminate the plate temperature T_p in Equation 6.15. After some simplification it can be shown that the thermal efficiency can be written

$$\eta_{\text{thermal}} = \frac{J_C}{F_{\text{abs}}} = \left[1 + \frac{\bar{U}_C}{\dot{\sigma}C_f[1 - \exp(-H/\dot{\sigma}C_f)]}\right]^{-1}\left[1 - \frac{\bar{U}_c(T_{f,i} - T_a)}{F_{\text{abs}}}\right] \tag{6.16}$$

A similar calculation using Equation 6.15 shows that

$$\frac{\Delta T_f}{F_{\text{abs}}} = \frac{(T_{f,e} - T_{f,i})}{F_{\text{abs}}} = \frac{1}{\dot{\sigma}C_f}\left[1 + \frac{\bar{U}_c}{\dot{\sigma}C_f[1 - \exp(-H/\dot{\sigma}C_f]}\right]^{-1}\left[1 - \frac{\bar{U}_c(T_{f,i} - T_a)}{F_{\text{abs}}}\right] \tag{6.17}$$

When the fluid flow rate is held fixed, the thermal efficiency in Equation 6.16 is a linear function of the variable $(T_{f,i} - T_a)/F_{\text{abs}}$. This is plotted in Figure 6.4*a*. The function

$$\frac{\Delta T_f}{F_{\text{abs}}} = \frac{T_{f,e} - T_{f,i}}{F_{\text{abs}}}$$

in Equation 6.17 is also a linear function of the same variable and is plotted in Figure 6.4*b*. Equations 6.16 and 6.17 are particularly simple to apply when the flow rate through the panel is held fixed. The efficiency and the temperature output of the panel can be obtained directly from plots such as those in Figures 6.4*a* and 6.4*b* once $T_{f,i}$, T_a, and F_{abs} are known.

EXAMPLE

A solar heating panel is characterized by the parameters $\bar{U}_c = 6\ \text{W/m}^2\text{-°C}$ and $H = 15\ \text{W/m}^2\text{-°C}$. The transfer fluid, water ($C_f =$

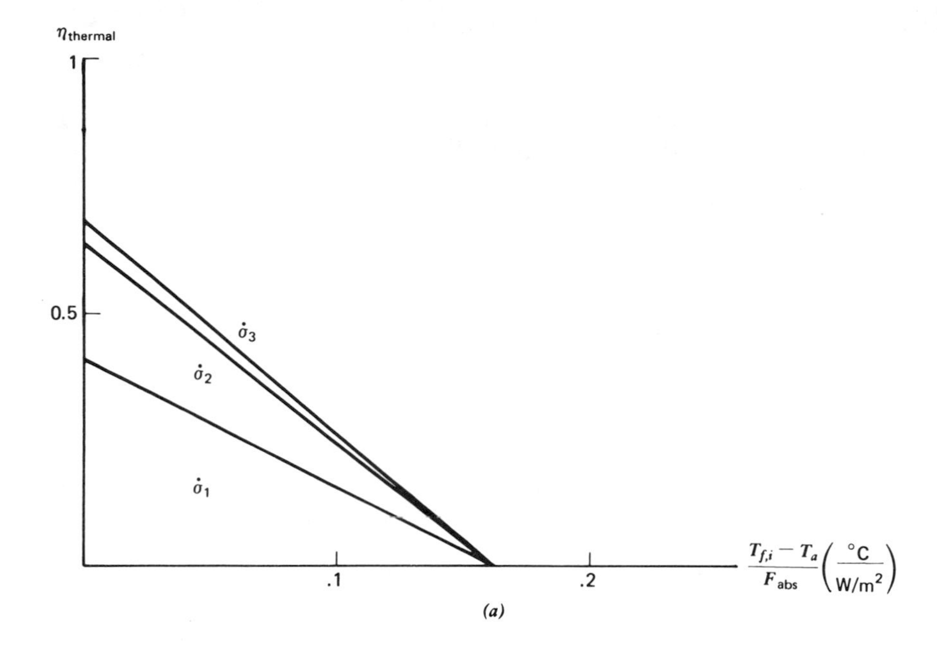

(a)

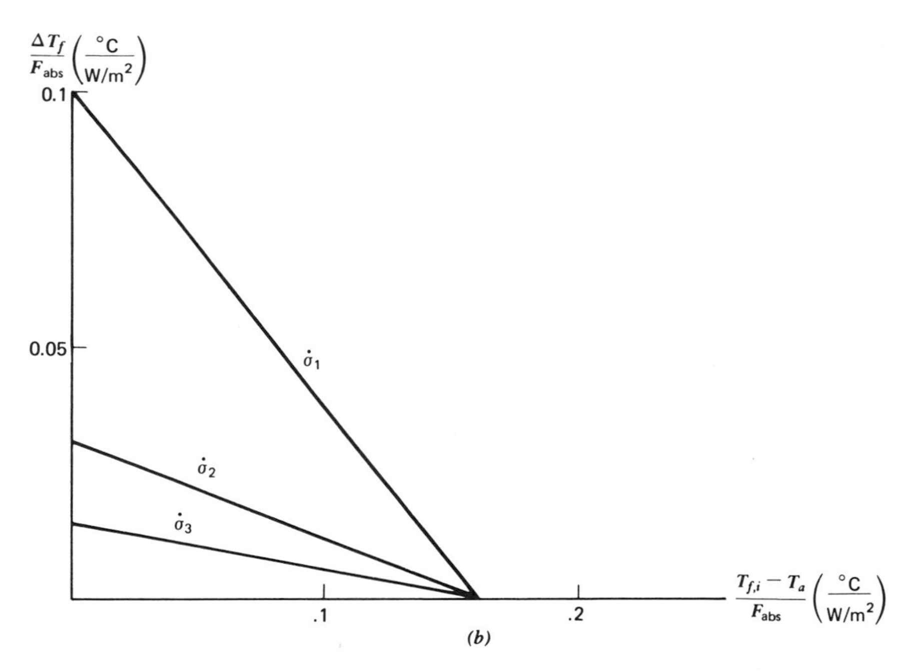

(b)

FIGURE 6.4 *(a) Plots of the thermal efficiency of a typical flat plate versus the variable* $(T_{f,i} - T_a)/F_{abs}$. *(b) Plots of the* $(T_{f,e} - T_{f,i})/F_{abs}$ *versus* $(T_{f,i} - T_a)/F_{abs}$ *for the same collector. These plots are obtained from Equations 6.16 and 6.17 using* $H = 15\ W/m^2 - °C$ *and* $\bar{U}_c = 6\ W/m^2 - °C$. *The flow rates are* $\dot{\sigma}_1 = 0.001\ kg/sec - m^2$, $\dot{\sigma}_2 = 0.005\ kg/sec - m^2$, *and* $\dot{\sigma}_3 = 0.01\ kg/sec - m^2$. *The transfer fluid is water* $(C_f = 4186\ J/kg - °C)$.

4186 J/kg-°C), is flowing at a rate $\dot{\sigma} = 0.001$ kg/sec-m^2 and enters at $T_{f,i} = 25°C$. Find the thermal efficiency of the panel and the exit temperature of the fluid when the ambient temperature is 10°C and the absorbed solar flux is 510 W/m^2.

Equations 6.16 and 6.17 take the form

$$\eta_{\text{thermal}} = 0.404\left[1 - \frac{6(T_{f,i} - T_a)}{F_{abs}}\right]$$

and

$$\frac{\Delta T_f}{F_{abs}} = 0.097\left[1 - \frac{6(T_{f,i} - T_a)}{F_{abs}}\right]$$

Since $F_{abs} = 510$ W/m^2, $T_{f,i} = 25°C$, and $T_a = 10°C$, we find

$$\eta_{\text{thermal}} = 0.33 = 33\%$$

and

$$\frac{\Delta T_f}{F_{abs}} = 0.08$$

Therefore

$$T_{f,e} = (0.08)(510) + 25 = 66°C$$

EXAMPLE

Referring to the preceding example, find the efficiency and the exit temperature of the water when the flow rate is increased to $\dot{\sigma} = 0.01$ kg/sec-m^2.

For the new flow rate, Equations 6.16 and 6.17 take the form

$$\eta_{\text{thermal}} = 0.678\left[1 - \frac{6(T_{f,i} - T_a)}{F_{abs}}\right]$$

and

$$\frac{\Delta T_f}{F_{abs}} = 0.016\left[1 - \frac{6(T_{f,i} - T_a)}{F_{abs}}\right]$$

Using $F_{abs} = 510$ W/m^2, $T_{f,i} = 25°C$, and $T_a = 10°C$, we find

$$\eta_{\text{thermal}} = 0.56 = 56\%$$

and

$$\frac{\Delta T_f}{F_{abs}} = 0.013$$

Therefore

$$T_{f,e} = (0.013)(510) + 25 = 32°C$$

Comparing these examples, we observe that the tenfold increase in the flow rate increases the thermal efficiency from 33 to 56 percent but decreases the output temperature from 66 to 32°C.

The results presented here have been derived from a rather simplified model of a heating panel. In a real operational collector the absorber is not a perfect thermal conductor; hence there is no single well-defined plate temperature. In some applications it is convenient to use some *average* or *effective* value to represent the plate temperature. As the flow rate of the transfer fluid is increased, both the average temperature of the plate and the exit temperature of the fluid decrease. Although this reduces the grade of the heat, it also reduces the heat loss so that the collector efficiency increases.

The actual operational characteristics of heating panels are generally determined by using experimental procedures. In addition to insolation measurements discussed in Chapter 3, these procedures involve the measurement of temperature and flow rate in the panel. In the next few sections we will examine how these measurements are made.

Temperature Measurement

The measurement of temperature is a particularly important procedure in determining the performance characteristics of a solar heating panel. The particular method used depends on the range of the temperature and on the accuracy and type of signal desired. We will restrict our discussion to electrical sensors.

For high temperature measurements two devices commonly used are *metallic resistance* thermometers and *junction thermocouples.* A resistance thermometer uses a metallic element such as platinum or copper as its sensor. The resistance of the element increases with increasing temperature. The temperature coefficient of resistance is defined as

$$\boxed{\beta = \frac{1}{R}\frac{dR}{dt}} \qquad (6.18)$$

where R is the resistance at the temperature T. For most metals β is a slowly varying function of temperature. Furthermore, the coefficient is usually small enough so that over restricted temperature ranges Equation 6.18 can be integrated as

$$\boxed{R = R_0[1 + \beta(T - T_0)]} \tag{6.19}$$

where R_0 is the resistance at some reference temperature, T_0, in the operating range. The temperature can be determined by measuring the resistance of the element (Figure 6.5*a*). The temperature coefficients of resistance of some metals are shown in Table 6.1.

EXAMPLE

The element of a platinum thermometer has a resistance of 0.018 Ω at 0°C. The element's resistance is measured as 0.029 Ω. Find the temperature of the element.

Using Equation 6.19 and setting $\beta_{Pt.} = 0.003°C^{-1}$, we find

$$0.029 = 0.018(1 + 0.003T)$$

so that

$$T = 204°C$$

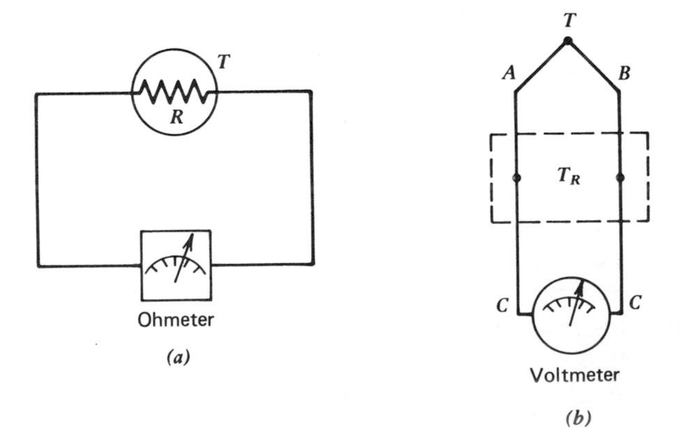

FIGURE 6.5 (*a*) *A simple circuit using a metallic resistance element to measure temperature.* (*b*) *A simple junction thermocouple. Junction AB is the test junction. The reference junctions AC and BC are maintained at a temperature* T_R.

TABLE 6.1 *The Temperature Coefficients of Resistance of Various Metals at* $T_0 = 0°C$

Metal	β(°C^{-1})
Aluminum	0.0039
Copper	0.00393
Gold	0.0034
Iron	0.0050
Lead	0.0043
Nichrome	0.0004
Nickel	0.006
Platinum	0.003
Silver	0.0038
Tungsten	0.0045

A second device for high temperature measurement is the metal-junction thermocouple. When two dissimilar metals are joined to form a junction, a voltage is produced whose magnitude increases with increasing temperature. The mechanism behind this phenomenon will be discussed in Chapter 8. A thermocouple thermometer can be fabricated from two dissimilar wires made from metals A and B, along with a pair of lead-in wires of metal C. Metals A and B could be elements, such as iron and nickel, or alloys, such as constantan (copper and nickel) and chromel (chromium and aluminum). The lead-in wires are typically copper. A thermocouple arrangement is shown in Figure 6.5*b*. The junction AB is the *test* junction used to measure temperature. The junctions AC and AB are *reference* junctions and are maintained at a fixed reference temperature T_R. As the temperature of the test junction increases, the voltage across the lead-in wires also increases. The rate of increase of the voltage with temperature is called the thermocouple or *Seebeck* coefficient

$$\boxed{\mathcal{S}_{AB} = \frac{dV_{AB}}{dT}} \tag{6.20}$$

This coefficient depends on the metals making up the junction and varies somewhat with the temperature of the test junction. However, it is independent of the reference temperature and of the metal used for the lead-in wires. Because the Seebeck coefficient depends on the temperature of the test junction, Equation 6.20 does not lead to a linear relationship between voltage and temperature. For most junctions $\mathcal{S}_{AB}$ is a slowly varying function of temperature so that Equation 6.20 can be integrated as a power series of the form

$$\boxed{V_{AB} \simeq a(T - T_R) + b(T - T_R)^2 + c(T - T_R)^3 + \cdots} \tag{6.21}$$

where a, b, c, and so on are constants for the operating temperature range. If the temperature difference $T - T_R$ is small, a linear approximation involving only the first term in Equation 6.21 is usually sufficient.

It is conventional to use the ice point, 0°C, as the reference temperature. The reference junctions can be immersed in an ice–water mixture or they can be affixed to a surface whose temperature is maintained at the ice point. More recently, electronic ice-point references have been developed consisting of temperature-compensating circuits that automatically simulate ice-point conditions regardless of the ambient temperature. The variation of V_{AB} with T for three different junctions is shown in Figure 6.6.

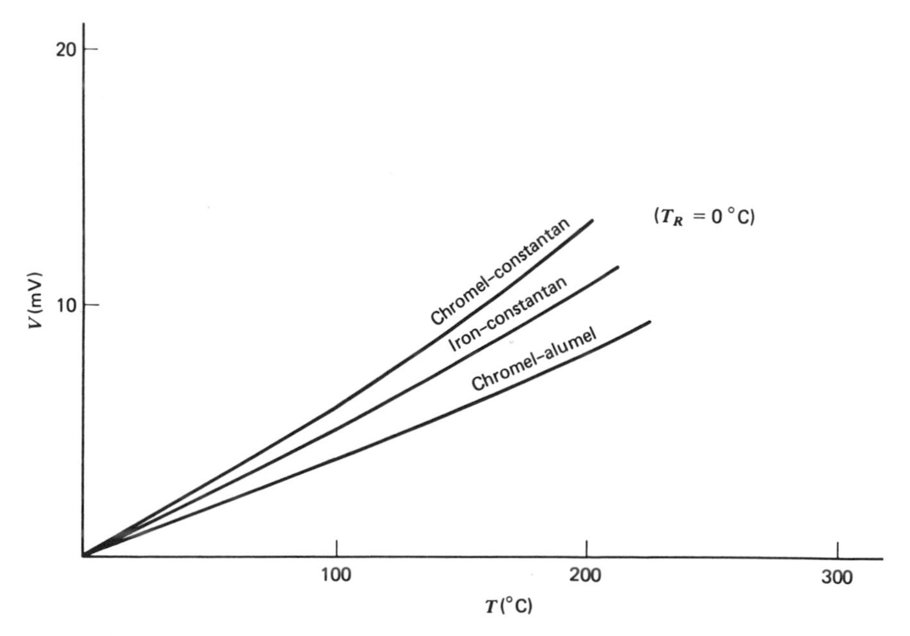

FIGURE 6.6 *The thermocouple voltage versus temperature for various materials. The reference junction in each case is $T_R = 0$°C.*

EXAMPLE

An iron-constantan thermocouple whose reference junction is fixed at $T_R = 0°C$ is used to measure temperatures of 62, 91, and 163°C, respectively. Using the equation $V_{AB} = (4.75 \times 10^{-2})T + (8.52 \times 10^{-5})T^2 - (3.01 \times 10^{-7})T^3$, ($V$ in millivolts), find the corresponding voltages produced by the test junction. Using a linearized approximation of the form $V_{AB} = 0.052T$, find the temperatures corresponding to these voltages and the percent error in each case.

From the iron-constantan equation, we find $V = 3.2$, 4.8, and 8.7 mv, respectively. Using the linearized form, we find $T = 61.5$, 92.3, and 167°C, respectively. The percent errors in the temperatures are 0.8, 1, and 2 percent, respectively.

Thermocouples are difficult to use when the temperature to be measured is close to the reference temperature because the output signal is so small that high-gain dc amplifiers must be used to obtain a reading on a conventional voltmeter. Problems can arise because of both amplifier drift and secondary thermoelectric voltages produced by temperature differences in the circuitry. It is sometimes possible to use a series of thermocouples called a *thermopile* to boost the signal. However, in most situations when temperatures between the ice point and boiling point of water are to be measured, it is more convenient to use semiconductor probes. The type most often used is the *thermistor.*

A thermistor is made from a semiconductor material whose temperature coefficient of resistance, as defined in Equation 6.18, is large and negative. The coefficient is typically 10 times as large as that of a metal, making the thermistor very sensitive to temperature changes. Unlike a metal, the resistance of a thermistor *decreases* markedly with increasing temperature. The relationship between resistance and temperature is quite nonlinear. By using a specially designed thermistor *pair*, one can, nevertheless, obtain a voltage signal that is approximately linear in temperature. A composite probe consisting of a YSI 44018 thermistor pair (Yellow Springs Instrument Co.) is shown in Figure 6.7*a*. An external network required to produce a signal that is linear in the Celsius temperature is shown in Figure 6.7*b*. It consists of a properly chosen pair of resistors, R_1 and R_2, and a battery of emf ε. Auxiliary circuits (Figure 6.7*c*) can be used to produce a voltage signal strictly proportional to the Celsius temperature, that is, a signal that vanishes at $T = 0°C$ and is set to some convenient value at

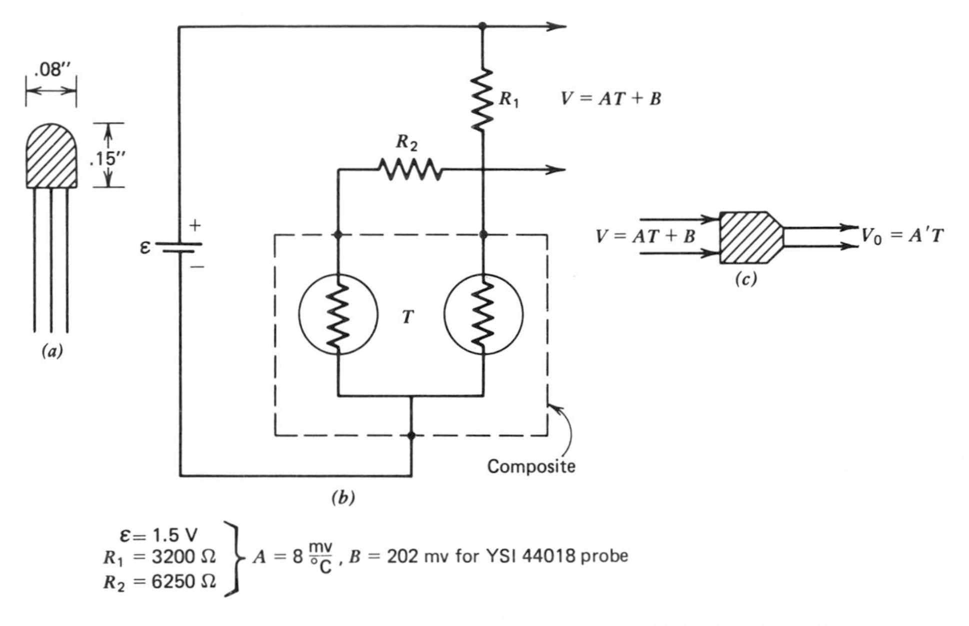

FIGURE 6.7 (*a*) *A thermistor-composite probe, YSI 44018 (Yellow Springs Instrument Company).* (*b*) *The thermilinear network for the composite. The resistor values shown produce linear outputs for* $0°C \leq T \leq 100°C$. (*c*) *An auxiliary network producing an output voltage strictly proportional to the Celsius temperature.*

$T = 100°C$. For example, if this value were 10 V, the output signal would have a scale of 10°C/V.

Integrated circuit sensors are also available to measure the temperature. These devices consist of a complete circuit network enclosed in a probe. An example of a temperature-sensing integrated circuit is the AD 590 (Analog Devices), which is the two-terminal device shown in Figure 6.8*a*. This device is a temperature-dependent current source that produces 1 $\mu a/K$. External components can be added to produce a voltage linear in the Celsius temperature (Figure 6.8*b*). The value of R determines the constants A and B in the output signal. Auxiliary networks can be used to produce a voltage signal that is strictly proportional to the Celsius temperature (Figure 6.8*c*).

For low temperature solar heating panels, that is, those operating between 0 and 100°C, the semiconductor probes are the most versatile. They are very sensitive to temperature changes, require no reference junction, and produce relatively large voltage signals that are linear in the temperature. Once such a voltage is obtained, signal processing can be achieved quite easily.

In testing a solar panel, we often need to obtain a signal cor-

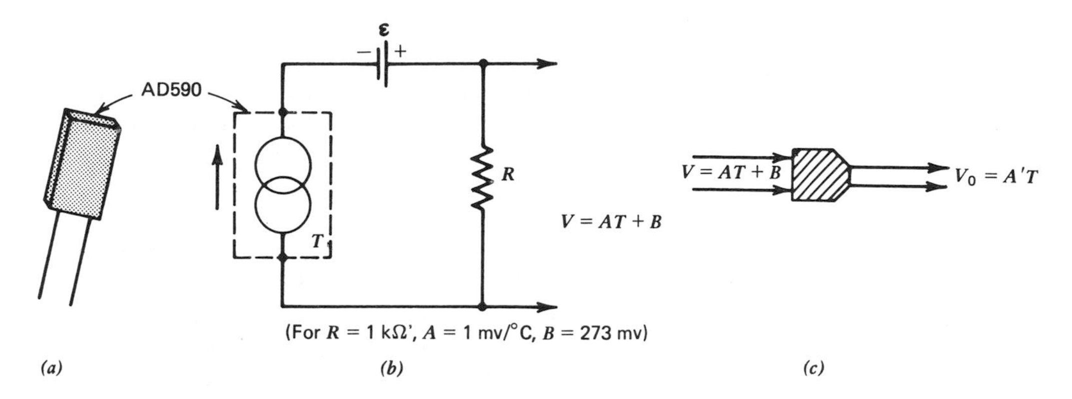

FIGURE 6.8 *(a) A two terminal, integrated circuit, thermal sensor AD 590 (Analog Devices Inc.). (b) A simple circuit using this device to produce a voltage linear in the temperature. (c) An auxiliary network producing an output voltage strictly proportional to the Celsius temperature.*

responding to the difference between two temperatures (e.g., $T_{f,e} - T_{f,i}$). This is easily achieved using a circuit known as a differencing amplifier (Figure 6.9). The basic amplifier used for differencing is called an *operational* amplifier or op-amp. Op-amps are readily available in a variety of integrated circuits. External resistors allow the op-amp to be used as a differencing amplifier so that it produces an output voltage equal to the difference between two input signals. By using differencing amplifiers, we can record temperatures and temperature differences for various points in the solar heating system.

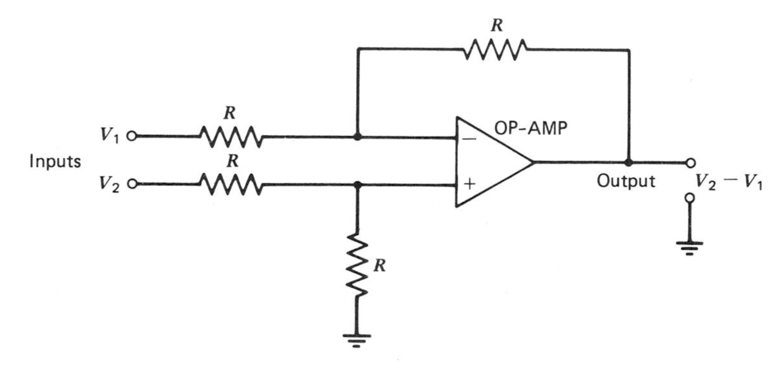

FIGURE 6.9 *Using an operational amplifier (op-amp) to produce an output voltage equal to the difference between two input voltages.*

Flow Measurement

It is somewhat more difficult to measure the mass flow of a fluid than its temperature. The flow rate is best determined using an *in-line* sensor, that is, a sensor that is placed in the stream of flow and directly measures the flow rate. We will consider here only those sensors that can be used to produce electrical signals.

One basic type of in-line or direct sensor is the paddle wheel or anemometer arrangement (Figure 6.10*a*). The paddle wheel is placed directly in the stream of flow. The angular velocity of the wheel is proportional to the fluid flow velocity, which in turn can be related to the volume flow rate (e.g. m^3/sec). The angular velocity is measured by a tachometer that produces a signal proportional to the volume flow rate. The mass flow rate is determined from the volume flow rate $\dot{V}$ using the relation

$$\dot{m} = \rho\dot{V}$$

where ρ is the mass density of the fluid.

A second kind of in-line sensor is the magnetic type. It has no moving parts and is useful for measuring flow rates of conducting liquids. Even water with small amounts of impurities is slightly conductive. Whenever a conducting fluid flows across a magnetic field, a voltage, directly proportional to the flow velocity, is induced in the fluid (Figure 6.10*b*). This voltage signal can be used to indicate the flow rate.

It is also possible to measure the flow rate of a fluid indirectly. For example, in the case of liquids there exists a type of pump known as a tubing or *peristaltic* pump (Figure 6.10*c*). This pump forces liquid through a collector by alternately compressing and expanding adjacent sections of a flexible tube, forcing the liquid from one section to another along the tube. The tube is compressed and expanded by an odd-shaped rotor. During each revolution a fixed volume of fluid is transferred along the tube. The angular velocity of the rotor, measured with a tachometer, is directly related to the volume flow rate. The signal produced is approximately proportional to the volume flow rate.

There are a number of other devices available for measuring flow rates. One such device consists of a pair of temperature probes situated on opposite sides of a heating element. The element and the probes are placed directly in the stream flow. The fluid leaving the heating element is necessarily warmer than the fluid entering. The temperature difference in the fluid decreases as the flow rate increases and vice versa. Thus the temperature difference signal produced by the probes can be used to represent the flow rate.

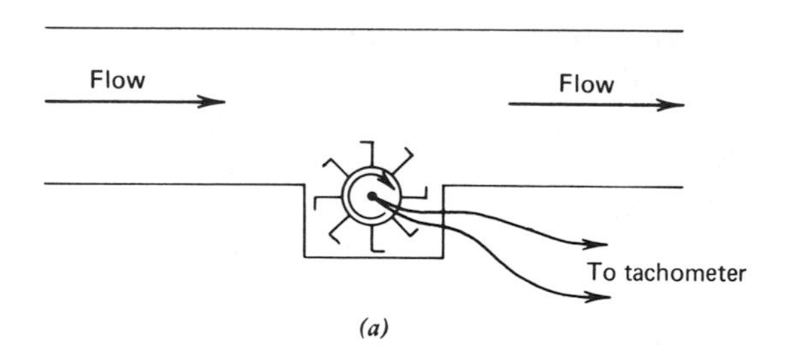

(a)

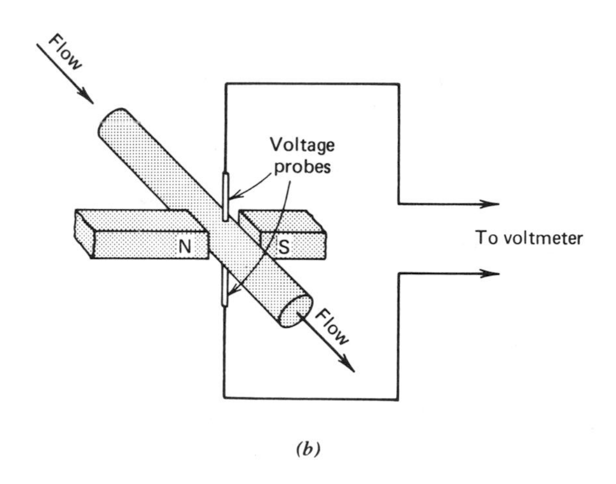

(b)

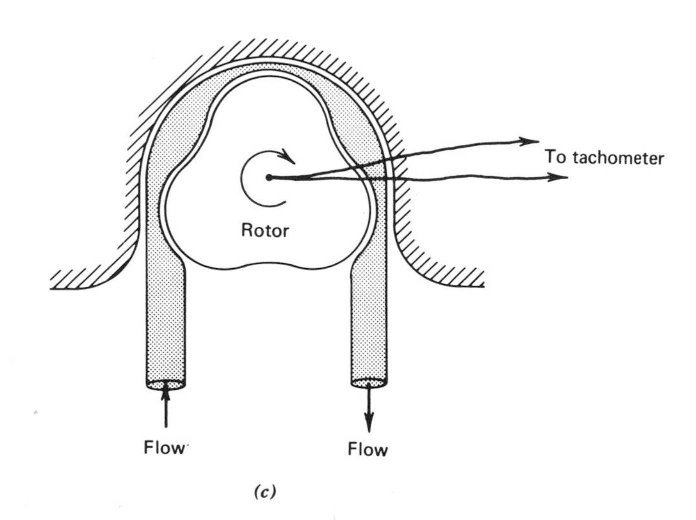

(c)

FIGURE 6.10 *(a) A paddle wheel or anemometer type flow meter. (b) A magnetic type flow meter. (c) Using a peristaltic pump to measure flow rate.*

Another type of device used to determine flow rates measures pressures. When a fluid flows from a wider to a narrower channel, the flow velocity increases, resulting in a drop in pressure. The pressure difference between a point in the wider channel and one in the narrower channel increases as the flow rate increases. Hence the pressure difference signal can be used to represent flow rate.

The problem with the thermal and pressure devices described is that the output signals are not linear in the flow rates, which makes signal processing difficult.

Measurement of Heat Collection Rate

To determine the rate of heat collection, we must make three measurements—the inlet fluid temperature $T_{f,i}$, the exit fluid temperature $T_{f,e}$, and the mass flow rate $\dot{m}$. It is assumed that the specific heat of the fluid C_f is known and is constant. According to Equation

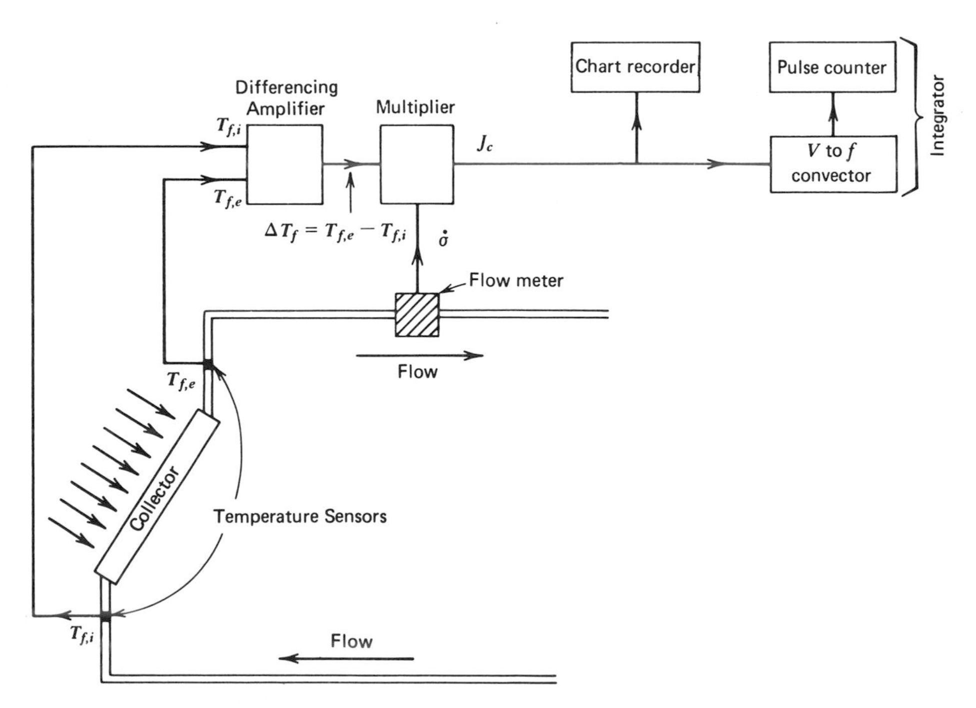

FIGURE 6.11 *A flow chart illustrating how the performance of a solar heating panel is monitored.*

6.13, the heat collected per unit area of the collector is

$$J_C = \dot{\sigma} C_f (T_{f,e} - T_{f,i})$$

where $\dot{\sigma} = \dot{m}/A$. Consequently, the collected flux is proportional to $\dot{m}(T_{f,e} - T_{f,i})$. To obtain a direct readout of the collected flux (e.g., in W/m^2), the signal corresponding to $\Delta T_f = T_{f,e} - T_{f,i}$ must be multiplied by the signal corresponding to the mass flow rate using an appropriate scale factor. (There are a number of integrated circuits available that can multiply two analog signals.) The resulting signal can be displayed on a chart recorder. It can also be applied to a digital integrator, consisting of a voltage-to-frequency convertor and a pulse counter, to give the total heat per unit area of the panel collected during a given period. The arrangement is shown in Figure 6.11.

Temperature Controllers

In many applications it is desirable to fix the operating temperature of a solar heating panel or the temperature of the fluid leaving the panel. To achieve this as the insolation level varies, we need to vary the mass flow rate of the fluid. For example, if the collectors were operating at a fixed mass flow rate $\dot{\sigma}$, then a large increase in insolation could result in high operating temperatures, excessive thermal losses, and low efficiency. On the other hand, with low insolation levels, the same flow rate might produce unacceptably low output temperatures. A controller adjusts the flow rate to bring about temperature stability in the system. We will consider three classes of devices—the proportional, the on–off, and the ramp controllers.

A proportional controller is a device that sets the pump speed in proportion to some temperature signal. For example, the flow rate can be set as proportional to the temperature difference, $\Delta T_f = T_{f,e} - T_{f,i}$. As the insolation level increases, the resulting increase in ΔT_f produces an increase in the flow rate. The increased flow rate in turn tends to minimize the increase in ΔT_f. The opposite effect occurs when the insolation decreases. Using this type of negative feedback, the proportional controller helps to stabilize the operating temperature of the system. However, such a controller cannot fix the operating temperature to any preset value. The actual ΔT_f of the panel is determined by the intersection between two curves—the controller curve and the heating panel curve (Figure 6.12). The controller curve is a plot of the $\dot{\sigma}$ versus ΔT_f produced by the proportional controller. Each heating panel curve is a plot of the flow

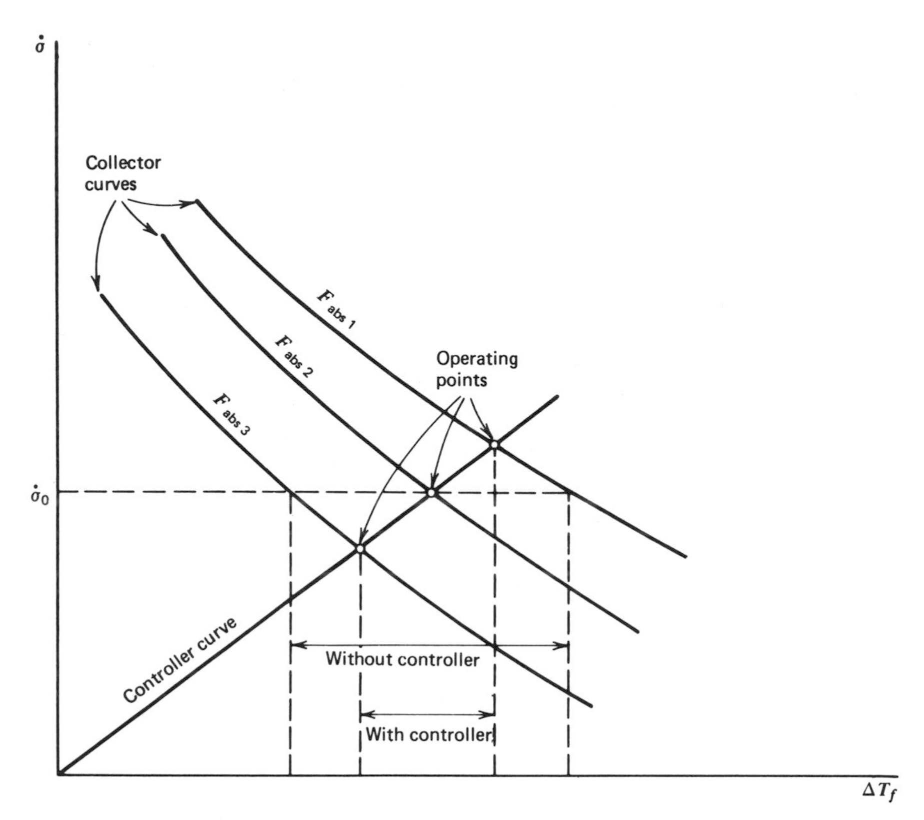

FIGURE 6.12 *Illustrating how a proportional controller reduces temperature variation in a solar heating panel. The variation of temperature at some fixed flow rate, $\dot{\sigma}_0$, is shown by the horizontal dashed line. The (absorbed) solar flux is varying from $F_{abs,1}$ to $F_{abs,3}$.*

rate $\dot{\sigma}$ versus ΔT_f for a fixed insolation level. As the flow rate increases, the temperature difference decreases. For a fixed insolation level the steady-state flow rate $\dot{\sigma}$ and temperature difference ΔT_f are obtained from the intersection of the two curves. As the insolation level changes, the operating point must move along the curve determined by the controller. Hence it can be seen that as the insolation changes, ΔT_f also changes, but by an amount smaller than if $\dot{\sigma}$ were kept fixed.

EXAMPLE

Assume that in a certain range of operation, a heating panel's curves can be approximated by the lines

$$\dot{\sigma} = -20\Delta T_f + 0.7F_{abs}$$

where $\dot{\sigma}$ is the flow rate in g/min-m^2 and F_{abs} is the absorbed solar flux in W/m^2. A proportional controller is used that adjusts the flow rate according to

$$\dot{\sigma} = 15\Delta T_f$$

Find the flow rates produced by the controller and the temperature differences produced by the panel when the absorbed solar fluxes are $F_{abs} = 800$, 600, and 400 W/m^2. Compare these values of ΔT_f with those obtained when the flow rate is held fixed at $\dot{\sigma} = 240$ g/min-m^2.

Eliminating $\dot{\sigma}$ from the panel and controller equations, we find

$$\Delta T_f = \frac{0.7F_{abs}}{35}$$

Thus for $F_{abs} = 800$, 600, and 400 W/m^2 the respective temperature differences are $\Delta T_f = 16$, 12, and 8°C. Substituting back into the controller equation, we find the respective flow rates to be $\dot{\sigma} = 240$, 180, and 120 g/min-m^2. If the flow rate is fixed at 240 g/min-m^2, the solar panel equations give $\Delta T_f = 16$, 9, and 2°C, respectively.

Thus although the proportional controller does not keep ΔT_f constant, it does reduce the magnitude of its variation when compared with the situation of a fixed flow rate.

An on–off controller is a device capable, in principle, of maintaining a preset temperature or a temperature difference in a solar collector. When a certain preset temperature is reached, the controller turns the pumps on. The large fluid flow causes the temperature to fall. When this happens, the pumps turn off and the cycle begins again. The two flow states may be high–low instead of on-off. Two-state controllers are often used to open valves that drain a liquid-type collector when subfreezing conditions exist, thus preventing freeze-up when water is used as a transfer fluid. If the solar heating system responds too rapidly when the flow turns on and off, frequent cycling will result. To prevent this, we can use a *hysteresis* or *dead-band* circuit. This circuit allows the pumps to be turned on at a temperature T_{on} and off at a lower temperature T_{off}. Within the dead band, that is, for $T_{off} < T < T_{on}$, the pumps may be either on or off. Although hysteresis prevents frequent cycling, it limits the accuracy to which the temperature can be regulated.

A third type of controller used to regulate temperature is a ramp controller. This device adjusts or modulates the pump speed continuously in a manner analogous to a proportional controller. However, the ramp controller can, in principle, be set to maintain a fixed preset output temperature $T_{f,e}^{(set)}$. If the sensed output temperature is above this preset value, the pump speed is gradually increased or ramped up until the preset value is reached. As $T_{f,e}$ approaches the preset value, the ramp levels off and the required fluid flow remains constant. Should the insolation level drop and $T_{f,e}$ fall below the preset value, the pump speed is ramped down until the temperature again returns to the preset value. The rate of the ramp must be set so that it is compatible with the time constant of the particular heating panel used. A ramp rate that is too large causes overshoot whereas a rate that is too small produces too slow a response to changing conditions. The choice of controller to be used depends on the particular collector system and on the tolerance of temperature control required.

Overall Performance of Heating Panels

The criteria for rating the performance of solar heating panels are not always easy to establish. One difficulty is the large number of variable parameters involved. For example, some panels are intended to heat cold fluids whereas others are designed to heat warm fluids. Some are required to achieve the heating process alone whereas others are designed to be preheaters, allowing secondary devices such as auxiliary heaters to complete the heating process. Furthermore, some operate at a fixed flow rate so that the output temperature varies as the solar flux varies whereas others use a controller to maintain a fixed output temperature.

If the ambient conditions T_a and the fluid inlet temperature $T_{f,i}$ remain fixed throughout the day, the efficiency for a fixed flow rate can be obtained using plots similar to those of Figure 6.4. As F_{inc} (and consequently F_{abs}) varies throughout the day, the efficiency changes. If a controller is used to keep the operating temperature low, the panel will be more efficient. Thermal losses decrease as the operating temperature of the panel approaches the ambient temperature. Thus a panel can produce relatively high temperatures on a warm summer day; to produce such temperatures on a cold winter day, the collector would have to operate at near-zero efficiency.

We conclude this chapter with the following points regarding solar heating panels.

For a flat plate heating panel to be more efficient:

1. Apply highly absorptive coatings on the plate and use highly transmissive glazings to increase optical efficiency.
2. Use selective absorber coatings (i.e., highly reflective in thermal infrared region) with good insulation on back and sides of absorber plate. If possible, partially evacuate the enclosure or at least make it airtight to reduce convective losses. Use more than one glazing where feasible.
3. Arrange the channels or ducts as uniformly as possible across the absorber plate with good thermal contact so that heat exchange to transfer fluid is efficient.

To obtain maximum efficiency from a given flat plate panel:

1. Allow the panel to track the sun so that it remains normal to the sun's rays for as long as possible during the day. If this is not feasible, fix the panel in a position so that the obliquity of the sun's rays is as small as possible throughout the day. This increases the intercepted flux as well as the optical efficiency.
2. Operate the panel at a temperature as close to the ambient temperature as possible. Under severe conditions use the panel as a preheater to supply a warm rather than a hot fluid. Use a secondary heater to complete the process. Also use well-insulated pipes or ducts to carry the fluid to and from the panel.

In Chapter 7 we will deal with the peripheral equipment that interfaces with the panel to make a complete heating system. The complete system must be designed to increase overall efficiency and to provide storage capability so that heat can be derived during sunless periods. We will place particular emphasis on systems for space heating and hot water supply.

PROBLEMS

6-1. A single-glazed flat plate has a plate-to-glazing transfer coefficient, $\bar{U}_{p-g} = 6$ W/m^2-°C, and a glazing-to-air coefficient, $\bar{U}_{g-a} = 12$ W/m^2-°C.
 (a) Find the overall transfer coefficient $\bar{U}_c$ for the panel. Neglect back losses.
 (b) Find the stagnation temperature and the glazing tem-

perature when the absorbed solar flux is $F_{abs} = 400\ W/m^2$ and when $T_a = 0°C$.

(c) Plot $(T_{p,s} - T_a)$ and $(T_{g,s} - T_a)$ versus F_{abs}.

6-2. The plate of the collector in Problem 6-1 is coated with a selective absorber and the region above the plate is partially evacuated so that the plate-to-glazing transfer coefficient is reduced to $\bar{U}_{p-g} = 4\ W/m^2\text{-}°C$. Repeat (a), (b), and (c), showing that the glazing temperature is unaffected.

6-3. An absorber plate of a collector is made of copper ($C = 389\ J/kg\text{-}°C$) and has an area $2\ m^2$ and a mass of 20 kg. The coefficient for heat transfer to the surroundings is $\bar{U}_c = 8\ W/m^2$.

(a) Find the time constant of the collector, in minutes.

(b) If the insolation suddenly changes from zero to some constant value, how many minutes will it take for $(T_p - T_a)$ to reach 60 percent of the stagnation limit?

6-4. A collector of area $1.8\ m^2$ is characterized by heat transfer coefficients $\bar{U}_c = 7\ W/m^2\text{-}°C$ and $H = 12\ W/m^{2}°C$. Water ($C_f = 4186\ J/kg\text{-}°C$) is used as a transfer fluid. The water is flowing at $\dot{m} = 420\ g/min$.

(a) Find $\dot{\sigma}$ in $kg/sec\text{-}m^2$.

(b) Using Equation 6.16, plot the thermal efficiency $\eta_{thermal}$ versus $(T_{f,i} - T_a)/F_{abs}$.

(c) Plot the function $\Delta T_f / F_{abs}$ versus $(T_{f,i} - T_a)/F_{abs}$.

(d) Find $\eta_{thermal}$ and the exit temperature, $T_{f,e}$, when $F_{abs} = 600\ W/m^2$, $T_a = 10°C$, and $T_{f,i} = 15°C$.

(e) How much thermal power Q_c is being collected?

(f) Find T_p, assuming the plate to be at a uniform temperature.

(g) If the optical efficiency is 88 percent, how much flux is actually incident? What is the overall efficiency of the panel?

6-5. As shown in the text, increasing the flow rate results in a lower exit temperature but a higher thermal efficiency.

(a) Show, however, using Equation 6.16, that for very high flow rates (i.e., $\dot{\sigma} \rightarrow \infty$) the thermal efficiency becomes independent of $\dot{\sigma}$.

(b) Show, nevertheless, that $T_{f,e}$ continues to approach $T_{f,i}$ as $\dot{\sigma} \rightarrow \infty$.

(c) Show that in this high flow limit, the thermal efficiency can only be improved by increasing H. Explain why. (This problem illustrates that poor heat transfer from the plate to the transfer fluid limits the thermal efficiency even when high flow rates are used.)

6-6. A resistance thermometer uses a metallic element. Its resistance at 0 and 100°C is 0.5 and 0.695 Ω, respectively.
   (a) Find β for the metal, using 0°C as the reference temperature.
   (b) Find the temperature of the element when its resistance is 0.7 Ω, assuming that β is constant.

6-7. A thermocouple whose reference junctions are maintained at $T_R = 0°C$ produces a voltage given by

$$V \simeq 16.34T - 0.021T^2 \qquad \text{for} \qquad 0°\text{C} \leq T \leq 200°\text{C}$$

where V is in microvolts ($1\ \mu V = 10^{-6}$ V) and T is in °C.
   (a) If the voltage produced is 2000 μV, find T.
   (b) Compare the temperature errors for $V = 750$, 1000, and 1500 μV if the approximate equation $V \simeq 14.3T$ is used.

6-8. A system of sensors is designed to monitor the performance of a solar heating panel of area 1 m^2 that uses water as a transfer fluid. The temperature sensors produce voltage signals corresponding to 10°C/V. The flow meter produces a signal corresponding to 500 g/min-V. If the output to a recorder is to represent collected power corresponding to 100 W/V, what should the scale or gain factor of the multiplier in Figure 6.11 be?

6-9. Assume that the operating curves of a solar panel, such as those shown in Figure 6.12, can be represented by straight lines of the form

$$\dot{\sigma} = -25\Delta T_f + 0.6F_{abs}$$

where $\dot{\sigma}$ is in g/min-m^2 and F_{abs} is in W/m^2. A proportional controller is used that fixes the flow rate according to

$$\dot{\sigma} = 10\Delta T_f$$

   (a) Plot the controller curve and the collector curves for $F_{abs} =$ 400, 500, and 600 W/m^2.
   (b) Find the operating point (i.e., ΔT_f and $\dot{\sigma}$) when $F_{abs} =$ 500 W/m^2.
   (c) If F_{abs} varies between 400 and 600 W/m^2, find the variation of ΔT_f and $\dot{\sigma}$ for the controller in (a).
   (d) If the flow rate is held fixed at the value determined in (b), find the variation of ΔT_f.
   (e) If the temperature is held fixed (say, by a ramp controller) at the value found in (b), find the variation of $\dot{\sigma}$.

REFERENCES

1. Brinkworth, B. J., *Solar Energy for Man,* Wiley, New York (1972), Chapter 5.
2. Duffie, J. A. and W. A. Beckman, *Solar Engineering of Thermal Processes,* Wiley, New York (1980), Chapters 6 and 7.
3. Lunde, P. J., *Solar Thermal Engineering,* Wiley, New York (1980), Chapters 4, 5, 7, and 8.
4. McDaniels, D. K., *The Sun,* Wiley, New York (1979), Chapters VII and VIII.
5. Meinel, A. B. and M. P. Meinel, *Applied Solar Energy,* Addison-Wesley, Reading, Mass. (1976), Chapters 11–13.
6. Sheingold, D. H., *Transducer Interfacing Handbook,* Analog Devices, Norwood, Mass. (1980), Chapters 1–2, 7–10, and 13.
7. Szokolay, S. V., *Solar Energy and Building,* 2nd ed., Wiley, New York (1977), Chapters 2 and 3.
8. Zemansky, M. W., *Heat and Thermodynamics,* 4th ed., McGraw-Hill, New York (1957), Chapter 1.

CHAPTER 7

Solar Heating Systems

In the previous chapters we have considered both the theoretical and experimental aspects of individual solar heating panels. In this chapter we treat the solar heating system as a whole. In particular, we will consider how an array of panels is arranged and interfaced with other components to make a complete, efficient, and cost-effective heating system.

Array Orientation

Unlike concentrators that normally require daily tracking, flat plates can operate with a fixed orientation. Although tracking will improve the performance of a flat plate, the gains are usually more than offset by the increased costs of manufacture and maintenance of the tracking apparatus.

A fixed array of flat plates should be oriented so that the daily intercepted flux is largest during the operating season. Because it is difficult if not impossible to optimize tilt with respect to diffuse solar radiation, we will optimize with respect to the direct component only. Using Equations 2.10, 2.16, and 3.13, we obtain the daily direct flux intercepted by a fixed array with tilt coordinates Δ and ψ by integrating over the hours of available insolation,

$$F_{\text{dir}}^{\text{daily}}(\Delta, \psi) = S \int_{t_1}^{t_2} e^{-\tau/\cos Z} \cos \theta \, dt \tag{7.1}$$

where

$S = 1352$ W/m² (solar constant)
τ = optical thickness of the atmosphere
Z = solar zenith angle (varies with solar time, see Equation 2.10)
θ = obliquity angle of sun's rays to array (varies with solar time, see Equation 2.16)
t_2, t_1 = solar times in between which $\cos \theta$ is positive, that is, when the sun's rays strike the array on the front face
Δ, ψ = tilt and azimuth of the array

In Table 7.1 we have listed daily fluxes obtained from Equation 7.1. The values are for an array situated at a colatitude $L' = 49°$ and for an atmosphere of optical thickness $\tau = 0.3$. For south-facing arrays the daily flux during any season is largest when the tilt is set so that the array is approximately perpendicular to the sun's rays at solar noon, that is, when

$$\Delta \simeq Z_{\text{noon}} = |D' - L'|$$

At the winter solstice a vertical south-facing array is more effective than a horizontal one. The opposite is true at the summer solstice. As seen from Table 7.1, an east- (or west-) facing array is generally less effective than a southerly array. Note also that for an east- (or west-) facing array the intercepted flux increases as the array becomes more horizontal. If a solar array, situated at $L' = 49°$, is to be used to provide space heating during the winter solstice, a southerly tilt of 64.5° is optimum. This tilt still provides adequate heating at the equinoxes. However, if the array is used to supply hot water to a summer house, a tilt of ~17.5° is more effective.

The results presented in Table 7.1 are based on an oversimplified model and are for comparison purposes only. For one thing, diffuse radiation and the radiation reflected toward the array from the underlying terrain have not been included. Also the flux predicted by Equation 7.1 assumes that the insolation pattern is symmetric about solar noon. Morning and afternoon insolations are often different. Consequently, east- and west-facing panels do not in

TABLE 7.1 *Approximate Clear Day Direct Fluxes on Surfaces at Various Orientations*[a]

Season	Tilt (Δ) Degrees (°)	Flux (kw-hr/m^2-day) (south, $\psi = 0°$)	Flux (kw-hr/m^2-day) (east-west, $\psi = \pm 90°$)
	0 (horizontal)	1.4	1.4
Winter	17.5	2.3	1.3
solstice	41	3.2	1.2
($D' = 113.5°$)	64.5	3.5	0.98
	90 (vertical)	3.3	0.69
	0 (horizontal)	4.5	4.5
Equinox	17.5	5.5	4.3
($D' = 90°$)	41	6.0	3.8
	64.5	5.5	3.0
	90 (vertical)	3.9	2.0
	0 (horizontal)	7.6	7.6
Summer	17.5	7.6	7.3
solstice	41	6.4	6.2
($D' = 66.5°$)	64.5	4.3	4.7
	90 (vertical)	1.5	3.0

[a]The results, derived from Equation 7.1, apply to surfaces situated at a colatitude $L' = 49°$ when the atmospheric optical thickness is $\tau = 0.3$.

general receive equal amounts of sunlight. A more precise treatment would require experimental data regarding the daily insolation available to inclined surfaces.

Array Size

The size of the array is determined by such factors as ambient conditions, heating needs, array efficiency, and available insolation. Suppose, for example, the daily heating needs of a home during the heating season are 100 kw-hr/day ($\sim 3.4 \times 10^5$ Btu/day) and that the available daily insolation on the array is 4 kw-hr/m^2-day. Also assume that each panel has an area of 1.5 m^2, an efficiency of 50 percent, and that one-third of the heating will come from auxiliary heaters. The solar heating requirement is then 66.7 kw-hr/day. Since the array is 50 percent efficient, the required array area is

$$A = \frac{P^{(\text{daily})}}{F^{(\text{daily})} \times \eta} = \frac{66.7}{4 \times 0.5} = 33.3 \text{ m}^2$$

Since each panel has an area of 1.5 m^2, the number of panels required is ~22.

Series and Parallel Arrays

A solar array can consist of heating panels arranged in either series, parallel, or a combination of the two, as shown in Figure 7.1. A large array will not produce a higher temperature than is possible with a single collector. An array of N panels, however, does have the potential for collecting N times the amount of heat than is possible with a single panel. To collect this heat, the fluid flow rate supplied to the array must be increased by a factor of N. In a series array, the outlet of one panel is coupled directly to the inlet of the next one (Figure 7.1*a*). Consequently the increased flow must pass through each and every panel of the array. As the fluid velocity increases so does resistance to flow. Furthermore, the longer the overall length of the pipe through which the fluid is flowing, the greater the flow resistance. Therefore a long series array

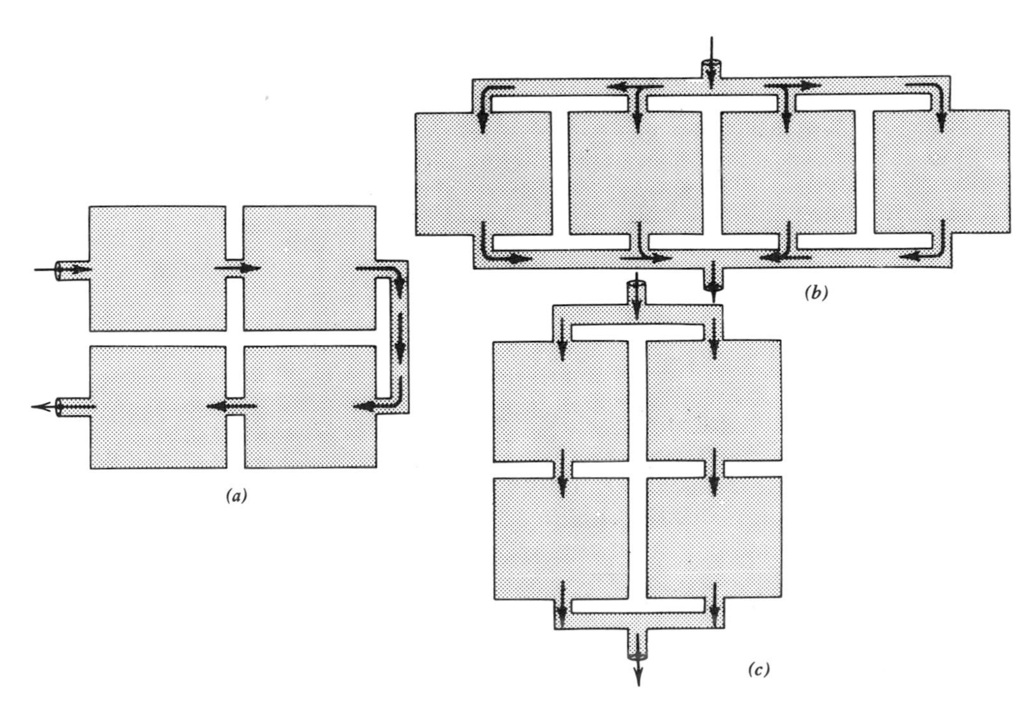

FIGURE 7.1 *Illustrating various arrays of four panels. (a) The series array. (b) The parallel array. (c) The combination array.*

offers a large resistance to the flow of the transfer fluid. To maintain the flow, the pumps must produce large pressures so that the pressure at the inlet is far greater than that at the outlet. This produces a strain on both the pump and on the panels of the array. Also, in a series array all the panels do not operate at the same efficiency. Those closer to the inlet operate at a lower temperature and are therefore more efficient. The opposite is true for those panels closer to the outlet.

In a parallel array (Figure 7.1*b*) the inlets of each panel are connected to a common feeder line. The outlets are similarly connected to a common drain. A parallel array, although more difficult to implement than a series array, offers a smaller resistance to fluid flow. Furthermore, if the total flow rate entering the array is equally divided into the individual panels, the performance characteristics of the array can be easily deduced from those of a single panel. The efficiency and temperature increase of such an N-panel parallel array are the same as those of an individual panel, but the flow rate and the useful heat collected are N times as large. In practice, a combination array is often used to facilitate installation (Figure 7.1*c*).

Once an array has been installed, the overall system's performance can be analyzed in a manner similar to that used for a single panel. Efficiency curves, that is, plots of η versus $(T_{f,i} - T_a)/F$, such as those shown in Figure 6.4, can be developed for an array. In addition, the outlet temperature can be determined once the flow rate is prescribed. If a controller is used, the flow rate can be adjusted to set the operating temperature of the array.

Pipe Losses

Exterior pipes that carry warm transfer fluids from the array will lose heat to the cooler surroundings. Because the ambient air acts as a heat reservoir at a temperature T_a, the heat transfer process can be approximated using the single-current, heat exchanger equation, Equation 4.31. If we apply Equation 4.31 to an exterior pipe carrying fluid from an array at a hot temperature T_H to a storage tank, we find that the temperature of the fluid reaching the tank is

$$\boxed{T_{\text{storage}} = T_a + (T_H - T_a)\exp(-\bar{U}_L L/\dot{m}C_f)} \tag{7.2}$$

where $\bar{U}_L$ is the overall coefficient per unit length of pipe for heat transfer from the fluid to the surrounding air and L is the length of the pipe. From Equation 4.32, the heat loss from the pipe is given by

$$\boxed{\dot{Q}_{\text{pipe loss}} = \dot{m}C_f(T_H - T_a)[1 - \exp(-\bar{U}_L L/\dot{m}C_f)]} \tag{7.3}$$

For well-insulated, short lengths of pipe, the product $H' = \bar{U}_L L$ will be small; in this case we would find $T_{\text{storage}} \simeq T_H$ and $\dot{Q}_{\text{pipe loss}} \simeq 0$. The effect of pipe losses on both small and large arrays is demonstrated by the next two examples.

EXAMPLE

A single solar heating panel uses water ($C_f = 4186$ J/kg-°C) as the transfer fluid. The water is flowing at $\dot{m} = 0.005$ kg/sec; it enters the panel at 20°C and leaves at 50°C. The fluid is carried to a storage tank by an exterior pipe 10 m long whose overall heat transfer coefficient per unit length is $\bar{U}_L = 0.2$ W/m-°C. The ambient temperature is $T_a = 15$°C. Find the temperature of the water entering the storage tank and the percent of the heat produced by the panel lost by the pipe.

The solar panel is collecting heat at a rate

$$\dot{Q}_c = \dot{m}C_f(T_H - T_C) = (0.005)(4186)(50\text{–}20) \\ = 628 \text{ W}$$

Using Equation 7.2, we find that the temperature at storage is

$$T_{\text{storage}} = 15 + (50 - 15)\exp[-(0.2)(10)/(0.005)(4186)] \\ = 46.8°\text{C}$$

According to Equation 7.3, the pipe loss is

$$\dot{Q}_{\text{pipe loss}} = (0.005)(4186)(50 - 15)\{1 - \exp[-(0.2)(10)/(0.005)(4186)]\} \\ = 67 \text{ W}$$

The percent loss is

$$\% \text{ loss} = \frac{67}{628} \simeq 11\%$$

EXAMPLE

Consider a parallel array of 10 panels, each of which is operating as the panel in the preceding example. Find the collection rate, the temperature entering storage, and the percent pipe loss.

The flow rate is increased by a factor of 10 to $\dot{m} = 0.05$ kg/sec. Assuming that the temperatures of the inlet and outlet remain at 20 and 50°C, respectively, the collection rate is increased by a factor of 10 to $\dot{Q}_c = 6280$ W. Using Equations 7.2 and 7.3 we find

$$\begin{aligned} T_{\text{storage}} &= 15 + (50 - 15)\exp[-(0.2)(10)/(0.05)(4186)] \\ &= 49.7°\text{C} \end{aligned}$$

and

$$\begin{aligned} \dot{Q}_{\text{pipe loss}} &= (0.05)(4186)(50 - 15)\{1 - \exp[-(0.2)(10)/(0.05)(4186)]\} \\ &= 70 \text{ W} \end{aligned}$$

The percent loss is now

$$\% \text{ loss} = \frac{70}{6280} \simeq 1.1\%$$

From the preceding examples, we see that pipe losses represent a smaller fraction of the heat collected for large arrays than for small arrays. Thus the smaller the array, the more important it is to keep the exterior pipes short and well insulated.

Heat Exchangers

A basic use of solar heating is to supply hot water. Tap water heated by a solar array can be used directly, or if higher temperatures are required, the water can be heated further by an auxiliary heater (Figure 7.2*a*). In many applications, however, it is necessary to transfer the heat from a solar-heated transfer fluid to a cold water supply. For example, the transfer fluid might be water to which antifreeze has been added to prevent freeze up and corrosion within the array. The transfer fluid is circulated in a closed loop and the heat transferred through a

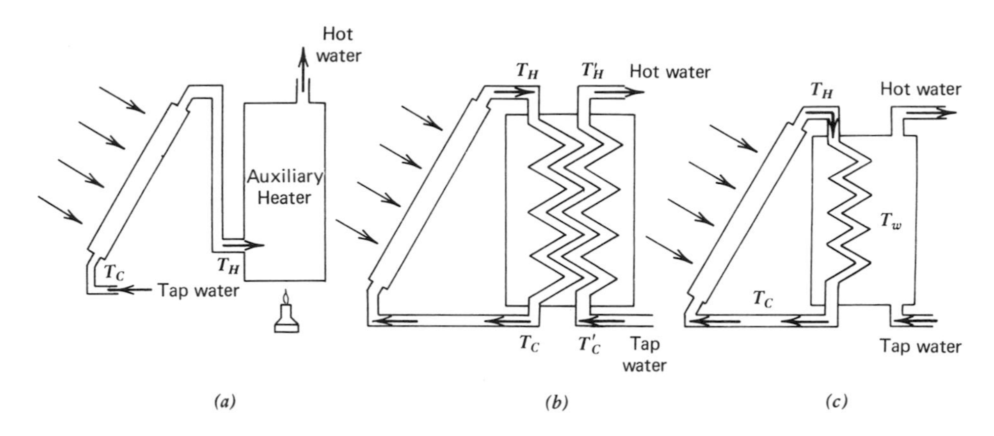

FIGURE 7.2 *Various solar hot-water supply systems. (a) An open loop system in which tap water is preheated by solar energy and then heated by an auxiliary heater. (b) A closed loop system in which heat is transferred from a solar-heated transfer fluid to a countercurrent of water. (c) A closed loop system in which heat is transferred to a holding tank of water by a single-current heat exchanger.*

countercurrent heat exchanger to a stream of tap water (Figure 7.2*b*). The heat may also be transferred through a single-current exchanger to a storage tank of water (Figure 7.2*c*).

As shown in Chapter 4, the transfer rate of heat from a warmer to a cooler fluid in a countercurrent heat exchanger is

$$\boxed{\dot{Q} = \bar{U}_L L \Lambda} \tag{7.4}$$

where $\bar{U}_L$ is the heat transfer coefficient and L is the length of the exchanger. The log mean temperature difference between the hot and cold fluids is

$$\boxed{\Lambda = \frac{(T_\text{H} - T'_\text{H}) - (T_\text{C} - T'_\text{C})}{\ln[(T_\text{H} - T'_\text{H})/(T_\text{C} - T'_\text{C})]}} \tag{7.5}$$

where T_H and T_C refer to the hot (inlet) and cool (outlet) temperatures of the primary (solar heated) fluid and T'_H and T'_C refer to the warm (outlet) and cold (inlet) temperatures of the secondary fluid (e.g., water). The heat transfer rate is given by

$$\dot{Q} = \dot{m}_\text{f} C_\text{f}(T_\text{H} - T_\text{C})$$

or equivalently by

$$\dot{Q} = \dot{m}_f' C_f' (T_H' - T_C') \tag{7.6}$$

EXAMPLE

A small solar array is used to heat tap water through a countercurrent heat exchanger. The primary transfer fluid ($C_f = 3700$ J/kg-°C) is circulated through the array at a rate of 0.015 kg/sec (~15 gal/hr). It enters the exchanger at 60°C (140°F) and leaves at 25°C (77°F). The tap water ($C_f' = 4186$ J/kg-°C) is flowing at a rate of 0.016 kg/sec (~16 gal/hr) and enters the exchanger at 10°C (50°F). The exchanger transfer coefficient is $\bar{U}_L = 20$ W/m-°C. Find the heat transfer rate of the system, the temperature of the water when it leaves the exchanger, and the required length of the exchanger.

The rate at which heat is extracted from the primary fluid and transferred to the tap water is, using Equation 7.6,

$$\dot{Q} = (0.015)(3700)(60 - 25) = 1943 \text{ W} = 1.943 \text{ kW}$$

The exit temperature of the water is found from

$$1943 = (0.016)(4186)(T_H' - 10)$$

so that

$$T_H' = 39°\text{C } (102°\text{F})$$

Using Equation 7.5, we find

$$\Lambda = \frac{(60 - 39) - (25 - 10)}{\ln \dfrac{(60 - 39)}{(25 - 10)}} = 17.8°\text{C}$$

Substituting into Equation 7.4, we have

$$L = \frac{\dot{Q}}{\bar{U}_L \Lambda} = \frac{1943}{(20)(17.8)} = 5.4 \text{ m}$$

This example shows that the final product, the heated tap water, is not as hot as the solar heated transfer fluid ($T_H' < T_H$). Consequently the heat supplied by the array is degraded somewhat by the exchanger. On the other hand, the transfer fluid reenters the array at a higher temperature than the tap water would if it were heated directly ($T'_C > T_C$).

The solar heat derived from the array can also be stored in a large reservoir using a single-current heat exchanger. The transfer fluid flows through the heat exchanger, which is immersed in a storage medium at a temperature T_B. If the fluid that enters at T_H is flowing at a rate $\dot{m}_f$, it will leave at a temperature

$$\boxed{T_C = T_B + (T_H - T_B)\exp(-\bar{U}_L L/\dot{m}_f C_f)} \tag{7.7}$$

Heat is stored in the medium at a rate

$$\boxed{\dot{Q}_{storage} = \dot{m}_f C_f (T_H - T_B)[1 - \exp(-\bar{U}_L L/\dot{m}_f C_f)]} \tag{7.8}$$

EXAMPLE

A transfer fluid ($C_f = 3500$ J/kg-°C) flows through a single-current heat exchanger immersed in a very large tank of water. The fluid enters at $T_H = 60$°C and is flowing at a rate $\dot{m}_f = 0.01$ kg/sec. If the temperature of the water tank is 50°C and if $U_L L = 50$ W/°C, find the rate at which heat is being stored in the water and determine the temperature of the fluid as it leaves the storage medium.

Using Equations 7.8 and 7.7, we have

$$\begin{aligned}\dot{Q}_{storage} &= (0.01)(3500)(60 - 50)\{1 - \exp[-(50)/(0.01)(3500)]\}\\ &= 266 \text{ W}\end{aligned}$$

and

$$\begin{aligned}T_C &= 50 + (60 - 50)\exp[-(50)/(0.01)(3500)]\\ &= 52°\text{C}\end{aligned}$$

The maximum possible transfer rate to storage is proportional to $(T_H - T_B)$. Showing that heat transfer to the storage medium becomes more efficient as $\bar{U}_L L$ and $(T_H - T_B)$ become large is straightforward. In fact, as $\bar{U}_L L \to \infty$ (an ideal exchanger), the transfer rate to storage becomes

$$\dot{Q}_{storage} = \dot{m}_f C_f (T_H - T_B)$$

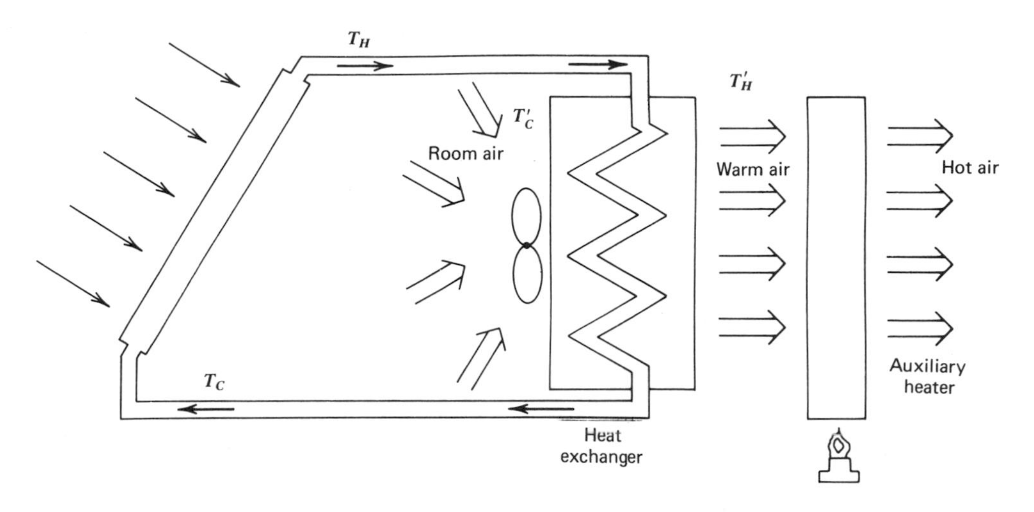

FIGURE 7.3 *A solar-assisted forced air system. Heat is transferred from a liquid in closed loop operation to a stream of air. An auxiliary heater raises the air temperature to the desired grade.*

and the exit temperature of the fluid becomes

$$T_C = T_B$$

If the temperature of the storage medium gradually increases as heat is being stored, the transfer process becomes less efficient. [See Problem 7-5.]

Exchangers may exchange heat between two liquids or between a liquid and air. For example, in forced-air space heating systems, heat may be transferred to a stream of forced air from a liquid in a thermal storage tank or from a countercurrent of a solar heated transfer liquid. The more efficient the exchanger is, the greater the amount of heat transferred. Furthermore, the air is heated to a higher temperature and the transfer fluid is cooled more effectively (Figure 7.3).

Storage

In a solar heating system, provision is usually made for heat storage. Energy derived during periods of abundant sunshine can be stored and used as needed during sunless periods. When heat is stored in a medium, the energy of that medium increases. The increase can be in the form of molecular kinetic energy, in which case the temperature of the medium will rise. It can also be in the form of potential energy, in which case the molecular structure changes as in a chemical change or in a phase transition. (e.g., melting or evaporation).

When the heat added produces only a temperature increase in the medium, we say that the thermal energy is being stored as *sensible* heat. As long as there is no phase change, the rise in temperature is approximately proportional to the heat stored and inversely proportional to the mass. The change in temperature can be written

$$\boxed{\Delta T = Q_s/m_s C_s} \qquad \text{(sensible heat storage)} \tag{7.9}$$

where Q_s is the heat stored and C_s and m_s are the specific heat and mass of the storage medium, respectively. When sensible heat is extracted from the medium, the temperature drops in accordance with Equation 7.9. An important parameter in a thermal storage system is the heat stored per unit volume. For sensible heat storage, Equation 7.9 may be written

$$\boxed{\begin{aligned} \frac{Q_s}{V_s} &= \frac{m_s}{V_s} C_s \Delta T \\ &= \rho_s C_s \Delta T \end{aligned}} \tag{7.10}$$

EXAMPLE

Find the sensible heat stored per unit volume for water ($\rho_s = 1000\ \text{kg/m}^3$, $C_s = 4186\ \text{J/kg-°C}$) and for methyl alcohol ($\rho_s = 810\ \text{kg/m}^3$, $C_s = 2512\ \text{J/kg-°C}$) when the medium's temperature increases by 10°C.

Using Equation 7.10, we have

$$\begin{aligned} \frac{Q_s}{V_s} &= (1000)(4186)(10) = 4.186 \times 10^7\ \text{J/m}^3 \\ &= 11.6\ \text{kw-hr/m}^3 \qquad \text{(water)} \end{aligned}$$

and

$$\begin{aligned} \frac{Q_s}{V_s} &= (810)(2512)(10) = 2.03 \times 10^7\ \text{J/m}^3 \\ &= 5.6\ \text{kw-hr/m}^3 \qquad \text{(methyl alcohol)} \end{aligned}$$

EXAMPLE

A pebble bed is used to store heat for an air-type solar heating system. The bed is to store 25 kw-hr when its temperature is raised by 20°C. If the average density of the bed is $\rho_s = 3000\ \text{kg/m}^3$ and its specific heat is $C_s = 800\ \text{J/kg-°C}$, find the stored energy per unit volume, the volume, and the mass of the bed.

Using Equation 7.10, we have

$$\frac{Q_s}{V_s} = (3000)(800)(20) = 4.8 \times 10^7\ \text{J/m}^3$$
$$= 13.3\ \text{kw-hr/m}^3$$

The required storage volume is

$$V_s = \frac{25}{13.3} = 1.88\ \text{m}^3$$

and the mass is

$$m_s = \rho_s V_s = (3000)(1.88) = 5639\ \text{kg} = 6.2\ \text{tons}$$

In certain applications it is desirable to keep the variation of the temperature of the storage medium to a minimum. According to Equation 7.9, this variation can be reduced by increasing the mass of the storage medium. However a more effective way of stabilizing the temperature is by using *latent* heat instead of sensible heat storage. When a solid is heated, its temperature continues to rise until its melting temperature T_m is reached. As more heat is added, its temperature remains constant at T_m but it begins to undergo a phase transition to the liquid state. No increase in temperature will occur until the melting process is completed. The heat per unit mass absorbed by and stored in the medium during the phase transition is termed *latent* heat, l. The latent heat storage per unit volume is

$$\boxed{\frac{Q_s}{V_s} = \rho_s l} \qquad \text{(latent heat)} \qquad (7.11)$$

where ρ_s is the average density of the medium. For example, the latent heat of ice is $l = 3.35 \times 10^5\ \text{J/kg} = 0.093\ \text{kw-hr/kg}$. If we take the density of the ice–water medium to be approximately that of water, namely, $\rho_s = 1000\ \text{kg/m}^3$, Equation 7.11 gives

$$\frac{Q_s}{V_s} = (1000)(0.093) = 93\ \text{kw-hr/m}^3$$

Comparing this with water in Equation 7.10, we observe that the same storage density can be accomplished with sensible heat, provided the water is allowed to change by 80°C! The latent heat storage occurs with no change in the storage temperature. The problem with storing latent heat in ice water is that the melting temperature is far too low to be useful. The objective then is to find a solid whose melting temperature is below what the solar array can supply and yet above the required level for the heating application.

Most materials that are solid at room temperature have a melting point that is too high to be useful in conventional solar storage applications. There are, however, some hydrated salts that have low enough transition temperatures for storing solar energy (see Table 7.2). In fact, certain mixtures of salts melt at lower temperatures than do their individual components. Such systems are called *eutectic* mixtures. The melting temperature of such a mixture can be reduced to a minimum value by the proper choice of the mixing ratios.

EXAMPLE

Glauber salt (hydrated sodium sulphate $Na_2SO_4 \cdot 10\,H_2O$) has the following characteristics.

$$T_m = 32°\text{C}\ (90°\text{F})$$
$$C_s^{\text{solid}} = 1926\ \text{J/kg-°C}$$
$$C_s^{\text{liquid}} = 2846\ \text{J/kg-°C}$$
$$l = 2.4 \times 10^5\ \text{J/kg}\ (104\ \text{Btu/lb})$$
$$\rho_s = 1600\ \text{kg/m}^3\ (100\ \text{lb/ft}^3)$$

Find the heat storage per unit volume if the Glauber salt is raised from 27 to 37°C. Find the volume required to store 25 kw-hr and determine the percent stored by latent heat.

Sensible heat is being stored when the solid is raised from 27 to 32°C and when the liquid is warmed from 32 to 37°C. The total sensible heat stored per volume is

$$\frac{Q_s}{V_s} = (1600)(1926)(32-27) + (1600)(2846)(37-32)$$
$$= 3.8 \times 10^7\ \text{J/m}^3 = 10.6\ \text{kw-hr/m}^3$$

TABLE 7.2 *Materials for Latent Heat Storage and Their Properties*

Material	Melting Temperature (°C)	Latent Heat (J/kg)
H_3PO_2	26	1.5×10^5
$CaCl_2 \cdot 6H_2O$	30	1.7×10^5
$Na_2SO_4 \cdot 10H_2O$	32	2.4×10^5
$Na_2CO_3 \cdot 10H_2O$	34	2.7×10^5
$Na_2HPO_4 \cdot 12H_2O$	36	2.6×10^5
$Ca(NO_3)_2 \cdot 4H_2O$	41	2.1×10^5
$Na_2S_2O_3 \cdot 5H_2O$	50	1.8×10^5
Hitec ($NaNO_3 + NaNO_2 + KNO_3$)	143	0.807×10^5

The latent heat storage is, from Equation 7.11,

$$\frac{Q_s}{V_s} = (1600)(2.4 \times 10^5) = 3.84 \times 10^8 \text{ J/m}^3$$
$$= 107 \text{ kw-hr/m}^3$$

The total storage is 117.6 kw-hr/m^3, with 90 percent stored in latent heat. The volume required to store 25 kw-hr is

$$V_s = \frac{25}{117.6} = 0.21 \text{ m}^3$$

Clearly, from this example there are some distinct advantages to storing thermal energy in latent rather than in sensible heat. The storage takes place over a narrower temperature range and the energy per unit volume is considerably larger. However, some technological problems exist. Chemical deterioration of the storage medium and the surface with which it comes into contact can be a problem. Furthermore, heat transfer to the solid phase is not always an efficient process, especially if the medium is a poor thermal conductor. The storage medium must also melt at a temperature compatible with the particular solar heating application.

In any storage system the primary transfer fluid must enter the tank at a higher temperature than that of the tank; otherwise the fluid will extract rather than deposit heat. A flow controller can be used to regulate the flow of the transfer fluid. When the fluid temperature falls below that of the storage medium, the flow can be made to bypass the tank or it may be halted entirely.

Solar-Assisted Systems

As discussed in Chapter 6, the efficiency of a solar heating panel decreases as its operating temperature increases, because thermal losses increase as the difference in temperature between the absorber plate and the surroundings increases. If we take the average operating temperature of the panel to be $\bar{T} \simeq (T_{f,e} + T_{f,i})/2$, we require that $\bar{T} - T_a$ be as small as possible for maximum efficiency. For $\bar{T}$ to be low it is necessary for the fluid inlet and exit temperatures to be as low as possible. Let us assume for the moment that the inlet temperature $T_{f,i}$ is fixed. For example, in a hot water supply system, $T_{f,i}$ is fixed by the temperature of the cold water supply, whereas in a simple space heating system, $T_{f,i}$ is fixed by the air temperature in the room. Thus to reduce $\bar{T}$ and increase efficiency, we need to reduce the output temperature $T_{f,e}$ by increasing the flow rate of the transfer fluid. If this temperature turns out to be too low to be useful, it can be raised to an acceptable level by an auxiliary heater. The increased operating efficiency of the array means that fewer panels will be necessary. In fact, under severe environmental conditions, the stagnation temperature may itself be below the level required to produce useful heat. In this situation the array would then be completely incapable of operating without an auxiliary system.

EXAMPLE

A collector array has an efficiency of 60 percent when heating cold tap water from 10 (50) to 45°C (110°F). When the same water is heated to 60°C (140°F), the array efficiency decreases to 40 percent. The system is designed to supply hot water at 60°C. Compare the size of the array for an unassisted system with that of an assisted system in which an auxiliary heater is used to raise the water from 45 to 60°C. Assume that both systems heat water at equal rates (Figure 7.4).

Let us assume that 100 units of heating power are to be supplied by each system. At an efficiency of 40 percent the unassisted system must intercept 250 units of radiant power. In the assisted system the array raises the temperature by 35°C and

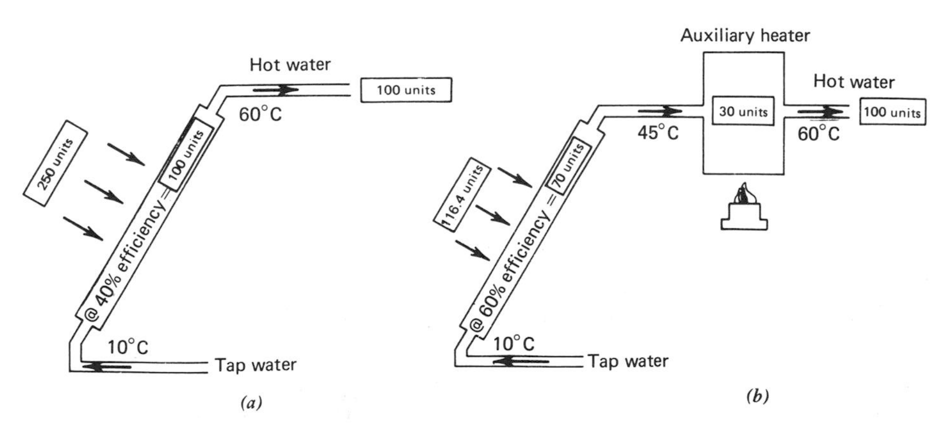

FIGURE 7.4 *Illustrating how an auxiliary heater increases array efficiency and reduces array size. (a) The array is producing heat at 60°C and is operating at 40 percent efficiency. (b) The array is producing heat at 45°C and is operating at 60 percent efficiency.*

the auxiliary heater raises it by 15°C. Thus the array supplies 70 percent of the heating needs or 70 units. Since the assisted array is 60% efficient, the intercepted radiant power must be 116.4 units. Consequently, the unassisted array requires more than twice the number of panels than the assisted one. In the assisted system approximately one-third of the heating comes from the auxiliary heater.

The Solar-Assisted Heat Pump

In unassisted space-heating applications the collector's operating temperature is limited by the fact that $T_{f,i}$ can be no lower than the temperature of the space to be heated. For example, if a solar-heated transfer fluid such as air transfers its heat to a room, its temperature when it returns to the collector can be no colder than the temperature of the room. In fact, $T_{f,i}$ will likely be substantially higher than the temperature of the space to be heated. This limits the operating efficiency of the array. By using a device known as a *heat pump*, we can extract heat from the transfer fluid and return the fluid to the array at a temperature *below* that of the space being heated. This will increase the array's efficiency significantly.

With a heat pump the array can actually be operated at a temperature below that of the space to be heated. Heat can be transferred from a *colder* region (the array) to a *warmer* region (the heated space). To appreciate how it is possible for heat to flow from cold to hot, we need to consider the second law of thermodynamics, which states:

> *It is impossible to construct a device operating in a cycle that produces no other effect than the transfer of heat from a cold reservoir to a hot one.*

We will discuss this law in some detail in Chapter 8. For now let us consider its implications with regard to a heat pump. The second law prohibits heat from flowing from cold to hot unless there is another "effect," say, the utilization of work. Thus a device can be made that, when work is supplied to it, will transfer heat from a cooler to a warmer region. An example of such a device is an air conditioner. On a warm summer day the temperature in a room might be 20°C (68°F) when the outside air is at 30°C (86°F). If the air conditioner is not operating, heat flows from the outside to the inside, gradually raising the room temperature. When the air conditioner is turned on and electrical work supplied, heat is extracted from the cooler room and rejected to the warmer outside air, thereby keeping the room cool.

Consider next a cool day when the outside air is 10°C (50°F) and the air in a room is 20°C (68°F). If the room is unheated, it will eventually cool to 10°C. Suppose that an air conditioner is installed in a window but in a reverse manner—with the side normally outside now inside. The unit now extracts heat from the cooler exterior and transfers it to the warmer interior so that the room's temperature is maintained at 20°C. A device that supplies heat in this manner is called a *heat pump* (Figure 7.5). The heat supplied to the room by the heat pump is equal to the heat extracted from the outside air plus the heat equivalent of the work supplied to the device. Thus we may set

$$Q_H = Q_C + W \tag{7.12}$$

where

Q_H = heat supplied to hot reservoir
Q_C = heat extracted from cold reservoir

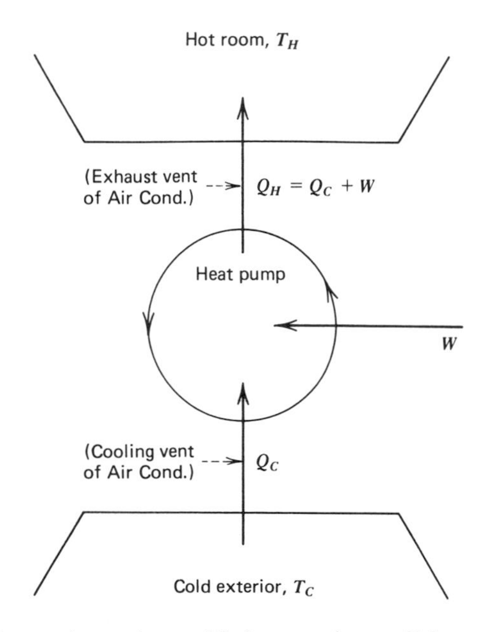

FIGURE 7.5 *The thermodynamic model for an air conditioner being operated as a heat pump. When electrical work is supplied, the unit extracts heat from the colder exterior and transfers it to the warmer interior.*

and

W = work supplied to heat pump.

The coefficient of performance (COP) of the heat pump is defined as

$$\boxed{\text{COP} = \frac{Q_H}{W} = \frac{\text{heat supplied}}{\text{work supplied}}} \qquad \text{(heat pump)} \qquad (7.13)$$

The higher the COP is, the more effective the heat pump. The lowest possible COP is unity. This corresponds to the situation where work such as electrical energy is simply converted to heat and no heat is extracted from a cold reservoir. A solar-assisted heat pump arrangement is shown in Figure 7.6. The heat pump extracts heat from a low temperature fluid and supplies it to a high temperature environment. As it extracts heat, the heat pump actually chills the transfer fluid before it is returned to the heating panel. Thus the heating panel is being operated at reduced temperatures, with lower heat losses to the surroundings. The heat pump raises the grade of the heat to a level useful for the specific application.

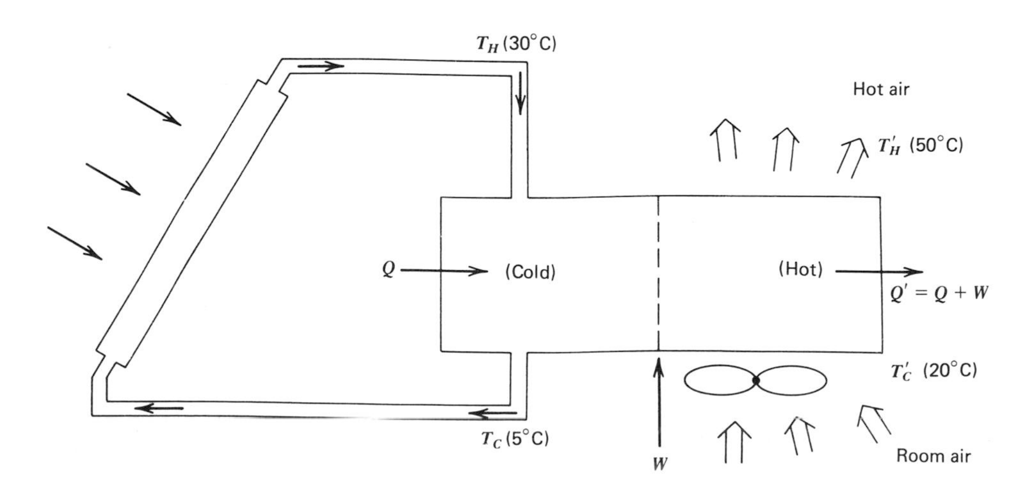

FIGURE 7.6 *A solar-assisted heat pump. The pump maintains a cold and a hot region by using work to extract heat from the former and supply it to the latter. Low-grade heat is supplied by the array. The grade is raised by the heat pump and supplied to the room. Note that* $T'_H > T_H$ *and* $T'_C > T_C$.

As shown, the total heat supplied is equal to the heat extracted from the solar transfer fluid plus the heat equivalent of the work required by the heat pump. The less work required by the heat pump, the higher its COP. It will be shown in Chapter 8 that there exists a theoretical upper limit to the COP of a heat pump operating between cold and hot reservoirs at temperatures T_C and T_H, respectively. The limit is given by *Carnot's* formula

$$\boxed{COP_{actual} \leq COP_{Carnot} = \frac{T_H}{T_H - T_C}} \qquad \text{(heat pump)} \qquad (7.14)$$

where T_H and T_C are in kelvins. A heat pump operates at the Carnot limit if it uses a reversible thermodynamic cycle called a Carnot cycle. We defer a discussion of this cycle to Chapter 8; suffice to say, a real heat pump has a much smaller COP than does a Carnot heat pump. The Carnot heat pump is an idealization and cannot be built for practical reasons. However, the Carnot limit does give a qualitative criterion for heat pump performance. From Equation 7.14 we see that as the temperature difference between the reservoirs becomes large, the heat pump becomes less effective. Although a heat pump can be used to extract heat from outside air at 10°C (50°F) and supply it to a room at 21°C (70°F), it becomes less effective as the temperature of the outside air drops toward the freezing point. In subfreezing conditions a solar array can be used to boost the temperature to an acceptable level for the heat pump.

EXAMPLE

A heat pump is operating with its cold reservoir at $T_C = 4$°C (39°F). A transfer fluid (water) from a solar array enters the reservoir at 30°C (86°F) and leaves at 5°C (41°F). The heat pump extracts 700 W of heat from this reservoir and supplies it to a hot reservoir at $T_H = 55$°C (131°F). The hot reservoir supplies heat to a stream of air entering at 20°C (68°F) and leaving at 50°C (122°F). The pump uses electrical work at the rate of 800 W. Find the flow rates of the two fluids. Find the actual COP of the heat pump and verify that it is below Carnot's limit. ($C_{water} =$ 4186 J/kg-°C, $C_{air} = 1047$ J/kg-°C.)

The flow rate of the transfer fluid (water) is found from

$$700 = (4186)\dot{m}_f(30 - 5)$$

so that

$$\dot{m}_f = 0.0066 \text{ kg/sec} = 6.6 \text{ gal/hr}$$

Since the heat supplied to the air is $Q_H = Q_C + W = 700 + 800 = 1500$, we have

$$1500 = (1047)\dot{m}_f(50 - 20)$$

or

$$\dot{m}'_f = 0.048 \text{ kg/sec} \simeq 4600 \text{ ft}^3/\text{hr}$$

The COP of the heat pump is

$$\text{COP} = \frac{Q_H}{W} = \frac{1500}{800} = 1.875$$

The COP of an ideal Carnot heat pump is

$$\text{COP}_{\text{Carnot}} = \frac{T_H}{T_H - T_C} = \frac{328}{328 - 277} \simeq 6.43$$

Clearly, the heat pump's COP is well below the Carnot limit. Note that the output temperature of the solar array is 30°C and that the array is supplying 700 W. Without the heat pump, the array would have to produce an output temperature in excess of 50°C and supply 1500 W. The increased output and the decreased efficiency would necessitate using a larger array. Of course, the heat pump does require electrical work. However, the higher the COP of the heat pump, the lower the cost of operation.

It is interesting that there is an optimum operating temperature for a solar-assisted heat pump. If the array is operated at a low temperature, its efficiency is high but the COP of the heat pump begins to drop. On the other hand, if the array temperature is too high, the COP of the heat pump increases but the array efficiency approaches zero. A temperature controller is useful here to fix an optimum output temperature for the array. A solar-assisted system using a heat pump is shown in Figure 7.7. The array supplies heat to a transfer fluid that is stored in a hot holding tank. When heat is not called for but insolation is available, controller 1 gradually fills the hot storage tank with hot transfer fluid at a preset temperature $T_{f,e}$. The process

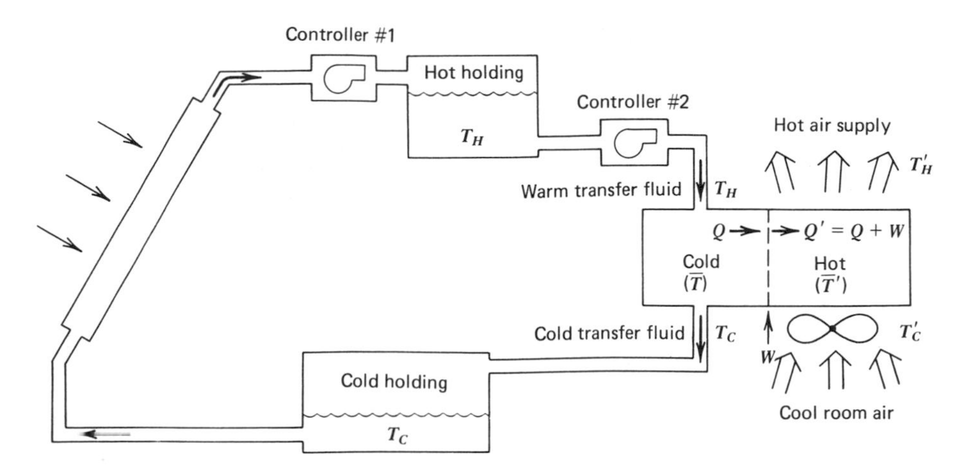

FIGURE 7.7 *A solar-assisted heat pump system for space heating applications. The system stores heat in a holding tank and uses controllers to maintain array and heat pump temperatures.*

stops when the cold holding tank is empty and the hot tank full. When heat is called for, controller 2 extracts fluid from the hot tank and returns it to the cold tank. Heat is extracted by the heat pump and the transfer fluid is chilled before it returns to the cold holding tank. Controller 2 regulates the flow in order to establish and maintain the proper temperature of the cold reservoir of the heat pump.

Comparing Figures 7.6 and 7.7 with Figure 7.3, we observe that the heat pump serves as a "super" heat exchanger which transfers heat from the transfer fluid of the solar system to the air stream. The heat pump enables heat to be transferred from a *cooler* transfer fluid to a *warmer* stream of air. Consequently, the temperature of the air supplied to the room is actually warmer than the output temperature of the solar array. This mode of heat transfer is possible only if some work is supplied. The greater the temperature difference between the heat reservoirs of the heat pump, the more work is required.

Radiative Coolers

We conclude this chapter with a brief discussion of the application of solar heating panels to radiative cooling. As discussed in the previous section, an air conditioner requires work to extract heat from a cold reservoir and exhaust it to a hot reservoir. As the temperature of the hot reservoir increases, the cooling process becomes less efficient. It is increasingly difficult to cool a room as the air surrounding the exhaust ducts (i.e., the hot reservoir) of the air conditioner warms. If the condenser coils (that is, the coils that exhaust heat to the warmer surroundings) are immersed in a cooler fluid, the cooling process becomes more efficient. Suppose a transfer liquid is circulated through solar panels at night when ambient temperatures are relatively low. On clear nights thermal losses produced by radiative cooling could drop the temperature of the fluid to a value below that of the ambient air. The cooler fluid could be stored at night and circulated, when needed, past the condenser coils of the air conditioner. The air conditioning process would then become more efficient.

Unfortunately the better the design of a solar heating panel, the more poorly it performs as a radiative cooler at night, since good solar heating panels are designed to minimize thermal losses. For example, a panel with a selective coating on its absorber plate does not radiate heat effectively. A good radiative cooler requires a highly emissive surface. It would also be necessary to remove the glazing at night in order to increase convection losses.

Until now we have dealt with the production of heat from solar energy. A significant fraction of the energy used by an industrial society involves nonheating applications, which fall into a category called work. Electricity, for example, is an energy form that allows electric motors to perform work. Similarly, the chemical energy stored in fossil fuels is used by internal combustion and steam engines to do work. In the next two chapters we will consider how solar energy can be converted to various forms of useful work.

PROBLEMS

7-1. Estimate the number of panels in an array required to supply hot water to a summer home if 100 gal of tap water per day are to be heated from 10 to 50°C. Assume that the average efficiency of the array is 60 percent and that the average daily insolation intercepted by the array is 5 kw-hr/m^2-day. Each panel has an area of 1.5 m^2.

7-2. Consider an array of nine panels arranged in three parallel strings. Each string has three panels in series. Water flows into the array at 0.03 kg/sec and divides equally into each string. The water enters the array at 10°C and leaves at 50°C. The first panel of each string raises the temperature by 20°C; the second, by 15°C; and the third, by 5°C. The area of each panel is 1.5 m^2 and the intercepted insolation is 800 W/m^2.

(a) Find the efficiency of each panel of the array.
(b) Find the rate at which heat is collected by the array.
(c) Determine the overall efficiency of the array.

7-3. An exterior pipe 5 m long and 2 cm in diameter carries water at 60°C from an array to a storage tank at a rate of 0.005 kg/sec. The pipe is wrapped in cylindrical insulation of rock wool ($K = 0.04$ W/m-°C) to a diameter of 10 cm. The air temperature is 0°C.

(a) Assuming that the transfer to the surrounding air is primarily by conduction through the insulation, find $\bar{U}_L$ for the pipe. (*Hint.* See Equation 4.9.)
(b) Find the temperature of the fluid as it enters the storage tank.
(c) Find the rate at which heat is lost through the pipe.

7-4. A solar hot water heater uses a transfer fluid consisting of a water–antifreeze solution ($C_f = 3800$ J/kg-°C). It passes through a countercurrent heat exchanger transferring its heat to a current of water. The transfer fluid flows through the exchanger, entering at 70°C and leaving at 20°C. Water enters at 10°C and leaves at 50°C.

(a) Find the log mean temperature difference for the exchanger.
(b) If the exchanger constant is $\bar{U}_L L = 200$ W/°C, find the rate at which heat is transferred by the exchanger?
(c) Find the flow rate of the transfer fluid in the exchanger.

7-5. The transfer fluid of a solar array transfers heat to a mass of water in a holding tank using a single-current heat exchanger. The fluid enters the exchanger at a *fixed* temperature T_H and flow rate $\dot{m}_f$. The exchanger constant is $\bar{U}_L L$ and the mass of water in the tank is m_w. The water is initially at a temperature T_{w0}.

(a) Show that the water temperature in the tank increases with time according to

$$T_w = T_H - (T_H - T_{w0})e^{-kt}$$

where

$$k = \frac{\dot{m}_f C_f}{m_w C_w}[1 - \exp[-\bar{U}_L L/\dot{m}_f C_f)]$$

[Assume that the change in T_w is negligible during the time that it takes for an element of fluid to traverse the exchanger.]

(b) Find an expression for the variation of the temperature of the transfer fluid as it leaves the exchanger.

(c) Show that the sensible heat storage *rate* decreases with time according to

$$\dot{Q}_{storage} = km_w C_w (T_H - T_{w0}) e^{-kt}$$

(d) Show that the total sensible heat stored increases with time according to

$$Q_{storage} = m_w C_w (T_H - T_{w0})(1 - e^{-kt})$$

(e) Show from (d) that the total possible heat stored in the process is

$$Q_{storage} \xrightarrow[t \to \infty]{} m_w C_w (T_H - T_{w0})$$

Explain this limit qualitatively.

7-6. A storage tank containing 50 kg of a eutectic salt ($l = 2 \times 10^5$ J/kg) has a single-current heat exchanger, whose constant is $H' = \bar{U}_L L = 150$ W/°C, immersed in it. The salt is in the solid phase and is at its melting point $T_m = 50$°C. A transfer liquid ($C = 3500$ J/kg-°C) from a solar array enters the exchanger at a temperature of 70°C with a flow rate of 0.02 kg/sec.

(a) Find the temperature at which the fluid leaves the storage medium.

(b) Find the time it takes to completely melt the salt.

(c) Find the total latent heat stored after this time.

7-7. The salt used in the storage tank described in Problem 7-6 is 50 percent liquid; the tank is equipped with a second single-current exchanger, which is used to extract heat. A stream of air ($C_{air} =$ 1047 J/kg-°C) flowing at 0.04 kg/sec enters this exchanger at 18°C and leaves at 40°C. If heat is being added by the solar transfer fluid at the rate determined in Problem 7-6, is the salt solidifying or melting; if so, at what rate?

7-8. A heat pump is operating with its hot and cold reservoirs at $T_H = 50$°C and $T_C = 5$°C, respectively. A transfer fluid ($C_f =$ 4000 J/kg-°C) from a solar array enters the cold reservoir at 35°C and leaves at 7°C. The fluid is flowing at a rate of 0.01 kg/sec. A stream of air (C_{air} = 1047 J/kg-°C) flowing at 0.05 kg/sec enters the hot reservoir at 18°C and leaves at 48°C.

(a) Find the solar heating power supplied by the array.
(b) Find the total heating power supplied to the air stream.
(c) How much electrical power is required to operate the heat pump?
(d) What is the COP of the heat pump?
(e) What is Carnot's limit for a heat pump operating between the two reservoirs?

REFERENCES

1. Anderson, B., *Solar Energy: Fundamentals in Building Design*, McGraw-Hill, New York (1977).
2. Beckman, W. A., S.A. Klein, and J.A. Duffie, *Solar Heating Design by the f-Chart Method*, Wiley-Interscience, New York (1977).
3. Duffie, J. A. and W. A. Beckman, *Solar Engineering of Thermal Processes*, Wiley-Interscience, New York (1980). Chapters 9–14.
4. Halacy, D. S., *The Coming Age of Solar Energy*, Harper & Row, New York (1973).
5. Kreider, J. F. and F. Kreith, *Solar Heating and Cooling*, McGraw-Hill, New York (1977), Chapters 4 and 5.
6. Lunde, P. J., *Solar Thermal Engineering*, Wiley, New York (1980), Chapters 7 and 8.
7. McVeigh, J. C., *Sun Power*, Pergamon, Oxford (1977).
8. Patton, A. R., *Solar Energy for Heating and Cooling Buildings*, Noyes Data, Park Ridge, N.J. (1975), Chapter 7.
9. Szokolay, S. V., *Solar Energy and Building*, 2nd ed., Wiley, New York (1977).

CHAPTER 8

Thermodynamic Conversion of Solar Energy to Work

In the last few chapters we considered how solar energy can be converted to heat for hot water supply and space heating systems. Solar energy is electromagnetic radiation. When such radiation is incident on a perfectly absorbing black surface, it is completely converted to heat. The primary task of an efficient solar heating system is to prevent this heat from being lost to the surroundings before it can be applied to a useful purpose. In principle at least, a solar heating system can approach 100 percent efficiency if optical and thermal losses are negligible.

There are many situations where *work* rather than heat is the form of energy required. In fact, in addition to producing heat, natural processes continually transform solar energy to work. The solar heating of oceans produces evaporation that eventually results in precipitation. The net effect is that water is raised from sea level to higher altitudes; equivalently, solar energy is converted to gravitational potential energy. As the water returns to the oceans, its potential energy can be converted to electrical energy as is done by hydroelectric generators. Solar heating of the earth and its atmosphere is also responsible for winds. This process converts solar energy to atmospheric kinetic energy. In this form the energy can be used to turn windmills and to produce useful work.

By far the most significant form of useful work produced by nature has been the conversion of solar energy to chemical energy through the process of photosynthesis. By absorbing sunlight, green plants convert raw materials (e.g., H_2O and CO_2) from a state of low chemical energy to one of higher energy (e.g., carbohydrates). Animals and man consume these materials and produce useful work. Organic materials produced over millions of years from decaying plant matter have evolved into the so-called fossil fuels—oil, coal, and natural gas, all of which can be used to produce work. Unfortunately, nature's conversion of solar energy to work is much too slow to fulfill the needs of a complex industrial society. A significant fraction of the fossil fuels produced over millions of years has been consumed by man in the last hundred years. Mankind has only recently begun to consider a technology that can produce rather than consume work derived from solar energy.

There are two distinct methods for converting solar energy to work: *thermodynamic conversion* and *direct conversion.* In the first method, solar energy is first converted to heat by solar collectors. This heat is then partially converted to work. As we will see, it is not possible to convert all the heat to work; some *must* be rejected to the surroundings. Unlike the conversion of solar energy to heat, there is a well-defined theoretical upper limit to the efficiency of any process that converts heat to work.

In *direct conversion* solar energy is converted directly to work without first converting the sunlight to heat. Although this method does have its practical limitations, it does not have the same theoretical limit imposed on thermodynamic conversion. In this chapter we will consider thermodynamic conversion; we defer a discussion of direct conversion to the next chapter.

Heat and Work—The Second Law of Thermodynamics

Heat and work are both forms of energy; however, from a thermodynamic perspective, they are two distinct forms. To appreciate the difference, we consider some basic elements of thermodynamics.

Imagine for the moment that the universe is divided into a number of subsystems. Assume that each system is in a state of equilibrium and that each is characterized by a set of thermodynamic variables such as temperature, pressure, volume, and so on. One such state variable is called the *entropy* S_i and represents the disorder of the ith subsystem. The total entropy or disorder of the universe S_u is the sum of the entropies of the subsystems. Next, let the subsystems interact so that a *process* is performed. Work may be done and heat may be transferred from one subsystem to another. At the end of the process, the state of each subsystem will, in general, have changed; the entropy of some subsystems may have decreased, whereas that of others increased. However, as a result of the process, the net change in the entropy of the universe is restricted according to the following law.

The second law of thermodynamics (entropy statement)—

> *During any process, the entropy (disorder) of the universe may increase or remain constant, but it may never decrease.*

Mathematically, we have

$$\Delta S_u \geq 0 \tag{8.1}$$

Using the second law, we can classify a process as being either *reversible* or *irreversible.* A process is irreversible if it causes the entropy of the universe to increase. This becomes apparent if we consider that there is no way of bringing the universe back to its original state without decreasing its entropy and thereby violating the second law. A process is reversible if and only if it does not change the entropy of the universe. We therefore classify processes as follows:

$\Delta S_u > 0$	irreversible
$\Delta S_u = 0$	reversible
$\Delta S_u < 0$	impossible!

Now consider that the universe is divided into only two subsystems (Figure 8.1*a*). One system is called the "device" and the other the "surroundings." Imagine that the surroundings consist of a single reservoir that is so large and massive that any heat deposited in or extracted from it does not change its temperature. We designate the temperature of this so-called *heat reservoir* by T. Whenever an amount of heat Q is extracted from a heat reservoir its entropy decreases by Q/T, where T is in kelvins. The entropy increases by the same amount when the heat is added. The transfer of work produces no change in entropy.

We will assume that our device is one that, during a sequence of processes, performs work and transfers heat to and from the surroundings but at the end of the sequence returns to its original thermodynamic state. Such a device is said to operate in a *cycle.* Suppose that at the end of one cycle, an amount of heat Q has been extracted from the surroundings by the device and the device has in turn performed an

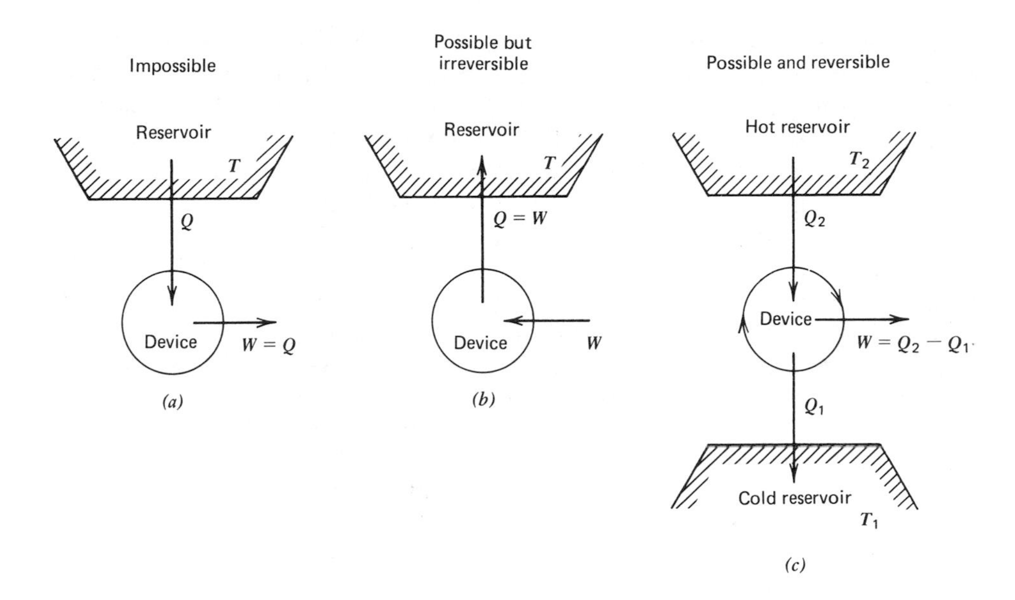

FIGURE 8.1 *Illustrating three thermodynamic cycles: (a) The conversion of heat to work using a single reservoir (impossible). (b) The conversion of work to heat using a single reservoir (irreversible). (c) The conversion of heat to work using two reservoirs (reversible).*

equal amount of work, W, on the surroundings. As a result of the heat extraction, the entropy of the surroundings decreased by Q/T. Because the device operates in a cycle, the net change in *its* entropy is *zero*. Thus at the end of the process, the net change in entropy of the universe is

$$\Delta S_{\mathrm{u}} = \Delta S_{\mathrm{surroundings}} = -\frac{Q}{T} < 0$$

Because this clearly violates the entropy statement of the second law of thermodynamics, the process is impossible. We are thus led to a second but equivalent version of the second law.

The second law of thermodynamics (Kelvin–Planck statement)—

> *It is impossible to construct a device, operating in a cycle, which achieves nothing else but the extraction of a quantity of heat from a single reservoir and the conversion of that heat to work.*

In this form the second law places severe limitations on the technology of thermodynamic processes. Were it not for this law, it would theoretically be possible to build an air conditioner that would neither require electricity nor reject heat but would instead *produce work* from the heat extracted from a room. It would also be possible for a ship to extract heat from the ocean and convert it completely to work for propelling itself. Neither of these processes are in fact possible according to the second law of thermodynamics.

Consider next the situation depicted in Figure 8.1*b*. In this case, work is being performed *on* the device and the device is in turn converting this work to heat and depositing it to the surroundings. At the end of the cycle the entropy of the universe is *increased* by Q/T. Consequently, a cyclic process that converts work to heat and deposits it in a single reservoir is certainly possible but irreversible. A heating appliance such as a toaster or an iron performs irreversible processes by converting electrical work into heat.

Finally, consider the case depicted in Figure 8.1*c*. Here, the universe is divided into a device and into surroundings that now consist of *two* heat reservoirs—one at a temperature T_2 and a second at a lower temperature T_1. The device operates in a cycle in such a way that at the end of the cycle, Q_2 units of heat are extracted from the hot reservoir and Q_1 are rejected to the cold reservoir. The net work done is $W = Q_2 - Q_1$. After one complete cycle the change in entropy of the device is zero and that of the universe is

$$\Delta S_u = \frac{Q_1}{T_1} - \frac{Q_2}{T_2} \geq 0 \tag{8.2}$$

As long as $Q_1/T_1 \geq Q_2/T_2$, the change in entropy of the universe is nonnegative and the process is possible. We therefore conclude that it is possible to construct a device, operating in a cycle, which extracts heat from one reservoir, converts some of this heat to work, and rejects the remainder to a second cooler reservoir. Such a device is called a *heat engine.* According to the second law of thermodynamics, heat engines require at least *two* heat reservoirs in order to convert heat to work. Furthermore, because some heat must be rejected to the cold reservoir, a 100 percent conversion efficiency is not possible. The efficiency of a heat engine operating between two reservoirs is

$$\boxed{\eta = \frac{W}{Q_2} = \frac{\text{work done per cycle}}{\text{heat absorbed per cycle}}} \tag{8.3}$$

where the work done per cycle is

$$\boxed{W = Q_2 - Q_1} \tag{8.4}$$

We summarize the results for cyclical processes as follows.

1. Conversion of work to heat, which in turn is deposited in a single reservoir—possible with 100 percent efficiency but irreversible.
2. Conversion of heat, derived from a single reservoir, completely to work—impossible.
3. Conversion of heat to work using two reservoirs—possible but with a limited efficiency.

According to (1), it is possible to convert solar energy completely into heat; however, according to (3), once this heat is produced, only a fraction can be converted to work. We will subsequently show that the conversion efficiency of any heat engine operating between two heat reservoirs is limited by *Carnot's* formula.

Carnot's Limit and Heat Engines

The most efficient engine operating between two heat reservoirs is an engine using reversible processes. According to Equations 8.2, 8.3, and 8.4, we find that the efficiency of any heat engine operating between two heat reservoirs is limited by

$$\eta = \frac{W}{Q_2} = \frac{Q_2 - Q_1}{Q_2} \leq \frac{T_2 - T_1}{T_2} \tag{8.5}$$

A reversible engine operating between two heat reservoirs has a limiting efficiency given by

$$\boxed{\eta_{\text{Carnot}} = \frac{T_2 - T_1}{T_2}} \quad (T \text{ in kelvins}) \tag{8.6}$$

Such an engine is called a Carnot engine; Equation 8.6 is known as Carnot's formula. Equation 8.5 implies that *no engine operating between two heat reservoirs can be more efficient than a Carnot engine operating between the same two reservoirs.*

EXAMPLE

A steam engine derives its heat from a hot reservoir (the boiler) at a temperature of 400°F (477 K) and rejects heat to a cold reservoir (the condenser) at 100°F (311 K). Find the maximum possible efficiency of such an engine.

According to Carnot's formula, Equation 8.6, the efficiency is restricted by

$$\eta_{steam} \leq \frac{477-311}{477} = 35\%$$

(Actual steam engines operating at these temperatures typically have efficiencies below 15 percent)

The Carnot efficiency increases as the temperature of the hot reservoir increases and as that of the cold reservoir decreases. Carnot's formula, however, does not provide us with a realistic upper limit to an actual heat engine's performance. In fact, most practical engines have efficiencies far below that predicted by Carnot's formula. We will discuss Carnot's engine as well as more realistic models for other engines subsequently. However, we digress to consider one other form of the second law of thermodynamics as it relates to refrigerators.

The Second Law of Thermodynamics and Refrigerators

Consider the device shown in Figure 8.2*a*. The device is operating between two reservoirs and is extracting an amount of heat Q from the cold reservoir (T_1) and transferring it to the hot reservoir (T_2). The net change in entropy of the universe is

$$\Delta S_u = \frac{Q}{T_2} - \frac{Q}{T_1} = Q\left(\frac{1}{T_2} - \frac{1}{T_1}\right) \qquad (8.7)$$

Since by assumption $T_2 > T_1$, Equation 8.7 requires that $\Delta S_u < 0$. Consequently, Figure 8.2*a* shows an impossible process. We therefore are led to still another statement of the second law.

The second law of thermodynamics (Clausius statement)—

It is impossible to build a device, operating in a cycle, which achieves no other effect than the extraction of heat from a cold reservoir and the deposition of that heat into a hot reservoir.

The process shown in Figure 8.2*b* involves the transfer of Q units of

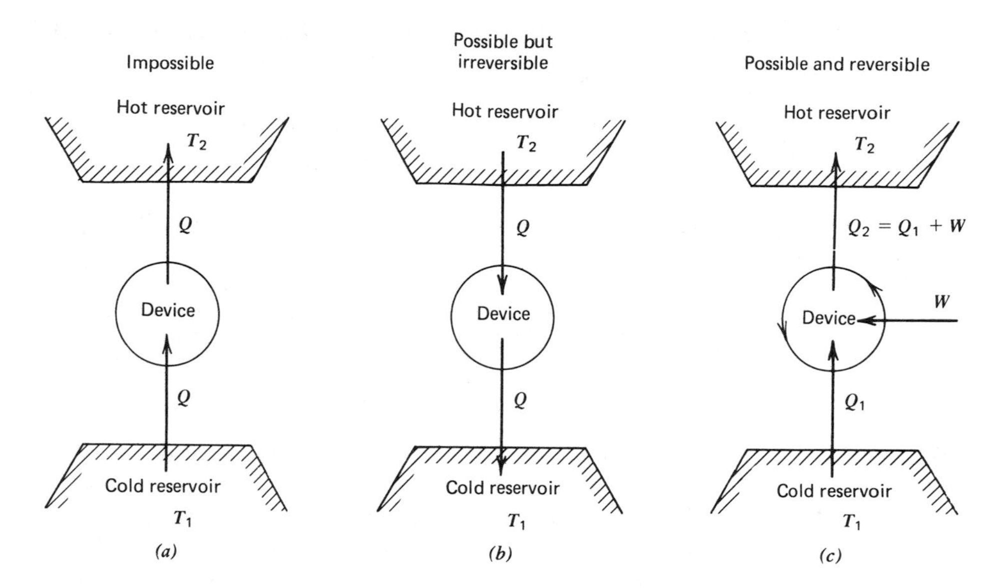

FIGURE 8.2 *Illustrating three thermodynamic cycles*: (*a*) *The transfer of heat from a cold to a hot reservoir with no other effect* (*impossible*). (*b*) *The transfer of heat from a hot to a cold reservoir with no other effect* (*irreversible*). (*c*) *Applying work to the transfer of heat from a cold to a hot reservoir* (*reversible*).

heat from a hot reservoir to a cold one. This process produces a net increase in the entropy of the universe and is therefore possible but irreversible. Consequently, if no other effect is involved, heat can flow irreversibly from a hot to a cold reservoir.

To transfer heat from a cold to a hot reservoir, another effect is required. As shown in Figure 8.2*c*, the effect can be the application of work. If work is applied to the device, then by conservation of energy, the heat deposited in the hot reservoir is

$$Q_2 = W + Q_1 \tag{8.8}$$

where Q_1 is the heat extracted from the cold reservoir. The net change in entropy of the universe is

$$\Delta S_u = \frac{Q_2}{T_2} - \frac{Q_1}{T_1} \tag{8.9}$$

Since $Q_1 < Q_2$, the change in entropy of the universe may be non-negative, making the process possible and even reversible.

A device that operates in a cycle and uses work to transfer heat from a cold to a hot reservoir is termed a *refrigerator.* Its coefficient of performance (COP) is defined as

$$\text{COP}_{(\text{refrigerator})} = \frac{Q_1}{W} = \frac{\text{heat extracted per cycle}}{\text{work required per cycle}} \tag{8.10}$$

Combining Equations 8.8, 8.9, and 8.10, we find that the COP of a refrigerator is limited by

$$\boxed{\text{COP}_{(\text{refrigerator})} \leq \frac{T_1}{T_2 - T_1} = \text{COP}_{\text{Carnot}}} \tag{8.11}$$

The equality holds for a refrigerator operating between two reservoirs and using reversible processes. Such a reversible refrigerator is called a Carnot refrigerator; consequently, the limit in Equation 8.11 is called the Carnot limit. Equation 8.11 can be expressed in words as follows. *No refrigerator operating between two heat reservoirs can have a higher COP than a Carnot refrigerator operating between the same two reservoirs.* The COP is usually but not necessarily greater than unity. Air conditioners typically have COPs between two and three.[1]

The preceding processes can be summarized as follows:

1. A process with no other effect than the transfer of heat from a hot to a cold reservoir—possible but irreversible.
2. A process with no other effect than the transfer of heat from a cold to a hot reservoir—impossible.
3. The application of work to transfer heat from a cold reservoir to a hot reservoir—possible but with a COP limited by Equation 8.11.

EXAMPLE

Find the maximum possible COP of an air conditioner extracting heat from a 70°F (294 K) room and rejecting it to outside air at 100°F (311 K). If the actual COP is 2.5 and if the unit requires 1000 W of electrical power, find the rate at which heat is removed from the room.

[1]When applied to air conditioners, the COP is often measured in Btu's/hr/watt and is called the *energy efficiency ratio* (EER).

Using Equation 8.11, we have

$$\mathrm{COP}_{\max} = \frac{294}{311 - 294} \simeq 17$$

The actual heat extraction rate is

$$\dot{Q}_1 = (\mathrm{COP}) \times P = (2.5)(1000) = 2500 \text{ W} \simeq 8500 \text{ Btu/hr}$$

Heat Engines

An actual heat engine takes a so-called *working substance* through a cycle composed of a sequence of thermodynamic processes. The substance can be a solid, liquid, or gas. Heat is extracted from hot reservoirs during certain processes and rejected to cold reservoirs during others. Work may be performed on or by the substance during parts of the cycle. The net effect after each cycle is that the difference between the heat absorbed and the heat rejected is equal to the work performed.

For a process to be reversible and therefore not increase the entropy of the universe, certain requirements must be satisfied. The working substance must always be in equilibrium with the surroundings. This means that the process must take place very slowly. Because most engines execute their cycles rapidly, the processes are irreversible. Reversibility also requires that the temperature of the working substance be virtually equal to that of the surroundings during heat transfer. This ensures that the heat transfer will occur slowly. In real engines, heat transfer occurs when the temperature difference between the working substance and the heat reservoirs is substantial.

To illustrate reversible cycles, we consider the following four processes and assume that the working substance is an ideal gas. The processes are shown in P-V diagrams in Figure 8.3.

Adiabatic The working substance is insulated from the surroundings so that heat neither enters nor leaves. During an adiabatic compression the volume decreases, the pressure and temperature increase, and work is done *on* the substance. The opposite is true for an adiabatic expansion. The adiabatic process requires no heat reservoirs.

Isothermal The system is exposed to a *single* heat reservoir. The

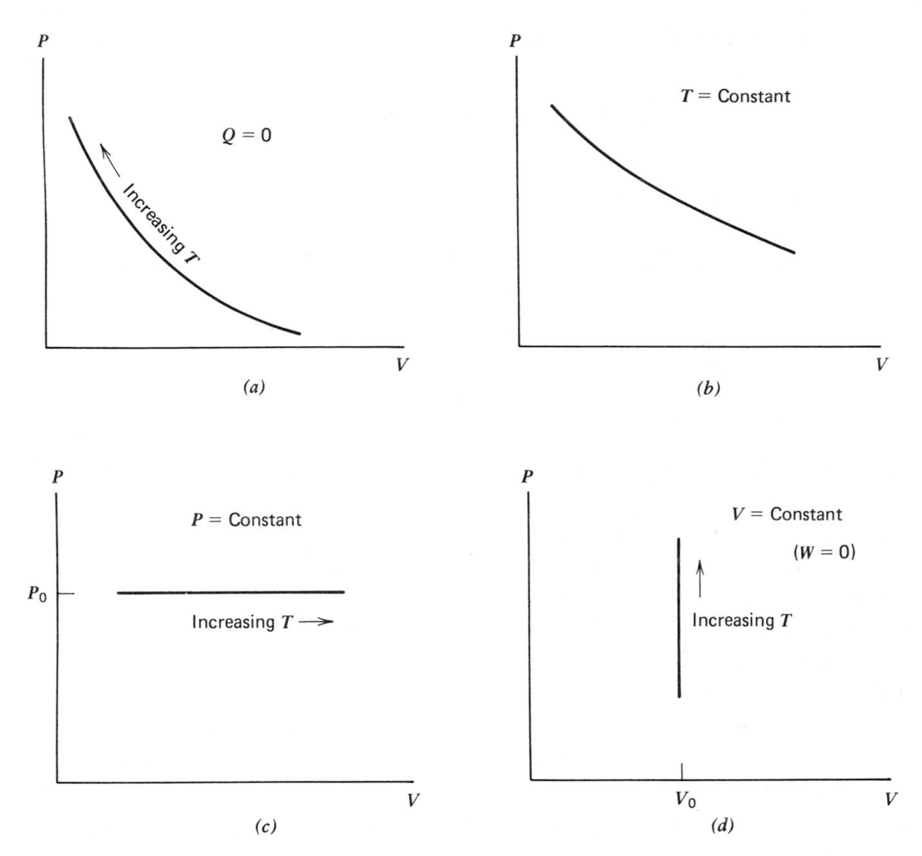

FIGURE 8.3 *The pressure versus volume variation for an ideal gas during four reversible processes*: (*a*) *An adiabatic process (no heat transfer*). (*b*) *An isothermal process (constant temperature*). (*c*) *An isobaric process (constant pressure*). (*d*) *An isochoric process (constant volume*).

temperature remains constant and equal to that of the reservoir. During an isothermal compression the pressure increases as the volume decreases. Heat is rejected to the reservoir and work is done *on* the substance. The opposite is true for an isothermal expansion.

Isobaric During the isobaric process the pressure remains constant. In an isobaric compression the volume and temperature both decrease. Heat is rejected and work is done *on* the substance. The opposite is true for an isobaric expansion. For an isobaric process to be reversible, the substance must be exposed to a *series* of reservoirs of different temperatures as heat is transferred. Thus a *reversible* isobaric process *cannot* be accomplished with a single heat reservoir. Any reversible engine cycle using an isobaric process requires a large number of heat reservoirs, each at a different temperature. Conversely, any ideal gas engine cycle using an isobaric process and operating between only two reservoirs is irreversible.

Isochoric During this process the volume is held fixed and no work is done by or on the substance. In isochoric heating, heat is absorbed and both the pressure and temperature increase. The opposite is true for isochoric cooling. Like a reversible isobaric process, a reversible isochoric process requires a series of heat reservoirs. Furthermore, any ideal gas cycle using an isochoric process and operating between only two reservoirs is necessarily irreversible.

Carnot Cycle

The Carnot cycle consists of four reversible processes—two adiabats alternated with two isotherms. Only two reservoirs are required, one for each isotherm. A Carnot cycle using an ideal gas as a working substance is shown in Figure 8.4. During the hot isotherm heat is absorbed reversibly from the reservoir at the temperature T_2. During the cold isotherm heat is rejected to the reservoir at the temperature T_1. All Carnot engines operating between the same two heat reservoirs have the same efficiency regardless of the working substance. The efficiency is given by

$$\eta_{\text{Carnot}} = \frac{T_2 - T_1}{T_2}$$

There is, however, no practical engine for which the Carnot cycle can serve as a model. Most reversible models on which practical heat engines are based neither extract heat from one reservoir at T_2 nor reject heat to a second reservoir at T_1. Equivalently, the heat transfer does not necessarily occur during an isothermal process. Instead, most engines absorb and reject heat while the temperature of the working substance is changing. Hence they must be simulated by reversible cycles in which a *series* of heat reservoirs are involved. These cycles give theoretical upper limits to the efficiencies of their practical counterparts which are lower and therefore more realistic than those given by Carnot's formula. For example, the Rankine cycle serves as a theoretical model for the steam engine, whereas the Stirling cycle serves as a model for one type of hot air engine. We begin with an analysis of the Rankine cycle and its applicability to the steam engine.

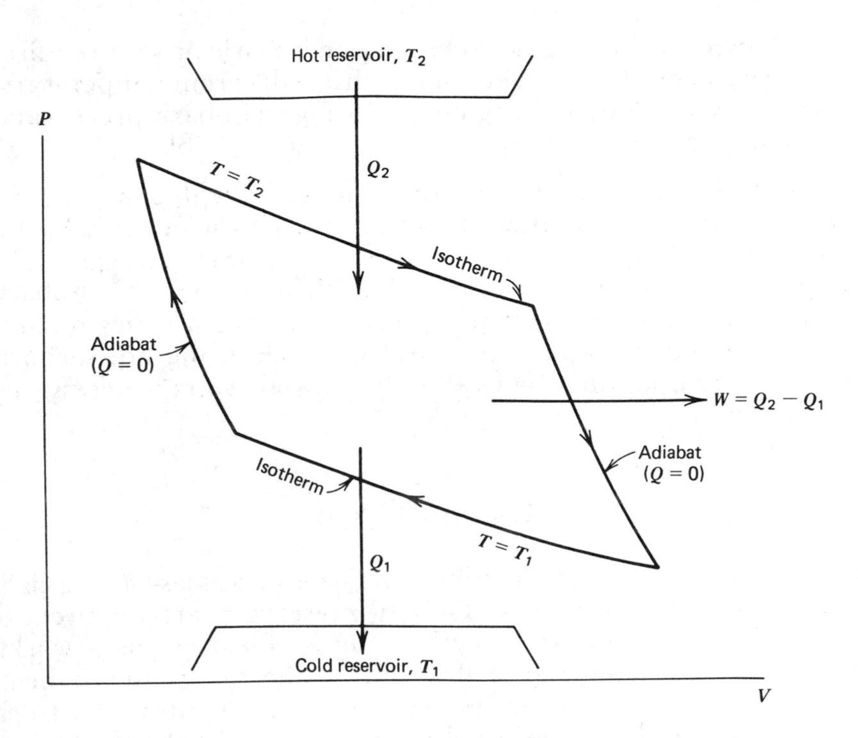

FIGURE 8.4 *A P-V diagram of a Carnot cycle using an ideal gas as its working substance.*

The Rankine Cycle and the Steam Engine

Although the typical steam engine is fossil fuel fired, the suggestion has been made that it could also be driven by solar energy. The working substance of a steam engine is a mixture of steam and water. Because two phases are involved, the theoretical model is somewhat more complicated than that for a simple gas. The actual steam engine processes are shown in Figure 8.5*a*.

Cool water is taken from a condenser at a low pressure P_C and a low temperature T_C. The water is compressed adiabatically by a pump to a high pressure called the boiler pressure P_B. Because water is highly incompressible, the volume decreases only slightly and the temperature increases nominally. For all practical purposes, the work performed by the pump on the water is negligible. The cool pressurized water enters the boiler where it is heated to the boiling point. The temperature at this point is usually well above the normal boiling point (212°F) because of the very high pressure in the boiler. This

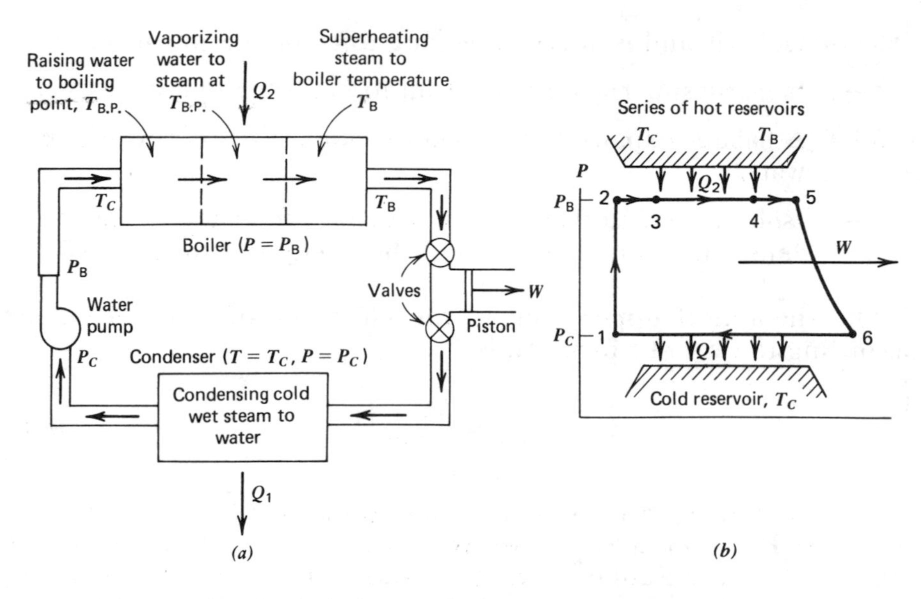

FIGURE 8.5 (*a*) *The schematic for a typical steam engine.* (*b*) *The P-V diagram of the reversible Rankine cycle used as a model for the steam engine.*

process is isobaric heating because heat is being supplied at constant pressure. However, because the temperature of the water is being raised, the boiler is acting as a series of reservoirs of increasing temperature. As the water continues to pass through the boiler, it is converted to steam and both the pressure and temperature remain constant. This process is both isobaric and isothermal so that the boiler is now acting as a single hot reservoir at the boiling point of water. As it approaches the exit of the boiler, the steam is superheated to the so-called boiler temperature T_B. This process is again isobaric but involves a series of reservoirs. Next, the superheated steam is adiabatically expanded against the blades of a turbine or against a moving piston. At the end of the process the steam has condensed to a low pressure and temperature mixture of steam and water. It enters the condenser where heat is extracted isothermally and the remaining steam is condensed. The cool, low-pressure water then begins the cycle over again. These six reversible processes which characterize the Rankine cycle are summarized here and shown on a *P-V* diagram in Figure 8.5*b*:

$1 \rightarrow 2$ Adiabatic (and virtually isochoric and isothermal) compression of liquid water to boiler pressure (P_B).

$2 \rightarrow 3$ Isobaric heating of pressurized water to boiling point.

3→4 Isobaric and isothermal heating to change water to steam.

4→5 Isobaric superheating of steam to boiler temperature (T_B).

5→6 Adiabatic expansion of steam to cool wet steam (steam plus water).

6→1 Isobaric and isothermal condensation to water at the condenser temperature (T_C) and the condenser pressure (P_C).

The theoretical upper limit to the efficiency of a steam engine, according to Carnot's formula, is

$$\eta_{\text{Carnot}} = \frac{T_B - T_C}{T_B} \tag{8.12}$$

where T_B and T_C are the boiler and condenser temperatures, respectively. However, a lower and more realistic theoretical limit can be derived from the Rankine cycle. It is shown in standard engineering texts on thermodynamics that the efficiency of the Rankine cycle can be expressed as (see Ref. 7, p. 223)

$$\boxed{\eta_{\text{Rankine}} = \frac{W}{Q_2} = \frac{h_5 - h_6 - (P_B - P_C)v_w}{h_5 - h_1 - (P_B - P_C)v_w}} \tag{8.13}$$

where v_w is the specific volume of water ($\sim 0.018\ \text{ft}^3/\text{lb}$) as it leaves the condenser and h is a property of the water known as *specific enthalpy*. If the boiling and condenser temperatures and pressures are known, it is possible to determine h at points 1, 5, 6 in the cycle (see Figure 8.5b) using so-called steam tables.[2]

We will illustrate the application of Equation 8.13 for two different boiler conditions using engineering units.

Case a: $T_B = 500°\text{F}$, $P_B = 300\ \text{lb/in}^2\ (43{,}200\ \text{lb/ft}^2)$

Case b: $T_B = 300°\text{F}$, $P_B = 30\ \text{lb/in}^2\ (4{,}320\ \text{lb/ft}^2)$

The condenser conditions in both cases are $T_C = 102°\text{F}$ and $P_C = 1\ \text{lb/in}^2\ (144\ \text{lb/ft}^2)$.

From steam tables, the specific enthalpies and efficiencies are

[2]See, for example, the *Handbook of Chemistry and Physics* published by the Chemical Rubber Co. (Cleveland, Ohio).

TABLE 8.1 *Comparison of the Carnot, Rankine, and Actual Efficiencies of a Steam Engine Operating at Two Different Boiler Temperatures and Pressures*

	T_B	P_B	T_C	P_C	Carnot (%)	Rankine (%)	Actual (%)
Case a	500°F	300 lb/in²	102°F	1 lb/in²	41	32	10–15
Case b	300°F	30 lb/in²	102°F	1 lb/in²	26	20	6–11

found to be:

Case a:

$$\begin{aligned} h_5 &= 1256 \text{ Btu/lb} \\ h_6 &= 876 \text{ Btu/lb} \\ h_1 &= 70 \text{ Btu/lb} \\ (P_B - P_C)v_w &= 1 \text{ Btu/lb} \\ \eta_{\text{Rankine}} &= \frac{1256 - 876 - 1}{1256 - 70 - 1} \\ &= 32\% \end{aligned}$$

Case b:

$$\begin{aligned} h_5 &= 1190 \text{ Btu/lb} \\ h_6 &= 966 \text{ Btu/lb} \\ h_1 &= 70 \text{ Btu/lb} \\ (P_B - P_C)v_w &= 0.1 \text{ Btu/lb} \\ \eta_{\text{Rankine}} &= \frac{1190 - 966 - 0.1}{1190 - 70 - 0.1} \\ &= 20\% \end{aligned}$$

As expected, a reduction in the boiler temperature reduces the efficiency of the Rankine cycle. Using kelvin temperatures in Equation 8.12, we find that the Carnot efficiencies of these cases are 41 and 26 percent, respectively. Actual steam engine efficiencies rarely exceed even half the corresponding Rankine limit. The results for these cases have been summarized in Table 8.1. In order to operate at an efficiency above 15 percent, a steam engine must have a boiler temperature considerably above the normal boiling point of water. Typical fossil-fuel fired steam plants operate with boiler temperatures well above 400°F.

A Solar-Powered Steam Plant—Overall Efficiency

A steam engine can also be driven by a solar-heated boiler, provided that temperatures considerably above the normal boiling point of water can be achieved. Because most flat plate solar collectors cannot produce such temperatures with any practical efficiency, concentrating arrays are generally required. The concentrated solar energy is directed toward a blackened receiver of a solar collector, which in turn is thermally coupled to the water in the boiler (Figure 8.6*a*). Some heat is lost to the surroundings by the collector so that the actual heat rate supplied to the boiler is

$$\dot{Q}_2 = \eta_C \dot{Q}_{inc} \tag{8.14}$$

where η_C is the collector efficiency and $\dot{Q}_{inc}$ the intercepted solar power. Actually, η_C includes the thermal efficiency as well as the optical efficiency. The overall efficiency of the solar driven engine is

$$\boxed{\eta = \frac{P}{\dot{Q}_{inc}} = \frac{\dot{Q}_2}{\dot{Q}_{inc}} \cdot \frac{P}{\dot{Q}_2} = \eta_C \cdot \eta_E} \tag{8.15}$$

where $\eta_E = P/\dot{Q}_2$ is the thermodynamic efficiency of the steam engine itself. As the operating temperature of the array is increased, the boiler temperature and consequently the engine efficiency increases; however, because of increased thermal losses the array efficiency is decreased. On the other hand, if the operating temperature is reduced, the array efficiency is increased but the engine efficiency is decreased. Consequently, there must be an operating temperature that yields an optimum *overall* efficiency. [See Problem 8-5.]

The total power produced by the solar driven engine is

$$P = \eta \dot{Q}_{inc} = \eta \cdot A \cdot F_{inc}$$

where F_{inc} is the intercepted flux and A is the effective area of the array. For a given overall efficiency the power output of the plant is proportional to the array size. A centralized solar power plant producing 5 MW of power and operating at an efficiency as high as 15 percent would still be required to intercept 33.3 MW. Assuming an incident flux of $F_{inc} = 1000\ \mathrm{W/m^2}$, the minimum interception area of the array would have to be $33{,}300\ \mathrm{m^2}$. It is impractical to fabricate a single concentrating element having so large an area. Instead, concentration could be accomplished using an array of flat mirrors, each arranged so that the incident solar energy is reflected toward the receiving element of the boiler. The receiving element is situated at the top of a tower and the mirrors are placed about its base (Figure 8.6*b*). Note that the mirrors must be spaced so that those closer to the tower do not obscure those behind them. Hence an open area even larger than the total area of the mirrors is required. Furthermore, because of the obliquity of the sun's rays (Figure 8.6*c*), the effective area of each mirror is smaller than its actual area. The

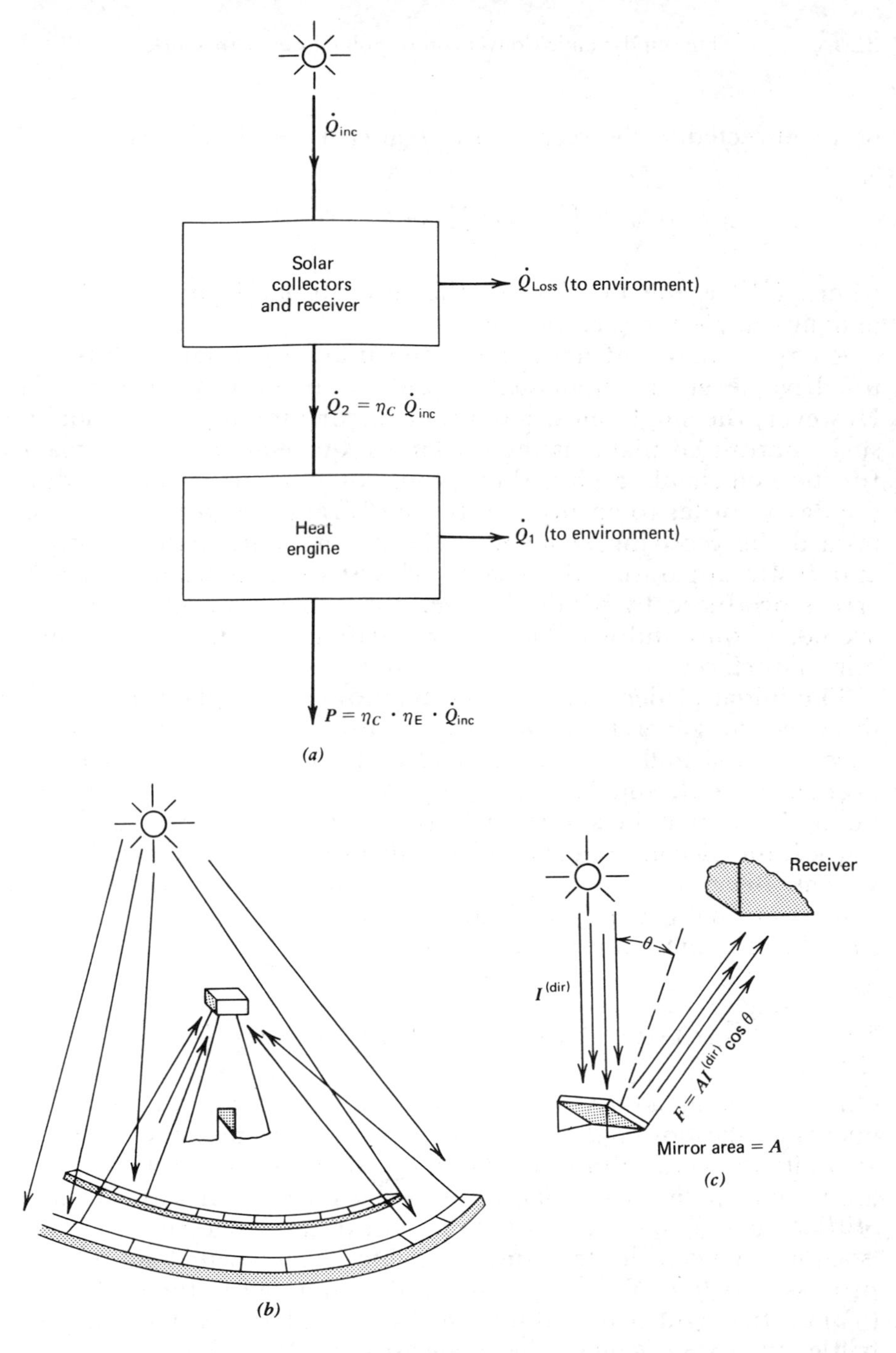

FIGURE 8.6 *(a) A block diagram showing the interception of solar power by a solar array and its conversion to mechanical power by a heat engine. (b) Two rows of an array of mirrors directing solar energy to a receiver at the top of a tower. (c) Illustrating the effect of obliquity on the flux directed toward the receiver.*

power directed to the receiver by N mirrors, each of area A, is

$$\dot{Q}_{inc} = I^{(dir)} \cdot A \cdot \sum_{i=1}^{N} \cos \theta_i \leq N \cdot A \cdot I^{(dir)}$$

where $I^{(dir)}$ is the intensity of the direct solar beam and θ_i is the obliquity angle for each mirror.

A large number of flat mirrors and much open land are required to drive even a moderately sized solar-powered steam plant. However, the single most important factor limiting the feasibility of such centralized plants is the tracking requirement. For the plant to function efficiently each and every mirror must track the sun during the day in order to ensure that the sun's rays are constantly directed toward the receiving element of the boiler. This requires elaborate and costly apparatus. Such large arrays are vulnerable to tracking errors produced by winds. Furthermore, unless proper precautions are taken, dust and windblown particulate matter will deteriorate the mirror surfaces.

The initial project costs for the Barstow power plant in California designed to generate 10 MW (peak) using this technique were in excess of 130 million dollars. Final costs could be even higher and average output, smaller than projected. Nevertheless, it should be pointed out that in a solar energy system costs are related to the fabrication of components and installation and maintenance of the system. Because the energy to operate the plant is supplied by the sun, continually rising fuel costs could conceivably even make a solar-fired steam plant cost effective.

Hot-Air Engines—The Stirling Cycle

One type of small engine that appears feasible for decentralized solar energy applications is the *hot-air engine.* A hot-air engine is one that uses air as its working substance. One such engine uses a cycle that can be approximated by a series of four reversible ideal gas processes called the *Stirling cycle.* The cycle consists of two isotherms alternated with two isochors. A two-cylinder Stirling engine is shown in Figure 8.7*a*. In process $(1 \rightarrow 2)$, cool air at a volume V_a contained in the cool cylinder (T_C) on the right is forced into the hot cylinder (T_H) via a large wire baffle called a *regenerator*. The air absorbs heat from the wire baffle. The regenerator thus acts as a series of heat reservoirs gradually raising the air temperature from T_C to T_H as it enters the hot cylinder. The volume of the air is unchanged and is equal to V_a. This process can therefore be simulated by reversible isochoric heating in which the temperature and

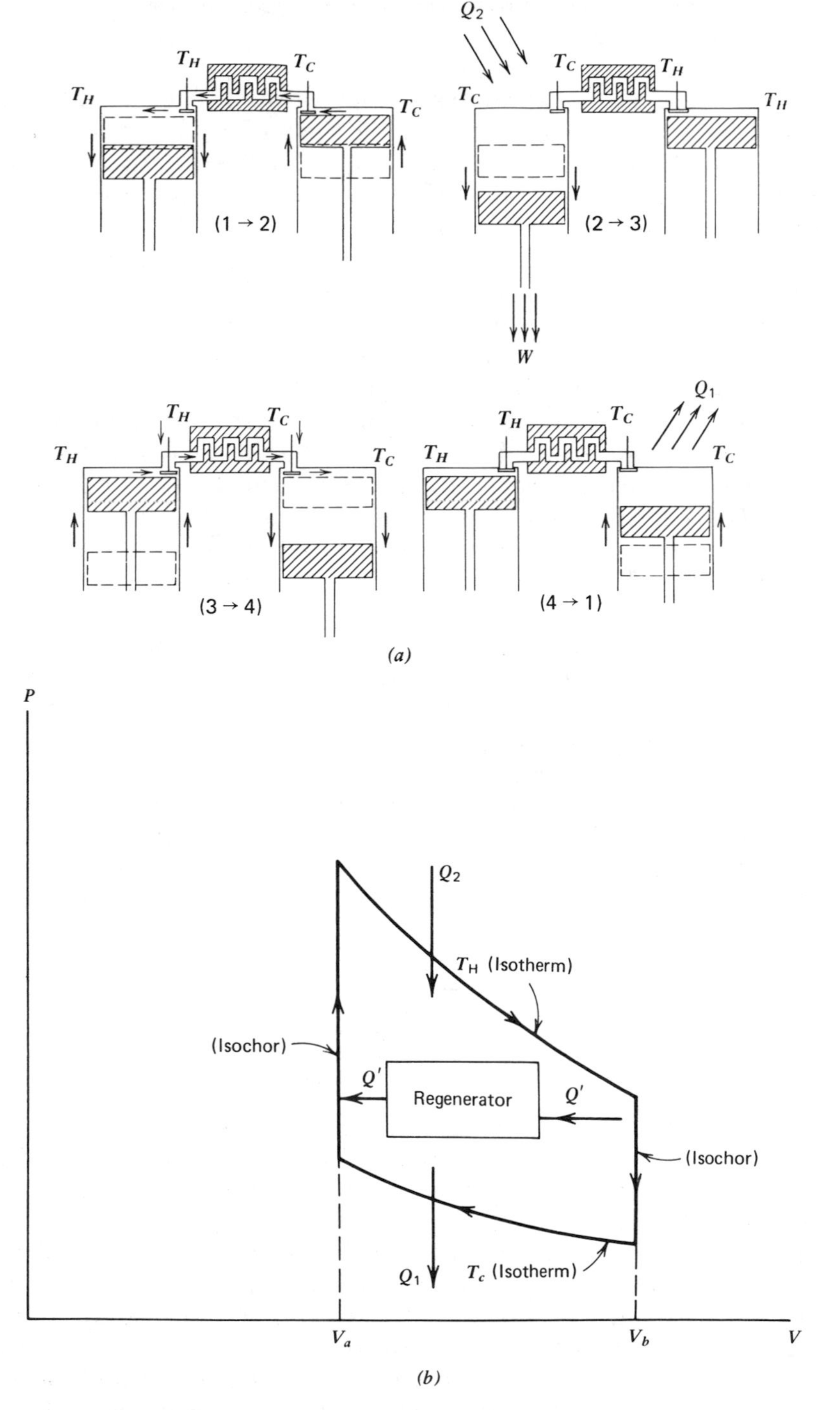

FIGURE 8.7 *(a) A schematic of the Stirling hot-air engine. (b) A P-V diagram for a reversible Stirling cycle using an ideal gas as its working substance.*

pressure increase but the volume remains constant. An amount of heat Q' is absorbed but no work is done. While the air is in the hot cylinder, it absorbs heat Q_2 at constant temperature T_H ($2 \rightarrow 3$). In doing so it expands isothermally to a value V_b as its pressure decreases. Work is done *by* the air during this isothermal expansion. The hot cylinder is maintained at a temperature T_H by solar energy which of course supplies the heat Q_2. The gas is then forced back into the cold cylinder ($3 \rightarrow 4$). As it passes the wire baffle, heat is returned to the regenerator and the gas is cooled to a temperature T_C before it enters the cold cylinder. The volume remains fixed at V_b. Hence the process can be simulated by reversible isochoric cooling. The pressure decreases with the temperature, but no work is done. During the final process ($4 \rightarrow 1$), the gas is compressed isothermally at a temperature T_C. The pressure increases as the volume returns to its original value V_a. Work is done *on* the air and heat Q_1 is rejected to a cold reservoir—usually the surrounding air. These processes are shown in a P-V diagram in Figure 8.7*b*.

It can be shown that if the air behaves as an ideal gas, the heat absorbed from the regenerator during isochoric heating is exactly equal to the heat rejected during isochoric cooling. In fact, after one complete cycle, the thermodynamic state of the regenerator reservoir is unchanged. Consequently, the heat transferred to and from the regenerator can be ignored in computing efficiency. (The temperature profile of the regenerator is actually produced and maintained by the hot air flowing back and forth.) A direct calculation on the reversible Stirling cycle (shown in Figure 8.7*b*) shows that the efficiency of the cycle is

$$\boxed{\eta_{\text{Stirling}} = \frac{T_H - T_C}{T_H}} \tag{8.16}$$

Hence the reversible Stirling cycle has the same efficiency as the Carnot cycle, provided the working substance behaves as an ideal gas. Because air does behave very much like an ideal gas, the theoretical limit for the efficiency of a Stirling engine is as high as the second law of thermodynamics will allow. The problem of course is whether or not the processes of the real hot-air engine can be performed in an approximately reversible manner. For example, because air is not a good thermal conductor, it is difficult to supply heat to it in the hot cylinder in any manner that remotely resembles a reversible isotherm. Other gases such as helium and hydrogen, which are better thermal conductors, have been used. Unfortunately, these gases tend to leak out of the engine.

There are other reversible cycles for hot-air engines whose efficiencies approach that of the Carnot cycle. One such cycle is the *Ericsson cycle,* which replaces the isochors of the Stirling cycle with isobars. Regeneration is used during these isobars as shown in Figure 8.8. The Ericsson engine is generally used to drive turbines rather than pistons.

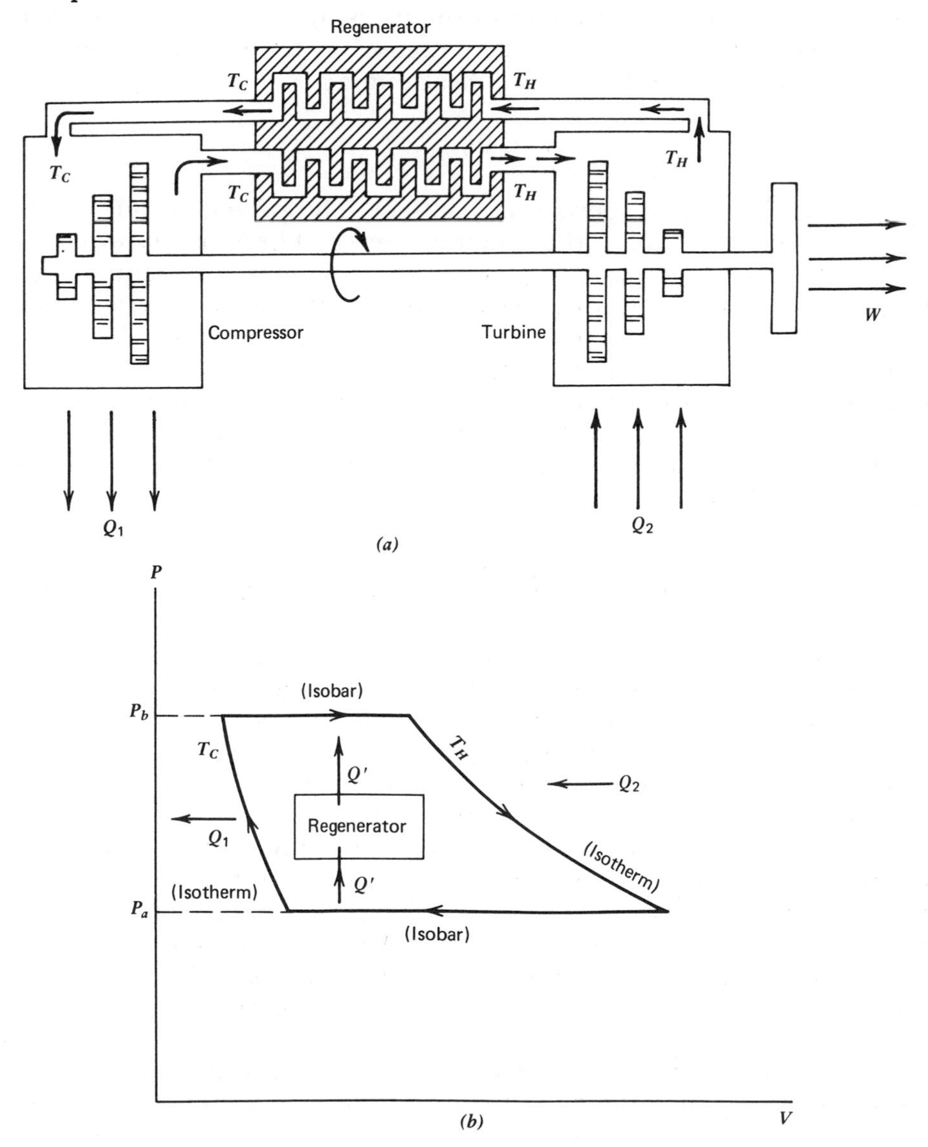

FIGURE 8.8 *(a) A schematic of the Ericsson hot-air engine. (b) A P-V diagram for a reversible Ericsson cycle using an ideal gas as its working substance.*

The work per cycle of a Stirling engine is found to be

$$\boxed{W_{\mathrm{cycle}} = nR(T_{\mathrm{H}} - T_{\mathrm{C}}) \ln \frac{V_{\mathrm{b}}}{V_{\mathrm{a}}}} \tag{8.17}$$

where n is the number of moles of the working gas and R is the ideal gas constant. The work per cycle increases as the size of the engine increases. The output power of a heat engine is

$$P = W_{\mathrm{cycle}} \times f$$

where f is the engine speed (in cycle/sec). The power output can be increased by increasing the engine speed. There is, however, a practical limit to the operating speed of an engine. Consequently, a large power output requires a large amount of work per cycle, which in turn means that the physical size of the engine must be large. Hot-air engines working at temperatures well below those of steam engines have been developed with reported operating efficiencies as high as 70 percent of the Carnot limit. However, these engines are relatively small-scale units and are most suitable for decentralized power production.

Thermoelectric Generation

One factor that reduces the efficiency of a mechanical heat engine is friction between its moving parts. It is possible to convert heat directly to a nonmechanical form of work such as electricity with a heat engine having no moving parts. One such engine is a *thermoelectric* generator. It is well known that when two dissimilar metals are joined at both ends and the junctions are maintained at different temperatures, an electric current will flow in the circuit. The existence of a current implies that there is an effective electromotive force (emf) in the circuit. The emf in turn means that the thermocouple acts as a source of electrical energy. As we will see, this work is derived from heat reservoirs. The process is therefore thermodynamic and is subject to the limitations of the second law of thermodynamics.

The theory behind thermoelectric generation is quite complicated; we will consider only the basic features. Consider a single metal wire whose ends are at temperatures T_2 and T_1 where $T_2 > T_1$ (Figure 8.9). As a result of the thermal gradient along the wire, some electrons migrate from higher to lower temperature regions. This

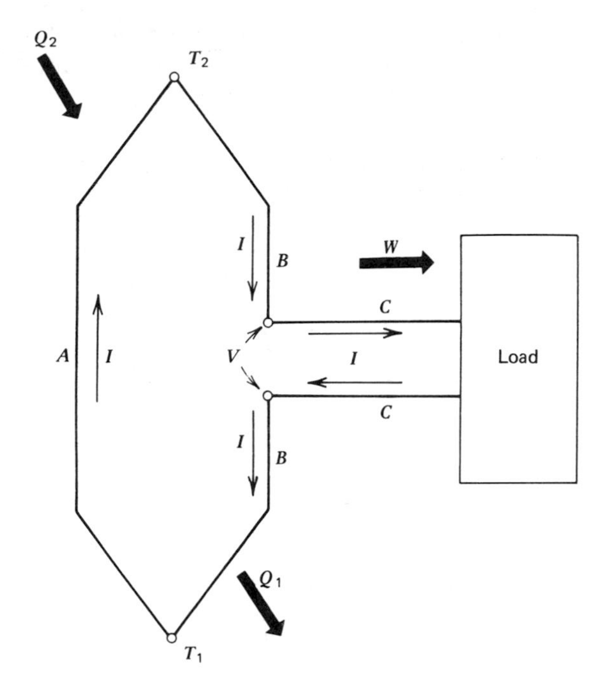

FIGURE 8.9 *A thermocouple of materials A and B extracting heat Q_2, rejecting heat Q_1, and supplying electrical work to a load. The power supplied is the product of the terminal voltage, V, and the current, I, delivered to the load.*

migration can be regarded as being caused by a "thermal" force on the electrons. The migration continues until the charge accumulation produces an opposing electrostatic force that balances the thermal force. The thermal force divided by the charge of the electron is called the thermoelectric field E_{thermal} and is related to the temperature gradient dT/dx by

$$E_{\text{thermal}} = \epsilon_A(T)\frac{dT}{dx}$$

The quantity $\epsilon_A(T)$ is a temperature-dependent parameter (of wire A) called the *absolute thermoelectric power.* By integrating over the length of wire A, we obtain

$$\boxed{\mathscr{E}_A = \int_0^L E_{\text{thermal}}\,dx = \int_0^L \epsilon_A \frac{dT}{dx}\,dx = \int_{T_1}^{T_2} \epsilon_A\,dT} \tag{8.18}$$

The quantity $\mathscr{E}_A$ in Equation 8.18 is called the *thermal* emf induced along the wire. Note that this emf is *independent* of both the length and the temperature profile of the wire. It is determined by the variation of ϵ_A with T and by the temperatures of the end points of the wire.

Suppose next that wires A and B are joined at the ends to form a thermocouple as in Figure 8.9. The net thermal emf around the circuit is

$$\boxed{\mathscr{E}_{AB} = \mathscr{E}_A - \mathscr{E}_B = \int_{T_1}^{T_2} \epsilon_{AB}\, dT} \tag{8.19}$$

where

$$\epsilon_{AB} = \epsilon_A - \epsilon_B$$

The parameter ϵ_{AB} is the *relative thermoelectric power* of the wires. It is also called the *Seebeck* coefficient, introduced and denoted by $\mathscr{S}_{AB}$ in Chapter 6. In many thermocouple systems, ϵ_{AB} varies slowly with temperature so that the Seebeck voltage in Equation 8.19 may be written

$$\boxed{\mathscr{E}_{AB} \simeq \epsilon_{AB}(T_2 - T_1)} \tag{8.20}$$

provided that $(T_2 - T_1)$ is not too large. For example, the Seebeck coefficient for an iron-constantan thermocouple is $\simeq 53\ \mu\text{V}/°\text{C}$ for $0°\text{C} \leq T \leq 100°\text{C}$.

A thermocouple performs useful work only if it is supplying a current to an external device. In short-circuit operation the electrical energy is dissipated as resistive heat within the wires. In the open circuit no work is done by the thermocouple because no current is flowing. When power is supplied to an external device as in Figure 8.9, heat is absorbed at certain points along the thermocouple wire and rejected at others. The difference is converted to electrical work and applied to the device.

The thermodynamic conversion of heat to work actually involves four distinct processes, only two of which are reversible. The reversible processes are the *Peltier* and *Thomson* effects. The irreversible ones are the *Fourier* effect (heat conduction) and the *Joule* effect (electrical resistance heating). In order to apply elementary thermodynamics to the thermoelectric system, we must ignore the irreversible processes. If the wires are assumed to have zero thermal conductivity, irreversible conduction of heat between the hot and cold reservoirs cannot occur. Furthermore, if the wires also have zero electrical resistivity, the irreversible process of joule heating is absent. We are thus left with the reversible Peltier and Thomson effects, which will be described below.

Equation 8.19 can be integrated by parts as

$$\mathscr{E}_{AB} = \int_{T_1}^{T_2} (\epsilon_A - \epsilon_B)\, dT = (\epsilon_A - \epsilon_B)T\Big|_{T_1}^{T_2} - \int_{T_1}^{T_2} T\left\{\frac{d}{dT}(\epsilon_A - \epsilon_B)\right\} dT \tag{8.21}$$

We will define the following functions of the Kelvin temperature:

$$\boxed{\pi_{AB}(T) = (\epsilon_A - \epsilon_B)T = \epsilon_{AB}T = \text{Peltier coefficient at junction } AB} \tag{8.22a}$$

and

$$\boxed{\begin{aligned} \sigma_A(T) &= -T\frac{d\epsilon_A}{dT} = \text{Thomson coefficient along wire } A \\ \sigma_B(T) &= -T\frac{d\epsilon_B}{dT} = \text{Thomson coefficient along wire } B \end{aligned}} \tag{8.22b}$$

In terms of these coefficients Equation 8.21 can be expressed as

$$\mathscr{E}_{AB} = \Pi_{AB}(T_2) - \Pi_{AB}(T_1) + \int_{T_1}^{T_2} \sigma_A\, dT - \int_{T_1}^{T_2} \sigma_B\, dT \tag{8.23}$$

The electrical power supplied to a device is found by multiplying Equation 8.23 by the current flowing, I, giving

$$\boxed{P = I\mathscr{E}_{AB} = I\,\Pi_{AB}(T_2) - I\,\Pi_{AB}(T_1) + I\int_{T_1}^{T_2} \sigma_A\, dT - I\int_{T_1}^{T_2} \sigma_B\, dT} \tag{8.24}$$

From Equation 8.24 it can be seen that the electrical power produced by the thermocouple is equal to the difference between two Peltier effects and two Thomson effects.

Whenever an electrical current traverses a junction immersed in a heat reservoir at temperature T, heat is either absorbed or rejected according to the direction of the current. This heat is called *Peltier heat* and is given by

$$\boxed{\dot{Q}_{\text{Peltier}} = \Pi_{AB}I} \quad \text{(Peltier heat)} \tag{8.25}$$

EXAMPLE

The Seebeck coefficient of a junction is 55 μV/K at 373 K and 50 μV/K at 273 K. Find the Peltier heats absorbed and rejected when the thermocouple is operating between these heat reservoirs and is supplying a current of 10 ma.

Using Equation 8.22a, the Peltier coefficients at these junctions are

$$\Pi_{AB}(373 \text{ K}) = (55 \times 10^{-6})(373) = 2.05 \times 10^{-2} \text{ W/amp}$$

and

$$\Pi_{AB}(273 \text{ K}) = (50 \times 10^{-6})(273) = 1.37 \times 10^{-2} \text{ W/amp}$$

From Equation 8.25 we find the Peltier heats absorbed and rejected to be

$$\dot{Q}_2 = (10 \times 10^{-3})(2.05 \times 10^{-2}) = 205 \ \mu\text{W}$$

and

$$\dot{Q}_1 = (10 \times 10^{-3})(1.37 \times 10^{-2}) = 137 \ \mu\text{W}$$

If no other heat transfer were involved, the difference between these values, 68 μW, would be supplied as electrical power.

In addition to Peltier heat transferred at the junctions, there is also *Thomson heat* transferred to and from the surroundings *along* the wires. This heat rate per unit length of wire is

$$\frac{d\dot{Q}_{\text{Thomson}}}{dx} = I\sigma_A \frac{dT}{dx}$$

so that the total heat transferred along wire A is

$$\boxed{\dot{Q}_{\text{Thomson}} = I\int_{T_1}^{T_2} \sigma_A \, dT} \qquad \text{(Thomson heat)} \qquad (8.26)$$

EXAMPLE

The absolute thermoelectric power of a wire increases linearly with temperature at a rate

$$\frac{d\epsilon_A}{dT} = 5.4 \times 10^{-9}\ \text{V/K}^2$$

The end points of the wire are maintained at 373 and 273 K. Find the Thomson heat transferred to the surroundings if a current of 10 ma is flowing in the wire.

According to Equation 8.22b, the Thomson coefficient varies with temperature as

$$\sigma_A = -T\frac{d\epsilon_A}{dT} = (-5.4 \times 10^{-9}T)\ \text{V/K}$$

Using Equation 8.26, we find the Thomson heat is

$$\dot{Q}_{\text{Thomson}} = (10 \times 10^{-3})(5.4 \times 10^{-9})\int_{273}^{373} T\,dT = 1.74\ \mu\text{W}$$

Returning to Equation 8.24, we may write the electrical power produced as

$$P = \dot{Q}_2 - \dot{Q}_1 + \dot{Q}' - \dot{Q}'' \tag{8.27}$$

where

$\dot{Q}_2$ is the Peltier heat absorbed at the hot junction
$\dot{Q}_1$ is the Peltier heat rejected at the cold junction
$\dot{Q}'$ is the Thomson heat absorbed along wire A and
$\dot{Q}''$ is the Thomson heat rejected along wire B

Since Thomson heat is reversibly transferred along a series of reservoirs varying in temperature from T_1 to T_2, the system will not necessarily have the *Carnot* efficiency. However, consider the highly idealized case in which the thermoelectric power of each wire is *independent* of temperature so that, according to Equation 8.22*b*, the

Thomson coefficient is zero and consequently the Thomson heat vanishes. In that case the system becomes a reversible heat engine operating between two reservoirs—one at each junction. The efficiency of this engine is, using Equations 8.20 and 8.24,

$$\eta = \frac{P}{\dot{Q}_2} = \frac{I\mathscr{E}_{AB}}{I\Pi_{AB}(T_2)} = \frac{I\epsilon_{AB}(T_2 - T_1)}{I\epsilon_{AB}T_2} = \frac{T_2 - T_1}{T_2} \tag{8.28}$$

Equation (8.28), as expected, leads to the Carnot efficiency. The thermodynamic processes for a reversible thermoelectric system are shown in Figure 8.10.

In practice, the irreversible processes of heat conduction along the

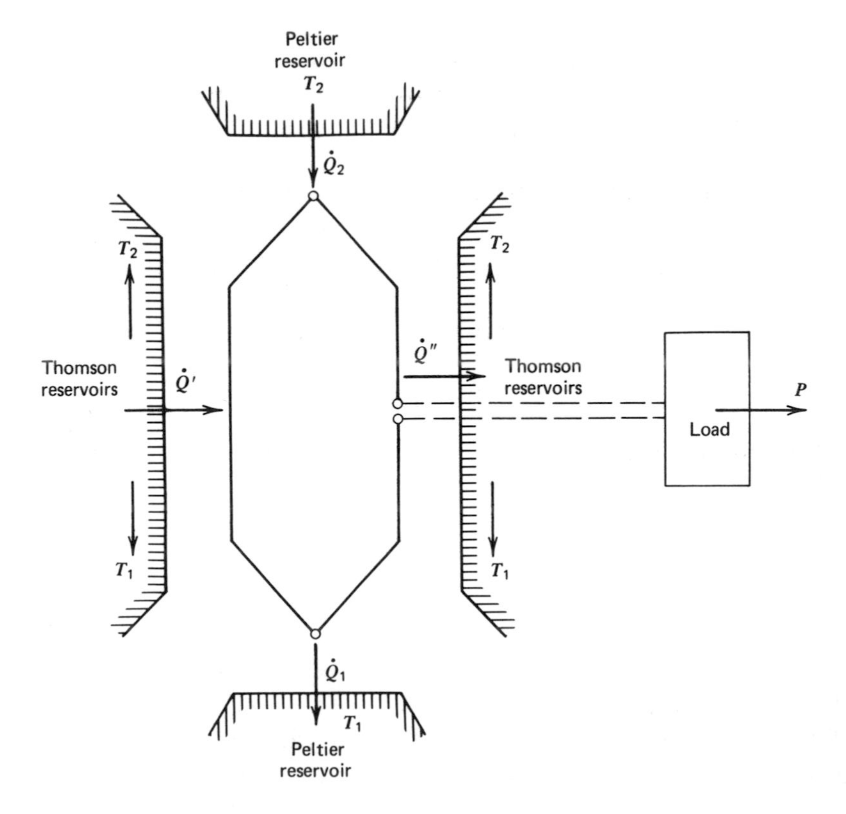

FIGURE 8.10 *Illustrating the Peltier and Thomson reservoirs in a thermoelectric system. Note although there are two Peltier reservoirs, there are two series of Thomson reservoirs. For a reversible system, the work supplied to the load is the difference between the Peltier heats* ($\dot{Q}_2 - \dot{Q}_1$) *plus the difference between the Thomson heats* ($\dot{Q}' - \dot{Q}''$).

wires and joule heating within the wires reduce efficiencies to levels well below the Carnot limit. The heat conduction process drains heat wastefully from the hot reservoir and deposits it in the cold reservoir without doing any work. Of the electrical power generated, an amount I^2R, where R is the total resistance of the wires, is dissipated as heat within the thermocouple wires themselves. If the wires are well insulated so that there is no transfer of joule heat to the surrounding air and if Thomson heat is negligible, half the joule heat produced in each wire will be conducted back to each reservoir. Thus the net heat rate actually extracted from the hot reservoir is

$$\dot{Q}_2 = \epsilon_{AB} T_2 I + \bar{U}_K (T_2 - T_1) - \tfrac{1}{2} I^2 R \tag{8.29}$$

where $\bar{U}_K$ is the net heat transfer coefficient for thermal conduction along the thermocouple wires. The terms on the right hand side of Equation 8.29 are the extracted Peltier heat, the heat leaving the hot reservoir by thermal conduction, and half the joule heat returned to the reservoir, respectively.

The actual electrical power supplied to an external device is

$$P = \mathscr{E}_{AB} I - I^2 R \tag{8.30}$$

This net power is the difference between the total power produced and the power wasted in joule heating within the thermocouple wires. Using both Equations 8.29 and 8.30, we find the actual efficiency to be

$$\boxed{\eta = \frac{P}{\dot{Q}_2} = \frac{\epsilon_{AB}(T_2 - T_1) I - I^2 R}{\epsilon_{AB} T_2 I + \bar{U}_K (T_2 - T_1) - \frac{1}{2} I^2 R}} \tag{8.31}$$

In obtaining Equation 8.31, we used the approximate relation, Equation 8.20, for $\mathscr{E}_{AB}$. Note that for a given thermocouple (i.e., for fixed ϵ_{AB}, $\bar{U}_K$, and R) operating between heat reservoirs at temperatures T_1 and T_2, the efficiency depends on the current drawn from the system. This current in turn depends on the resistance of the device being operated by the thermocouple. In terms of this external load resistance, the current can be expressed as

$$I = \frac{\mathscr{E}_{AB}}{R + R_{ext}} = \frac{\mathscr{E}}{R(1 + X)} \tag{8.32}$$

where

$$X = \frac{R_{ext}}{R}$$

The efficiency varies with the load resistance; it is possible to choose X to maximize the efficiency. Substituting Equation 8.32 into 8.31 and maximizing the efficiency (by setting $d\eta/dX = 0$), we find after simplifying that

$$\eta_{\text{max}} = \left\{\frac{(X_0 - 1)}{X_0 + (T_1/T_2)}\right\} \eta_{\text{Carnot}} \tag{8.33}$$

where

$$X_0 = \sqrt{1 + ZT_{\text{m}}}, \qquad T_{\text{m}} = \frac{T_1 + T_2}{2}, \qquad \text{and} \qquad Z = \frac{(\epsilon_{AB})^2}{\bar{U}_K R}$$

The quantity T_{m} is the mean operating temperature of the system and the parameter Z is the *figure of merit* of the thermocouple. The larger the value of Z is, the higher the quality of the thermocouple. The value of Z approaches infinity as U_K and R each approach zero. Furthermore, X_0 in Equation 8.33 also approaches infinity under these conditions so that

$$\eta_{\text{max}} \xrightarrow[X_0 \to \infty]{} \eta_{\text{Carnot}}$$

The effectiveness of a thermocouple depends on its figure of merit as defined in Equation 8.33. In order to have a large Z, a thermoelectric material must have a large Seebeck coefficient, a low electrical resistivity, and a low thermal conductivity. Although most metals do have a low electrical resistivity, their thermal conductivity is high and the Seebeck coefficient is only moderate. There are specially prepared semiconductor junctions that have high electrical conductivities and yet have relatively low thermal conductivities. In addition, they have larger Seebeck coefficients than metals. Semiconductor thermocouples are, however, more difficult to fabricate and more fragile than metallic thermocouples. The highest values of Z obtained are typically $Z \simeq 0.002\ \text{K}^{-1}$. For reservoirs of 373 and 273 K, even such a large figure of merit leads, according to Equation 8.33, to a thermodynamic efficiency of only 3 percent. The corresponding Carnot limit is 27 percent.

To produce a usable voltage, a series of thermocouple junctions known as a *thermopile* are required. Such a solar powered thermopile is shown in Figure 8.11. The overall efficiency of the thermopile is even smaller than its thermodynamic efficiency, because much of the solar

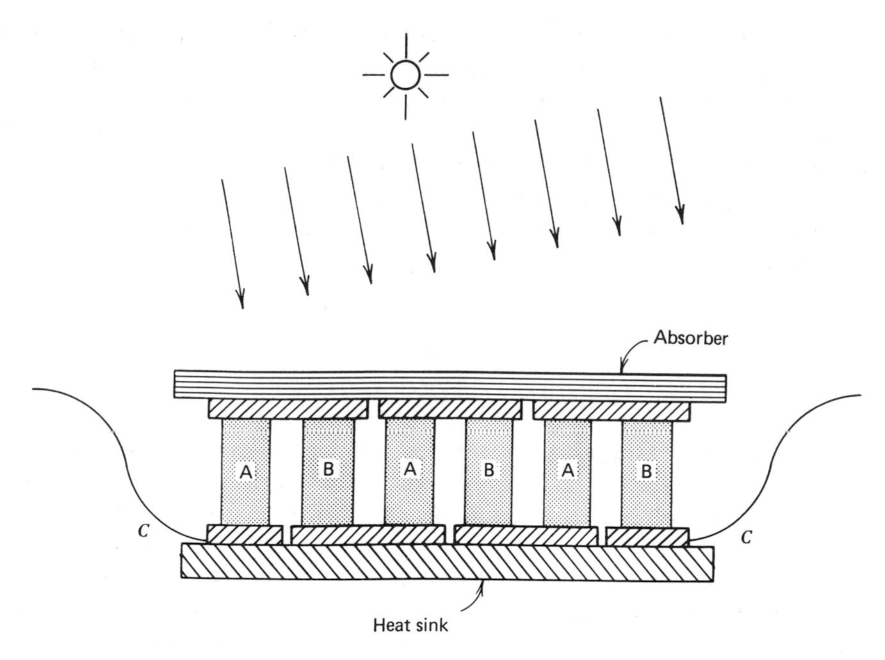

FIGURE 8.11 *Illustrating a thermopile consisting of three thermocouples used to convert solar heat to electricity.*

energy intercepted by the absorber plate $\dot{Q}_{inc}$ is lost to the surroundings; only a part $\dot{Q}_2$ is supplied to the thermoelectric system. A typical overall efficiency is only a few percent.

There are other thermodynamic systems that have been considered for solar energy systems. For example, in a *thermionic* converter, heat is supplied to an electrode (the cathode), resulting in electrons being emitted from the electrode's surface. These energetic electrons are collected by a second electrode called an anode where some heat is rejected. An emf appears across the electrodes; the system is therefore capable of doing electrical work. Because electrons cannot travel far in air, thermionic converters require that the electrodes be in a vacuum. This limits the size of the converter so that only small-scale power production is feasible.

In principle, it is possible to convert solar heat to work wherever solar energy makes available two heat reservoirs at different temperatures. For example, the solar heating of tropical waters maintains the temperature of the surface higher than that of the ocean depths. A thermodynamic engine can extract heat from the surface, convert some of it to work, and reject the rest to the ocean bottom. Ocean thermal energy is therefore a form of solar energy, with the ocean surface acting as both a giant collector and a massive heat reservoir. Unfortunately, the temperature difference between the ocean's surface and its depths is not very great. Hence the limiting Carnot efficiency is small to begin with. Furthermore, because the two heat reservoirs are so far apart physically, the technological problems are considerable. Nevertheless, because of the vast amounts of energy available from oceans, much interest has been shown in this resource.

Solar-Driven Coolers

As shown, the Clausius statement of the second law of thermodynamics states that it is impossible to construct a device operating in a cycle that involves no other effect than the transfer of heat from a cold to a hot reservoir. Refrigerators, air conditioners, and heat pumps do not violate this law because they involve another effect—they all require external work. It is possible to transfer heat from a cold reservoir to a hot one *without* doing external work and *without* violating the second law, provided a third and still *hotter* reservoir is available. This third reservoir provides the "effect" required by the second law and thus makes the heat transfer possible.

Consider three heat reservoirs—a hot reservoir at T_2, an intermediate reservoir at T', and a cold reservoir at T_1. One can imagine a heat engine extracting heat Q_2 from the hot reservoir, rejecting heat Q'' to the intermediate reservoir, and doing an amount of work $W = Q_2 - Q''$. This work in turn is used to drive a refrigerator that extracts heat Q_1 from the cold reservoir and deposits an amount of heat $Q''' = Q_1 + W$ to the intermediate reservoir (Figure 8.12a). The engine and refrigerator can be regarded as a composite device (Figure 8.12*b*). This device is a *heat-driven cooler*; it uses heat derived from a hot reservoir to extract heat from a cold reservoir. The sum of these heats, $Q' = Q'' + Q'''$, is rejected to an intermediate reservoir. Thus, in principle, it is possible to cool a region using solar heat instead of external work provided an intermediate reservoir is available.

A familiar example of a heat-driven cooler is a gas refrigerator. The hot reservoir is established by the gas flame, the intermediate reservoir by the room environment, and the cold reservoir by the interior space of the refrigerator. The heat rejected to the room is the heat derived from the flame plus the heat extracted from the interior of the refrigerator.

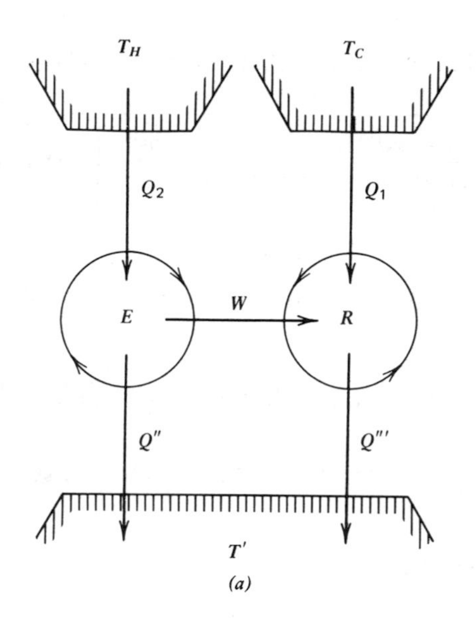

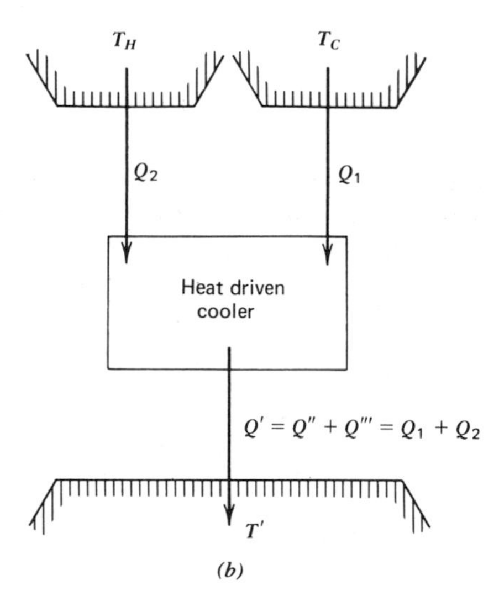

FIGURE 8.12 (*a*) *Illustrating how the work produced by a heat engine can be used to drive a refrigerator.* (*b*) *The composite system can be regarded as a three-reservoir heat-driven cooler using* Q_2 *units of heat from a hot reservoir, extracting* Q_1 *units from a cold reservoir, and rejecting* $Q_1 = Q' = Q'' + Q'''$ *units to an intermediate reservoir.*

The coefficient of performance of a heat-driven cooler is defined as

$$\boxed{\mathrm{COP} = \frac{Q_1}{Q_2} = \frac{\text{heat extracted from cold reservoir}}{\text{heat required from hot reservoir}}} \quad \text{(heat-driven cooler)} \tag{8.34}$$

If the overall device is composed of a Carnot engine and a Carnot refrigerator, the COP of the heat-driven cooler is, using Equations 8.6 and 8.11,

$$W = \eta_{\mathrm{Carnot}} Q_2 = \left(\frac{T_2 - T'}{T_2}\right) Q_2$$

$$Q_1 = \mathrm{COP}_{\mathrm{Carnot}}^{(\mathrm{refrig})} W = \left(\frac{T_1}{T' - T_1}\right) W$$

so that Equation 8.34 becomes

$$\boxed{\mathrm{COP}_{\mathrm{Carnot}} = \frac{Q_1}{Q_2} = \left(\frac{T_2 - T'}{T' - T_1}\right)\left(\frac{T_1}{T_2}\right)} \quad \text{(heat-driven cooler)} \tag{8.35}$$

EXAMPLE

A gas refrigerator has its working substance heated by a flame to 100°C (373 K). The interior space is to be kept at 0°C (273 K) when the room temperature is 20°C (293 K). Find the maximum possible COP of the refrigerator. If the actual COP is 0.8 and the heat is being extracted from the interior at 1000 Btu/hr, find both the rate at which heat is absorbed from the flame and the rate at which it is being rejected to the room.

The Carnot COP for the system is, from Equation 8.35,

$$\mathrm{COP}_{\mathrm{Carnot}} = \left(\frac{373 - 293}{293 - 273}\right)\left(\frac{273}{373}\right) = 2.9$$

From Equation 8.34, we find

$$\dot{Q}_2 = \frac{\dot{Q}_1}{\mathrm{COP}} = \frac{1000}{0.8} = 1250 \text{ Btu/hr}$$

and

$$\dot{Q}' = \dot{Q}_1 + \dot{Q}_2 = 2250 \text{ Btu/hr}$$

Note from Equation 8.35 that the COP increases with T_2; it also increases as T' approaches T_1. Hence for a heat-driven cooler it is desirable to produce a hot reservoir as hot as possible while providing an intermediate reservoir whose temperature is as close as possible to that of the region to be cooled.

In a solar-driven air conditioner the hot reservoir is supplied with heat energy derived from a solar array. The intermediate reservoir is generally a cooling tower situated outside the space to be cooled. Unfortunately, because of optical and thermal losses, only a fraction of the heat intercepted by the collector is actually available as heat, Q_2, for the cooling process. As with solar-driven heat engines, the overall efficiency of a solar-driven cooler can be written

$$\mathrm{COP}_{\text{overall}} = \frac{Q_1}{Q_{\text{inc}}} = \frac{\text{heat extracted from cold reservoir}}{\text{solar energy incident on collector}}$$

or

$$\boxed{\mathrm{COP}_{\text{overall}} = \frac{Q_1}{Q_2}\frac{Q_2}{Q_{\text{inc}}} = \mathrm{COP}_{\text{cooler}} \times \eta_{\text{collector}}} \qquad (8.36)$$

Hence the overall COP of a solar-driven cooler is equal to the product of the COP of the cooler and the efficiency of the collector. The higher the temperature produced by the collector is, the higher the COP of the cooler but the lower the efficiency of the collector itself. [See Problem 8-11.] Unless the solar panels can produce high enough temperatures with low thermal losses, a solar air conditioning system's overall COP may be too low to make the system cost effective. Although concentrating collectors do produce high temperatures with relatively low losses, the tracking requirement adds to the cost of the system. Recently developed solar-driven cooling systems use nontracking, evacuated collectors that operate at higher temperatures and have lower thermal losses than conventional flat plates. Nevertheless, when high temperatures are required for longer periods of the day tracking is essential.

Absorption Cooling

In a solar-driven cooler, work is transferred internally from the heat engine to the refrigerator. In principle, the heat engine provides mechanical work to a compressor, which in turn compresses the working substance (the refrigerant) of the refrigerator so that it can perform the cooling process. Because this type of work is mechanical, friction between moving parts reduces efficiency considerably. It is possible to devise a system in which *chemical* rather than mechanical work is transferred so that the number of moving parts is at a minimum. One such system is called an *absorption* cooler.

To understand this type of cooling, we first consider a conventional refrigeration cycle (Figure 8.13). The cycle consists of the following processes:

$1 \rightarrow 2$ The refrigerant (usually ammonia or freon) begins as a superheated, high-pressure vapor. It is isobarically cooled to the condensation point. Some heat is rejected to the surroundings (a series of reservoirs) through a device known as a *condenser.*

$2 \rightarrow 3$ The warm vapor is now condensed (isobarically and isothermally) to its liquid state. More heat (i.e., the latent heat of vaporization) is rejected to the surroundings (a single reservoir) through the condenser. The total heat rejected to the hot reservoirs in processes ($1 \rightarrow 2$) and ($2 \rightarrow 3$) is Q_2. The warm liquid is still under high pressure.

$3 \rightarrow 4$ The warm, high-pressure liquid is allowed to pass into a low pressure region through a small opening called a *throttling* valve. This throttling process is an intrinsically irreversible process regardless of how slowly and carefully it is performed. It is classified as an *irreversible adiabatic flow process.* At the end of the process the liquid's pressure and temperature fall markedly. The working substance is now a liquid at very low pressure and temperature in spite of the fact that no heat was extracted during the throttling process.

$4 \rightarrow 5$ Heat Q_1 is extracted from the surroundings through a device known as an *evaporator.* The cold, low-pressure liquid is transformed to a vapor isothermally and isobarically. The refrigerant is now a cool, low-pressure vapor.

$5 \rightarrow 1$ The cool vapor is adiabatically compressed to a superheated, high-pressure vapor by a compressor. It is now back to its original state so that the cycle can begin again. During the compression, an amount of work W is done by the compressor on the refrigerant.

The actual cooling process occurs in the refrigeration cycle during

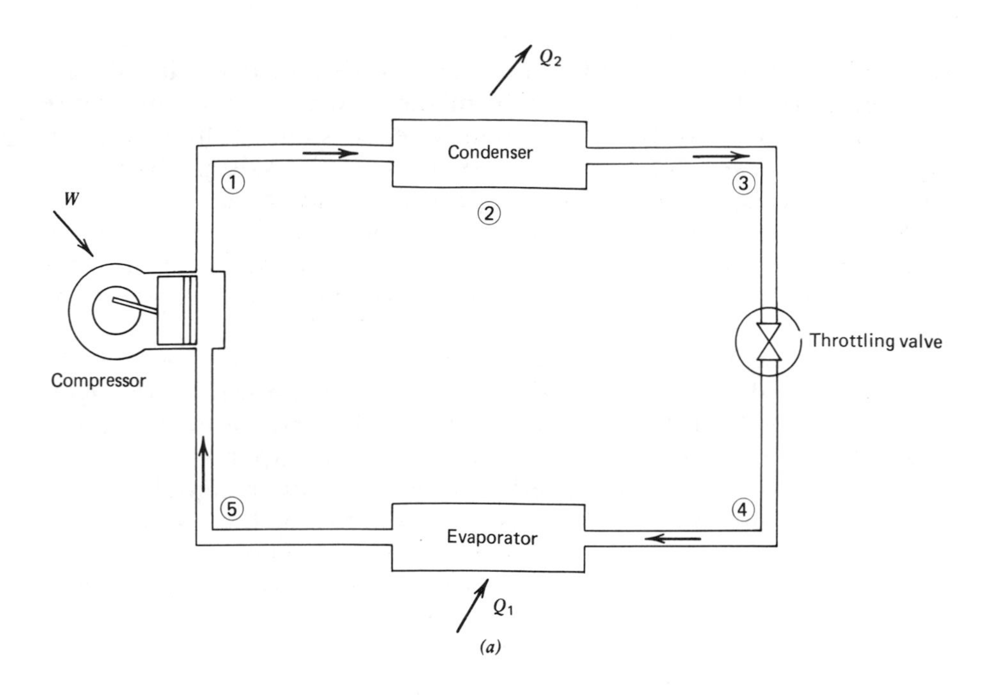

(a)

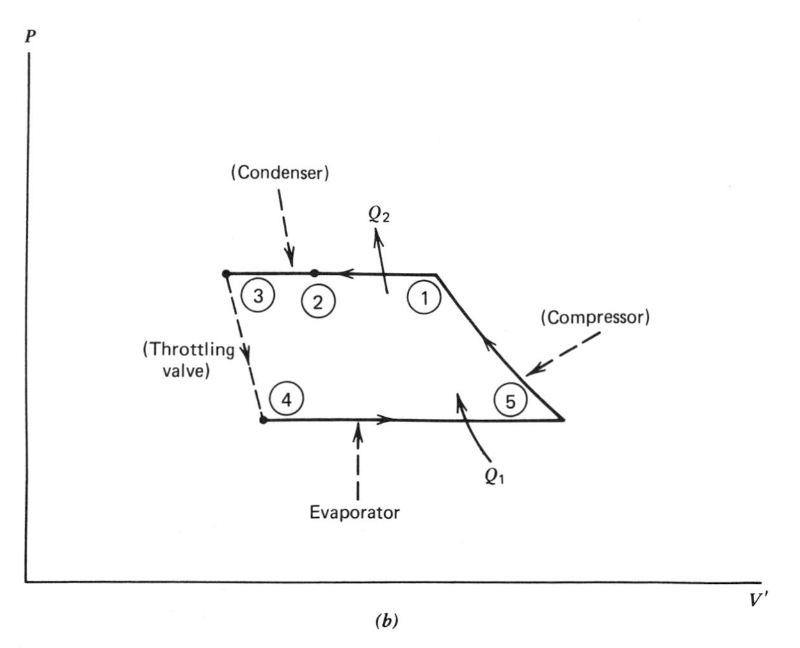

(b)

FIGURE 8.13 *(a) A schematic of a typical refrigeration cycle. (b) A P-V diagram for the refrigeration cycle. The broken line represents the throttling process, which is inherently irreversible.*

steps (3→4) and (4→5). The liquid is throttled to make it cold and then evaporated to extract heat from the cold reservoir. In a sense, the refrigerant falls from a higher energy state (a high-pressure liquid) to a lower one (a low-pressure vapor) as it performs the cooling process. The remaining processes—adiabatic compression (5→1), isobaric cooling (1→2), and condensation (2→3) serve to *regenerate* the refrigerant so that it returns to its high energy state. Note that to achieve regeneration of the refrigerant in conventional refrigeration, mechanical work is required (5→1).

It is also possible to regenerate the refrigerant from a low-pressure vapor to a high-pressure liquid using a process known as *absorption*. The process is heat driven and requires both a hot reservoir and an intermediate reservoir. The cool, low-pressure vapor enters an *absorber* where it mixes and combines readily with a liquid called a *carrier*. The absorption process releases energy in the form of heat that is transferred to an intermediate reservoir. The mixture is then pressurized by a pump as it passes into a very hot reservoir known as a generator or *separator* where heat is absorbed. The pump does not do any appreciable work because the mixture is in a liquid state and is therefore highly incompressible. At high temperature and pressure the mixture tends to separate into a liquid carrier and a vaporized refrigerant. The carrier is returned by the separator to the absorber where its pressure drops and it is cooled as heat is rejected. The carrier is then able to repeat its cycle by absorbing more refrigerant. The hot vapor continues through a condenser where it rejects more heat and condenses to a warm, high-pressure liquid. The refrigerant has thus been regenerated to its high energy state and can be throttled and evaporated to produce cooling. A solar-driven absorption cooler is shown in Figure 8.14. Typical absorption coolers use ammonia (refrigerant) in a water–ammonia solution (carrier) or water vapor (refrigerant) in a lithium–bromide solution (carrier).

EXAMPLE

A solar-driven air conditioning system has collectors that produce a temperature $T_2 = 90°C$ (363 K). The cooling towers provide an intermediate reservoir at $T' = 35°C$ (308 K). The system must maintain a cold reservoir at $T_1 = 15°C$ (288 K). Heat must be extracted at a rate of 20,000 Btu/hr to maintain this temperature. Find the maximum possible COP of this heat-driven system. If the actual COP of the cooler is 0.3 and the array efficiency 60 percent, find the minimum array area required when the intercepted solar flux is 1000 W/m^2

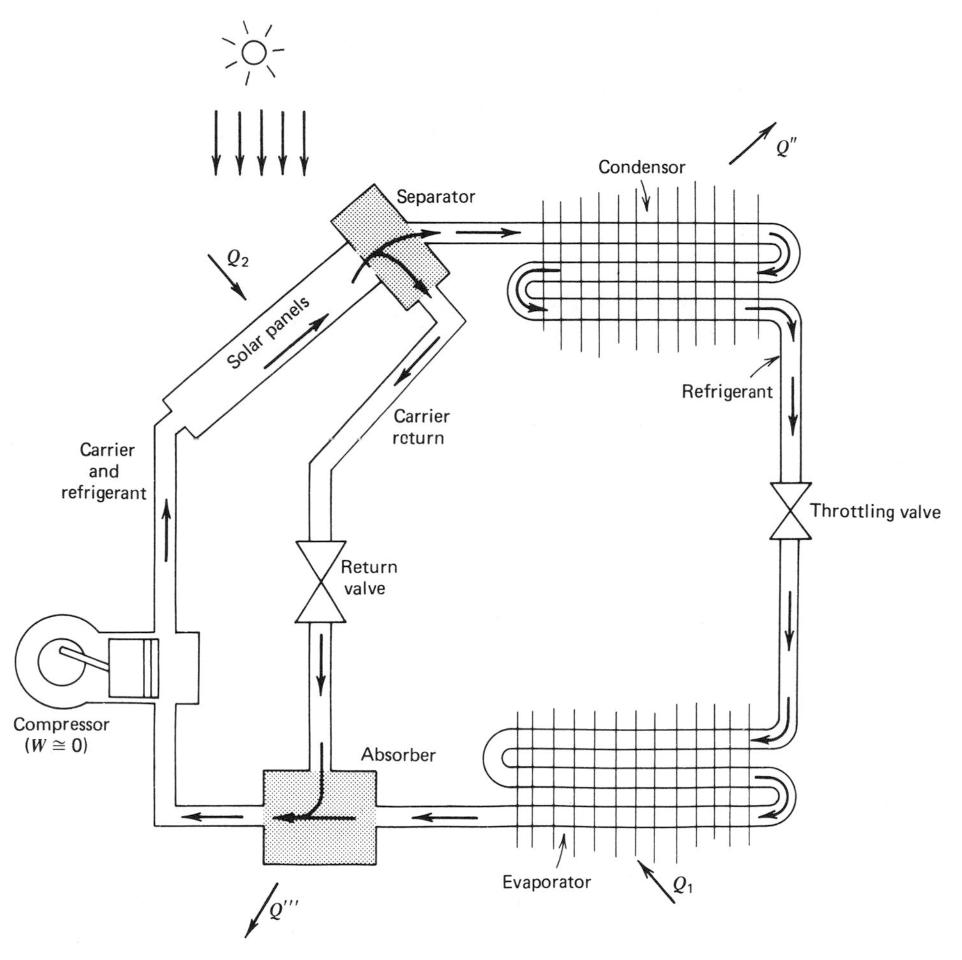

FIGURE 8.14 *A schematic of a typical solar-driven, absorption cooling cycle.*

(315 Btu/hr-ft^2). Also, find the heating power rejected to the cooling towers.

The maximum (Carnot) COP of the cooling system is, using Equation 8.35,

$$COP_{Carnot} = \left(\frac{363 - 308}{308 - 288}\right)\left(\frac{288}{363}\right) = 2.18$$

The overall COP of this system is, from Equation 8.36,

$$COP_{overall} = \frac{\dot{Q}_1}{\dot{Q}_{inc}} = (0.3)(0.6) = 0.18$$

Thus the intercepted power must be

$$\dot{Q}_{inc} = \frac{\dot{Q}_1}{COP_{overall}} = \frac{2 \times 10^4}{0.18} = 1.11 \times 10^5 \text{ Btu/hr}$$

The required array area is

$$A = \frac{\dot{Q}_{inc}}{F_{inc}} = \frac{1.11 \times 10^5 \text{ Btu/hr}}{315 \text{ Btu/hr-ft}^2} = 353 \text{ ft}^2$$

The actual thermal power supplied by the array to the cooling system is

$$\dot{Q}_2 = \eta_{coll}\dot{Q}_{inc} = (0.6)(1.11 \times 10^5) = 6.67 \times 10^4 \text{ Btu/hr}$$

The heat rate rejected is therefore

$$\dot{Q}' = \dot{Q}_1 + \dot{Q}_2 = 2 \times 10^4 + 6.67 \times 10^4 = 86{,}700 \text{ Btu/hr}$$

Solar-cooling systems require strong sunshine and specialized collectors (e.g., evacuated or concentrating) in order to produce the high temperatures necessary to establish an acceptable COP. As the technology continues to improve, it may eventually be possible to operate coolers using temperatures available from conventional but high quality flat plates.

All thermodynamic generators are limited in efficiency and performance by the temperatures that can be achieved by solar collectors. In fact, most real systems never come close to the theoretical limit imposed by Carnot's formula. It is therefore difficult and in some cases almost impossible to convert heat derived from solar energy into work efficiently. However, it is possible to produce work from solar radiation without the production of heat as an intermediate step. This process, called *direct conversion*, is not dependent on any reservoir temperature and its efficiency is not limited by Carnot's formula. The most common man-made method of direct conversion in use today is the production of electricity by photovoltaics. We consider this process in the following chapter.

PROBLEMS

8-1. A resistor is placed in a large tank of water at 50°C. Electrical power is supplied to the resistor at a rate of 10 W. Current is flowing for 5 min.
(a) How much electrical work is supplied to the resistor?
(b) How much heat is supplied to the water?
(c) If the temperature remains approximately constant during the interval, find the change in entropy of the water.
(d) What is ΔS for the resistor?
(e) Find ΔS_u and determine if the process is reversible, irreversible, or impossible.

8-2. An engine is operating between two heat reservoirs, one at $T = 100°C$ and the other at $T = 0°C$. Heat is being rejected at the rate of 300 W. Find the maximum possible power output of this engine.

8-3. An inventor has developed a window air conditioner that he claims requires 1000 W of electrical power to extract 2000 W of heat from a room at a temperature 70°F when the outside air temperature is 100°F.
(a) Assuming that the temperatures remain constant, compute the change in entropy of the universe after one hour of operation. Is the inventor correct in his claim?
(b) What is the maximum possible extraction rate when 1000 W of electrical power are supplied?

8-4. According to the first law of thermodynamics, the net heat supplied to a working substance during any process is equal to the increase in *internal* energy of the substance plus the work done by the substance on the surroundings.
(a) If the internal energy of an ideal gas depends only on its temperature, show that heat actually *enters* during an isothermal expansion even though there is no increase in temperature.
(b) If the equation of state of an ideal gas is $PV = KT$ (where K is a constant), show that heat enters during both an isobaric increase in volume and an isochoric increase in pressure.

8-5. Consider a heat engine driven by a solar array. Suppose the array efficiency varies as $\eta_{\text{collector}} = 1 - (T - T_a)/(T_s - T_a)$ where T_s and T are the stagnation and operating temperatures of the array, respectively, and T_a is the air temperature. Furthermore, suppose the engine is a Carnot engine operating between reservoirs T and T_a whose efficiency is $\eta_{engine} = (T - T_a)/T$.

(a) Show that the array efficiency decreases from unity to zero as the operating temperature T is increased from T_a to T_s.

(b) Show that the engine efficiency increases from zero as T is raised from T_a to T_s.

(c) Find an expression for the overall efficiency of the system as a function of T.

(d) Using (c), find the overall efficiency as a function of T for $T_s = 200°C$ and $T_a = 20°C$.

(e) Determine the maximum overall efficiency and the optimum operating temperature.

(f) Using the maximum efficiency, find the required array area if the incident solar flux is 1000 W/m^2 and if the engine is to produce 10 kw of power.

(g) How much heat is lost to the surroundings by the array?

(h) How much heat is rejected by the engine?

8-6. (a) Show that the efficiency of both the Stirling and Ericsson cycles is given by Carnot's formula provided an ideal gas is used as the working substance.

(b) Verify that the net heat transfer to the regenerator over a complete cycle is zero.

(c) Verify that the entropy change of the regenerator is zero over a cycle.

8-7. A reversible Stirling engine is operating between heat reservoirs at $T_H = 100°C$ and $T_C = 0°C$. The working substance is 1 mole of air. The maximum and minimum volumes of the air during the cycle are $V_b = 4000$ cc and $V_a = 1000$ cc, respectively.

(a) Using Equation 8.17, find the work done per cycle. (Use $R = 8.31$ J/mole-K).

(b) What should the engine rpm (i.e., cycles per minute) be in order to produce 1.5 kw of power?

(c) How much solar heating power must be supplied to the engine by a solar array to produce the output in (b)?

8-8. The absolute thermoelectric power of two materials (A and B) varies with the temperature as $\epsilon_A = 6 \times 10^{-9}T$ and $\epsilon_B = 4 \times 10^{-9}T$ where T is in kelvins and ϵ_A and ϵ_B are in volts/kelvin. The materials are joined to form a thermocouple whose hot and cold junctions are maintained at 300 and 10°C, respectively. The materials have a negligible electrical resistivity and thermal conductivity.

(a) Find the emf produced by the thermocouple.
(b) If a current of $I = 0.5$ A is supplied to an external electrical device, find the power produced.
(c) Find the Peltier heat rates absorbed and rejected at the junctions.
(d) Find the Thomson heat rates absorbed and rejected in materials A and B.
(e) Verify that the difference between the heat rates absorbed and those rejected is equal to the power produced in (b).
(f) Find the efficiency of the system and compare it with a Carnot system operating between the same two temperatures.

8-9. A thermocouple is composed of two materials. The thermoelectric power of each material is independent of temperature. The hot and cold junctions are at $T_H = 500$ K and $T_C = 300$ K, respectively. Using the values $\epsilon_A = 2 \times 10^{-4}$ V/K, $\epsilon_B = 4 \times 10^{-4}$ V/K, $R = 100\ \Omega$, and $\bar{U}_K = 3.33 \times 10^{-7}$ W/K,
(a) Find the open circuit voltage (i.e., the emf) of the thermocouple.
(b) Find the short-circuit current of the device.
(c) Find the figure of merit of the system.
(d) Determine the optimum load resistance required to achieve maximum efficiency and find this efficiency.

8-10. A heat-driven cooler consists of an engine whose efficiency is 20 percent. The engine's work is used to drive a conventional refrigerator whose COP is 2. The engine and refrigerator reject heat to the same reservoir. Find the effective COP of this heat-driven cooler.

8-11. A three-reservoir solar-driven air conditioner consists of a Carnot system whose efficiency is

$$\text{COP} = \left(\frac{T - T_a}{T_a - T_1}\right)\left(\frac{T_1}{T}\right)$$

where T, T_a, and T_1 are the temperatures of the hot, intermediate (ambient air), and cold reservoirs, respectively. The hot reservoir is maintained by a solar collector whose efficiency is

$$\eta_{\text{collector}} = 1 - \frac{T - T_a}{T_s - T_a}$$

where T_s and T are the stagnation and operating temperatures of the collector, respectively.

(a) Show that the COP of the cooler increases as T is increased but that the limiting COP, as $T \rightarrow \infty$, is

$$\mathrm{COP}_{\mathrm{limit}} = \frac{T_1}{T_a - T_1}$$

(b) Show that the efficiency of the collector approaches zero as T is increased toward T_s.
(c) Express the overall COP of the system as a function of T, using the following values: $T_s = 200$, $T_a = 30$, and $T_1 = 5$°C.
(d) Using (c), find the optimum operating temperature and the maximum overall COP.
(e) Using (d), and assuming the intercepted solar flux to be 1000 W/m^2, find the array area required to extract heat at a rate of 5 kw.
(f) Find the heat lost by the array to the surroundings.
(g) Find the heat rejected by the cooler to the surroundings.

REFERENCES

1. Angrist, S. W., *Direct Energy Conversion,* 3rd ed., Allyn & Bacon, Boston (1976), Chapter 4.
2. Brinkworth, B. J., *Solar Energy for Man,* Wiley, New York (1972), Chapters 6 and 7.
3. Daniels, F., *Direct Use of the Sun's Energy,* Ballantine, New York (1964), Chapters 13 and 14.
4. Levine, S. N., *Selected Papers on New Techniques for Energy Conversion,* Dover, New York (1961), pp. 294–346.
5. McDaniels, D. K., *The Sun,* Wiley, New York (1979), Chapters VIII and X.
6. Meinel, A. B. and M. P. Meinel, *Applied Solar Energy,* Addison-Wesley, Reading, Mass. (1976), Chapter 14.
7. Zemansky, M. W., *Heat and Thermodynamics,* 4th ed., McGraw-Hill, New York (1957), Chapters 7–10, 12, 14.

CHAPTER 9

Direct Conversion of Solar Energy to Work-Photovoltaics

In the previous chapter we considered how work can be derived from solar radiation using heat engines. The radiation is first converted to heat by some solar heating device (e.g., flat plate or concentrator); the heat is then partially converted to work. Although it is theoretically possible to convert all solar energy to heat, it is not possible to convert this heat completely to useful work. Some heat must be rejected to a cooler reservoir. The efficiency of conversion is limited by the ratio of the temperature of the cold to that of the hot reservoir. The smaller this ratio is, the higher the possible efficiency. In *direct* conversion, work is derived from solar energy without first converting the electromagnetic radiation to heat. The process does not require hot and cold reservoirs. Although the efficiency of conversion may have certain practical limitations, it is not restricted by Carnot's formula.

In order to illustrate how a beam of radiation can be converted directly to mechanical energy, we consider a device known as a *vacuum bulb radiometer*. This device consists of four vanes, each of which has one side silvered and the other blackened (Figure 9.1a). The vanes are enclosed in an evacuated glass bulb. As we will see, a beam of radiation is capable of exerting a pressure and consequently a force on any surface on which it is incident. The pressure exerted on a silvered surface is almost twice that exerted on a blackened one

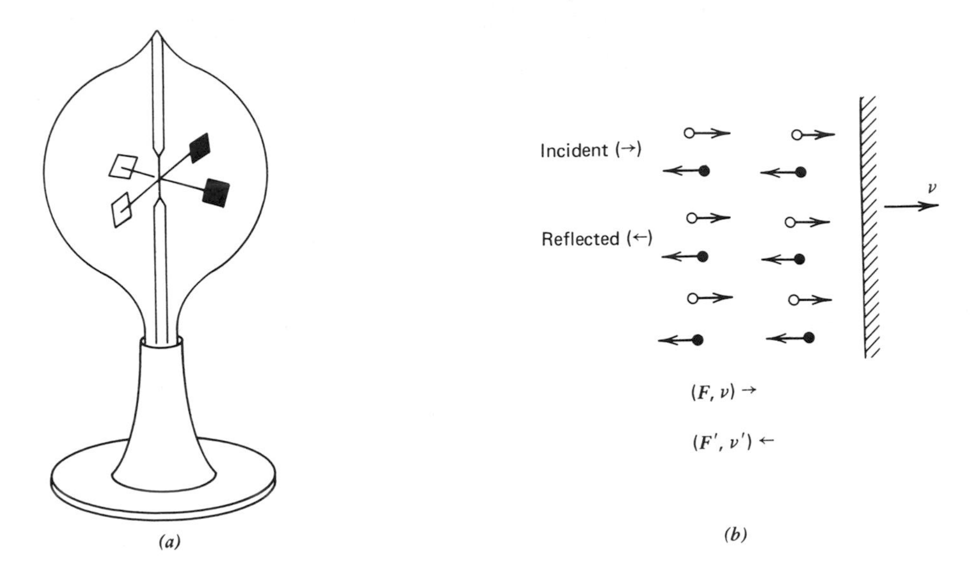

FIGURE 9.1 (*a*) *A vacuum bulb radiometer.* (*b*) *The photon beams incident and reflected from a moving mirror. The reflected photons experience a downward Doppler shift in frequency.*

because the silvered surface reflects radiation. Consequently, when one pair of opposite vanes is oriented toward a beam of radiation, a vane whose silvered surface is exposed to the beam experiences a greater force than an opposite vane whose blackened surface faces the beam. A net torque produced by these forces causes the vanes to rotate about a perpendicular axis (see Figure 9.1*a*). The mechanical power developed is equal to the torque multiplied by the angular velocity of the vanes. Of course in a typical radiometer this power is minuscule; it is just large enough to overcome the friction in the bearings of the system. Nevertheless, the vacuum bulb radiometer does illustrate that the production of heat is not a required intermediate step in the conversion of solar energy to work. Furthermore, the efficiency of conversion is not limited by the temperature that is produced by solar energy. To show this more clearly, we apply the principles of relativity and quantum theory to the conversion process.

Consider a parallel beam of radiation of frequency ν incident normally on a mirror that is perfectly flat and totally reflective. The mirror is moving at a speed v along the direction of propagation of the beam (Figure 9.1*b*). According to quantum theory, a beam of radiation can be regarded as a stream of particle-like quanta of energy called photons. The energy and momentum of each photon in the incident beam is

$$\epsilon = h\nu \tag{9.1}$$

and

$$p = \frac{h}{\lambda} = \frac{\epsilon}{c} \tag{9.2}$$

where λ is the wavelength and h is Planck's constant. If there are $\dot{N}$ photons crossing unit area per unit time in the beam, the incident flux is

$$F_{\text{inc}} = \dot{N}h\nu \tag{9.3}$$

According to the theory of relativity, the radiation that is reflected from the mirror experiences a *Doppler* shift. As a result, its frequency is decreased to a value given by

$$\nu' = \left[\frac{1-\beta}{1+\beta}\right]\nu \tag{9.4a}$$

where $\beta = v/c$, c being the speed of light. Furthermore, the theory of relativity predicts that the photon flux is reduced to a value

$$\dot{N}' = \left[\frac{1-\beta}{1+\beta}\right]\dot{N} \tag{9.4b}$$

The reflected flux is therefore

$$F_{\text{refl}} = \dot{N}'h\nu'$$

so that the net flux transferred by the incident beam is

$$F_{\text{net}} = F_{\text{inc}} - F_{\text{refl}} = h(\dot{N}\nu - \dot{N}'\nu') = \left[\frac{4\beta}{(1+\beta)^2}\right]F_{\text{inc}}$$

All this energy is not supplied to the mirror; some is used to "fill" the space in front of the mirror. To find the mechanical power actually supplied, we assume that the mirror is moving at a velocity that is much smaller than that of light (i.e., $\beta \ll 1$). A photon with momentum p will recoil with the same momentum magnitude so that the net momentum transferred to the mirror will be $2p$. The pressure exerted on the mirror is this momentum transfer multiplied by the photon flux

or

$$\text{Pressure} = 2p\dot{N}$$

Using Equations 9.1, 9.2, and 9.3, we find

$$\text{Pressure} = \frac{2F_{\text{inc}}}{c} \qquad (\beta \ll 1) \tag{9.5}$$

The mechanical power per unit area transferred to the mirror is

$$\frac{P}{A} = \text{Pressure} \cdot v = 2\beta F_{\text{inc}}$$

so that the transfer efficiency is

$$\eta = \frac{\frac{P}{A}}{F_{\text{inc}}} = 2\beta = 2\frac{v}{c} \qquad (\beta \ll 1) \tag{9.6}$$

Thus for small β, the conversion efficiency increases linearly with the mirror velocity.

For any practical mirror velocity, β is so small that no practical conversion efficiency could be expected. Furthermore, the maximum pressure exerted on a surface by strong sunlight is less than one-tenth of one-billioneth of an atmosphere. Although the minute radiation pressure may be sufficient to turn the small vanes of a vacuum bulb radiometer, this mode of conversion of solar energy to work is not practical. For example, if a frictionless solar "windmill" could be built whose blades each have an area of 100 ft^2 and move at 100 mph ($\beta \simeq 1.5 \times 10^{-7}$), the force per blade would be approximately 10^{-5} lb and the conversion efficiency would be less than 0.00003 percent. The important point is that direct conversion *is* possible and that this process does *not* involve a thermodynamic cycle; consequently, no heat reservoirs are required to supply or absorb heat.

There are processes that convert solar energy directly to chemical energy. Green plants use the sun's energy to convert water and carbon dioxide to complex carbohydrates. The process is a slow one and only a small fraction of the sunlight incident on the plant is converted to chemical energy. It has been suggested that chemical fuels could be made cost effectively from *biomass,* that is, from vegetation especially grown for that purpose. The mass production of alcohol from corn and sugar for gasohol is already a reality in some countries.

One of the most promising methods of direct conversion is the production of electricity using photovoltaics. A photovoltaic is a device that generates a voltage when sunlight is incident on it. We will consider a common class of such photovoltaics known as *p-n* junction devices. Devices made of silicon, for example, are capable of producing 0.5 V per cell in strong sunlight and have typical efficiencies of 10 to 12 percent. As we will see, the primary technological problem with *p-n* junction photovoltaics has been the high manufacturing costs. The theory of photovoltaics is rather complicated and so we will present only the fundamentals.

Intrinsic (Pure) Semiconductors

According to the quantum theory of matter, the electrons of an individual atom are allowed only certain states, each of which has a well-defined energy. There may be more than one state assigned the same energy. However, no more than one electron can be in any one state at a given time (Pauli's exclusion principle). When many identical atoms are arranged in a regular array, as in a pure crystalline solid, the outer levels tend to broaden into energy bands containing many closely spaced states. States within these bands are accessible to electrons, whereas energy levels between these bands are forbidden (Figure 9.2*a*).

Near absolute zero the electrons tend to occupy those states in the bands of lowest energy. According to the exclusion principle, only one electron can occupy a given state; consequently, the electrons are forced to seek levels of successively higher energy. Because the number of electrons in the material is finite, the occupied states reach only to a certain energy level. If this level is situated at the top of an energy band, the solid behaves as an insulator. The completely filled outermost band of the insulator is called the *valence* band. The empty band above the valence band is called the *conduction* band. For the insulator to exhibit electrical conduction, some electrons must acquire a small drift velocity in the presence of an electric field. In order to acquire this velocity, the electrons must increase their kinetic energy slightly. However, this is not possible because a slight increase in energy requires that the electrons near the top of the valence band move into the forbidden gap. Consequently, near absolute zero a solid with just enough electrons to completely fill the valence band behaves as a perfect insulator. When the temperature of the solid is raised substantially, some electrons near the top of the valence band are excited into the conduction band. The material then has a small

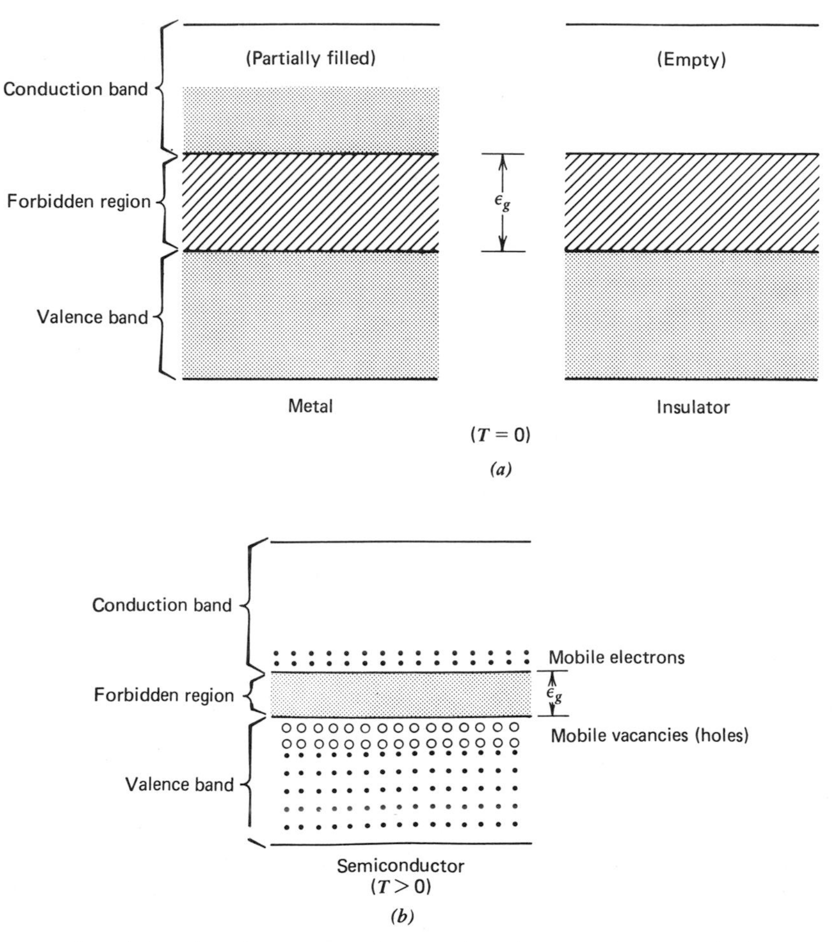

FIGURE 9.2 *(a) The band structure of a metal and an insulator at absolute zero. (b) The band structure of an intrinsic semiconductor at temperatures greater than absolute zero.*

number of electrons capable of acquiring a drift velocity and therefore exhibits a small electrical conductivity.

If the solid has a sufficient number of electrons to partially (e.g., half) fill the outermost (conduction) band, the electrons are excited to higher states within this band. As a result, a relatively large number of electrons acquire the necessary drift velocity for conduction and the solid behaves as a *metallic* conductor.

A semiconductor near absolute zero is quite similar to an insulator in that the semiconductor has a completely filled valence band and an empty conduction band. The distinction between the two is that in a semiconductor the energy gap between the conduction and valence bands is quite small. Thus at room temperature, a large number of valence band electrons are thermally excited into the conduction band (Figure 9.2*b*). Consequently, the electrical conductivity of a semiconductor increases dramatically as the temperature of the material is raised.

There is an interesting difference between electrical conduction in metals and in semiconductors. In metals, the current is due entirely to the flow of the free electrons in the conduction band. In semiconductors, however, the vacancies left behind in the valence band by the thermally excited electrons also contribute to the conduction process. Some electrons in the valence band can now be excited to the vacant states and thus acquire the necessary drift for conduction. The vacancies left in the valence band are said to produce positively charged *holes*. It is possible to describe the electrical conduction in the valence band as a flow of holes. When an electric field is applied, the holes produced in the valence band and the electrons produced in the conduction band flow in *opposite* directions but the net current is the *sum* of the hole and electron currents. The carrier density for an *intrinsic* or *pure* semiconductor is denoted by n_i for (negative) electrons and p_i for (positive) holes. Because carriers in an intrinsic semiconductor are generated in hole–electron pairs, we have $n_i = p_i$. It can be shown that the pair concentration varies with the kelvin temperature as

$$\boxed{n_i = p_i = AT^{3/2}e^{-\epsilon_g/2kT}} \tag{9.7}$$

where k is Boltzmann's constant and ϵ_g is the size of the energy gap, that is, the difference in energy between the bottom of the conduction band and the top of the valence band. The constant A is an empirical parameter of the semiconductor. The value of A is typically $\sim 10^{16}\ \text{cm}^{-3}\ \text{K}^{-3/2}$, whereas ϵ_g is of the order of an electron volt.[1] Boltzmann's constant can be written

$$k = \frac{1}{11{,}600} = 8.625 \times 10^{-5}\ \text{ev/K}.$$

According to Equation 9.7, the carrier concentration vanishes as T approaches zero so that the semiconductor begins to behave as an insulator. As T increases, both the carrier density and the electrical conductivity increase markedly. When an electric field E is applied to

[1]An electron volt is an amount of energy equal to 1.6×10^{-19} J.

the semiconductor, the resulting current density will be

$$J = E/\rho$$

where ρ is the *resistivity*. This resistivity can be expressed as

$$\rho = \frac{1}{en_i(\mu_n + \mu_p)} \tag{9.8}$$

where e is the electronic charge (1.6×10^{-19} C) and μ_n and μ_p are the electron and hole *mobilities*, respectively.

The resistivity of the material decreases as both the carrier concentration and carrier mobility increase. In metals n_i is temperature independent, $p_i = 0$ (i.e., no hole current), and μ_n decreases with temperature. Consequently, the resistivity of metals increases with increasing temperature. In semiconductors the increase in n_i more than compensates for any decrease in μ_n and μ_p so that the resistivity of semiconductors decreases with increasing temperature.

EXAMPLE

Find the resistivity of intrinsic silicon at both 0 and 50°C. Assume $A = 2.8 \times 10^{16}$ cm^3-K$^{3/2}$, $\epsilon_g = 1.12$ ev, and the mobilities remain approximately constant at $\mu_n = 1600$ cm^2/V-sec and $\mu_p = 400$ cm^2/V-sec.

Using Equation 9.8, we find the intrinsic carrier concentrations to be

$$n_i(0°\text{C}) = (2.8 \times 10^{16})(273)^{3/2}\exp[-(11{,}600)(1.12)/(2)(273)]$$
$$= 5.85 \times 10^9 \text{ particles/cm}^3$$

and

$$n_i(50°\text{C}) = (2.8 \times 10^{16})(323)^{3/2}\exp[-(11{,}600)(1.12)/(2)(323)]$$
$$= 3 \times 10^{11} \text{ particles/cm}^3$$

The corresponding resistivities are, from Equation 9.8,

$$\rho(0°\text{C}) = \frac{1}{(1.6 \times 10^{-19})(5.85 \times 10^9)(1600 + 400)}$$
$$= 5.34 \times 10^5 \ \Omega\text{-cm}$$

and

$$\rho(50°C) = \frac{1}{(1.6 \times 10^{-19})(3 \times 10^{11})(1600 + 400)}$$
$$= 1.04 \times 10^4 \ \Omega\text{-cm}$$

The values obtained for silicon in the preceding example are typical of intrinsic semiconductors. In contrast, a metal might have a resistivity of $\rho \sim 10^{-6}$ Ω-cm whereas an insulator such as rubber has $\rho \sim 10^{16}$ Ω-cm.

In summary, an intrinsic semiconductor can be regarded as having a uniform, temperature-dependent, concentration of hole–electron pairs. The equilibrium concentration is maintained by constant thermal production and subsequent recombination of these pairs. The resistivity of an intrinsic semiconductor is much smaller than that of an insulator but is still much larger than that of a metal. Furthermore, this resistivity is very sensitive to variations in temperature.

As we will see, when a small amount of impurity is added, the semiconductor is said to become *extrinsic* or *doped.* Even a minute amount of impurity changes the electrical characteristics of the semiconductor drastically.

Extrinsic (Doped) Semiconductors

Consider the effect on intrinsic silicon when a small amount of phosphorous is added. A phosphorous atom has a valence of +5 (one unit greater than that of a silicon atom whose valence is +4). The net effect of adding phosphorous atoms is twofold. First, an allowed energy level is introduced at the top of the forbidden gap, that is, just below the conduction band. Second, the phosphorous atoms contribute electrons to completely fill this level (Figure 9.3*a*). Even at room temperature, virtually all these electrons are excited into the conduction band so that a large number of electrons are available for electrical conduction. In fact, each phosphorous atom contributes a single electron to the conduction band so that these atoms are termed *donor* atoms. The electron carrier density supplied by the donor atom is

$$\boxed{n \simeq N_d} \qquad (n\text{-type material}) \tag{9.9}$$

where N_d is the donor atom concentration. It should be noted that these electrons do *not* leave a mobile hole behind. Instead, the

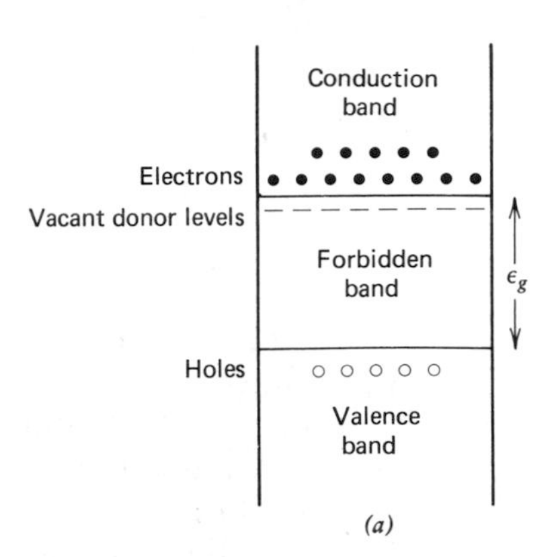

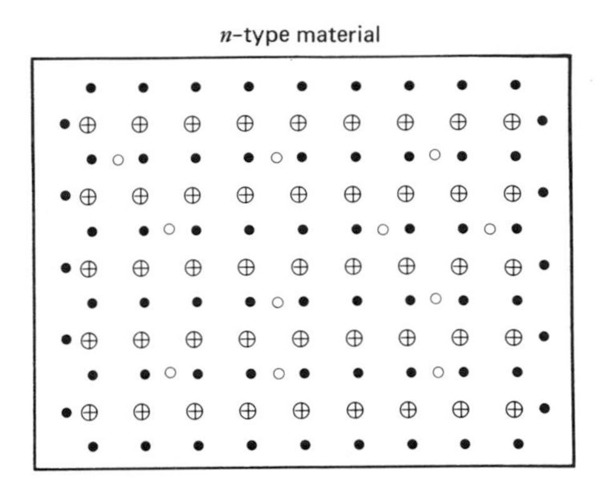

FIGURE 9.3 *(a) The band structure of an n-type semiconductor. (b) The distribution of majority carriers (electrons), minority carriers (holes), and fixed donor ions.* $(n = p_i + N_d,\ n \gg p_i)$.

positively charged donor ions remain *fixed* in the crystalline lattice. Although the donor impurity density is much smaller than the atomic density of the intrinsic lattice, the extrinsic carrier density is many orders of magnitude larger than the intrinsic hole–electron pair density produced by thermal excitation from the valence band. Because a donor impurity produces a large number of electrons for conduction, the doped material is said to be *n*-type and the electrons are called *majority* carriers.

The introduction of a large number of electrons affects the equilibrium concentration of the thermally induced *minority* carriers, namely, the holes. The reduction in the hole concentration is obtained using the *law of mass action*, which requires

$$np = n_i p_i = n_i^2$$

or

$$p = \frac{n_i^2}{n} \simeq \frac{n_i^2}{N_d}$$

where n_i is the equilibrium concentration of the intrinsic medium.

Using Equation 9.7, we find the minority concentration to be

$$\boxed{p \simeq \frac{A^2T^3e^{-\epsilon_g/kT}}{N_d}} \qquad (n\text{-type material}) \qquad (9.10)$$

Thus in an n-type material (e.g., silicon doped with phosphorous), the majority carriers consist of a large number of electrons contributed by the donor atoms plus a much smaller number produced by thermal excitation. The majority carrier concentration is approximately independent of temperature and is given by Equation 9.9. The minority carriers on the other hand are holes that are all produced by thermal excitation from the valence band. Their concentration varies considerably with temperature as given by Equation 9.10. In addition to majority electrons and minority holes, the n-type material consists of a large number of positively charged but immobile donor ions. Because the number of electrons is equal to the sum of the holes and the positive ions, the material is electrically neutral (Figure 9.3*b*).

When silicon is doped with arsenic atoms whose valence is +3, an allowed energy level is produced in the forbidden gap just above the valence band (Figure 9.4*a*). This level contains a number of vacancies equal to the number of impurity atoms. Even at low temperatures, electrons from the top of the valence band are easily excited to these *acceptor* levels. Once in these levels, the electrons bind to the arsenic atoms and become immobile, but the large number of holes left behind in the valence band become majority carriers. The impurities are called acceptor atoms and the material is classified as p-type. The majority carrier concentration is approximately temperature independent and equal to

$$\boxed{p \simeq N_a} \qquad (p\text{-type material}) \qquad (9.11)$$

where N_a is the acceptor atom concentration. Using reasoning similar to that used in deriving Equation 9.10, we find the minority carrier concentration (electrons) to be given by

$$\boxed{n \simeq \frac{A^2T^3e^{-\epsilon_g/kT}}{N_a}} \qquad (p\text{-type material}) \qquad (9.12)$$

The p-type material consists of a large (temperature-independent) concentration of free holes, a small (temperature-dependent) con-

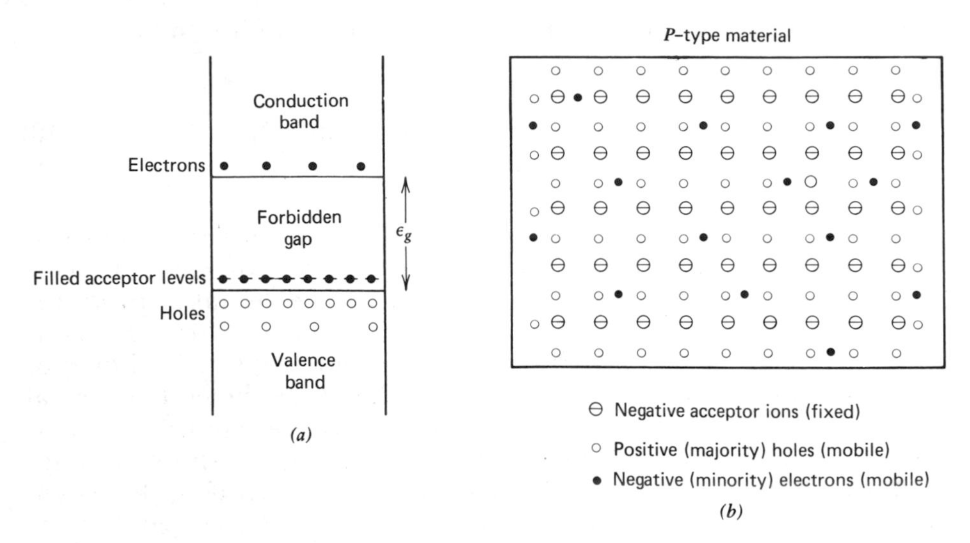

FIGURE 9.4 *(a) The band structure of a p-type semiconductor. (b) The distribution of majority carriers (holes), minority carriers (electrons), and fixed acceptor ions.* ($p = n_i + N_a$, $p \gg n_i$).

centration of free electrons, and a large number of negative but immobile acceptor ions (Figure 9.4*b*).

The *p*-*n* Junction

To understand the behavior of a photovoltaic device, we consider what happens when a junction of *p*-type materials is formed. When two such materials are brought into contact, majority carriers in each begin to diffuse across the junction in order to equalize their concentration. Holes near the junction diffuse from the *p*-type to the *n*-type materials, whereas electrons diffuse in a reverse manner. As a result, a large number of carriers recombine producing a thin region called the *depletion* layer that is essentially free of holes and electrons. This exposes the immobile ions in each region so that the depletion region in the *p*-type material becomes negatively charged while the region in the *n*-type material becomes positively charged. An electrostatic voltage and field appear across the depletion layer, inhibiting further diffusion of holes and electrons from outside the depletion layer. The *n*-type material is effectively at a higher electrostatic potential than the *p*-type material. This contact potential is not directly measurable; when metallic electrodes are affixed to the materials, secondary contact potentials appear so that the net voltage across the electrodes is

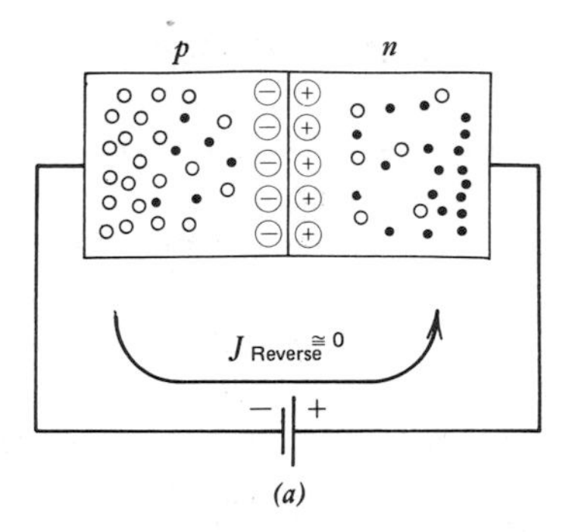

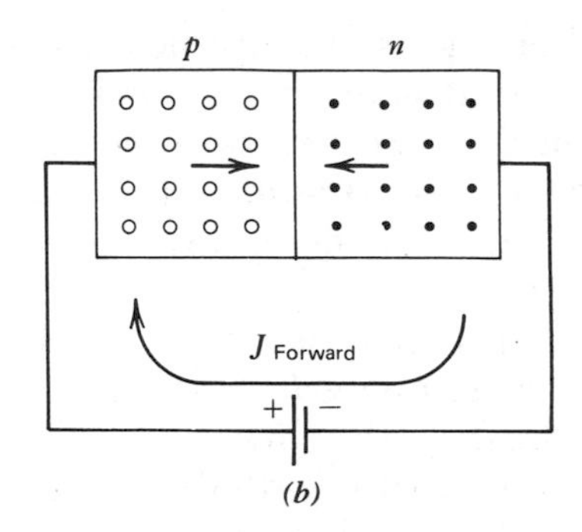

FIGURE 9.5 *The movement of carriers under (a) reverse bias and (b) forward bias.*

zero. (If the temperature of the p-n junction is different from that of the electrode junctions, a thermoelectric voltage will be generated. This effect is not being considered here.)

If an external dc voltage is applied across the junction with the positive pole of the battery on the n-type material (Figure 9.5a), current would tend to flow from the n-type to the p-type material. However, because majority carriers would be pulled away from this junction and because they cannot be thermally created in the depletion region, they could not contribute to a steady-state current. Only minority carriers that are created at the junction as hole–electron pairs by thermal excitation contribute to a small current. The maximum value of this current is called the *reverse saturation* or *dark* current; furthermore, the junction is said to be *reverse biased.* The reverse saturation current is proportional to the minority carrier concentration, as given by Equations 9.10 and 9.12. We may therefore write the reverse saturation current density as

$$\boxed{J_0 = DT^3 e^{-\epsilon_g/kT}} \tag{9.13}$$

where D is a characteristic constant of the junction. Note that the reverse saturation current is independent of voltage but increases rapidly with increasing temperature.

If we *forward bias* the junction by arranging the external voltage so that the positive pole is on the p-type material (Figure 9.5b), majority carriers will flow toward the junction. Thus a large forward current of majority carriers occurs. The forward current depends both on the applied voltage as well as on the temperature of the system. This current can be shown to be

$$\boxed{J = J_0(e^{V/kT} - 1)} \tag{9.14}$$

where V is the applied voltage expressed in volts and Boltzmann's constant is expressed as $k = 1/11{,}600$ ev/K. Equation 9.14 is called the *diode or rectifier equation* and actually describes *both* forward and reverse conduction. The plot in Figure 9.6 shows that the forward current can be many orders of magnitude larger than the reverse current.

A *p-n* junction used in electronics to rectify an alternating current is termed a *diode*. It is represented by the symbol ⊣◀⊢ where the arrow indicates the forward direction. Macroscopically, the electronic behavior of the diode is characterized by its reverse saturation current density J_0, which in turn is related to constants D and ϵ_g by Equation 9.13.

EXAMPLE

A given *p-n* junction diode is characterized by the constants $D = 0.2$ amp/cm^3-K^3 and $\epsilon_g = 1.12$ ev. Find its reverse saturation

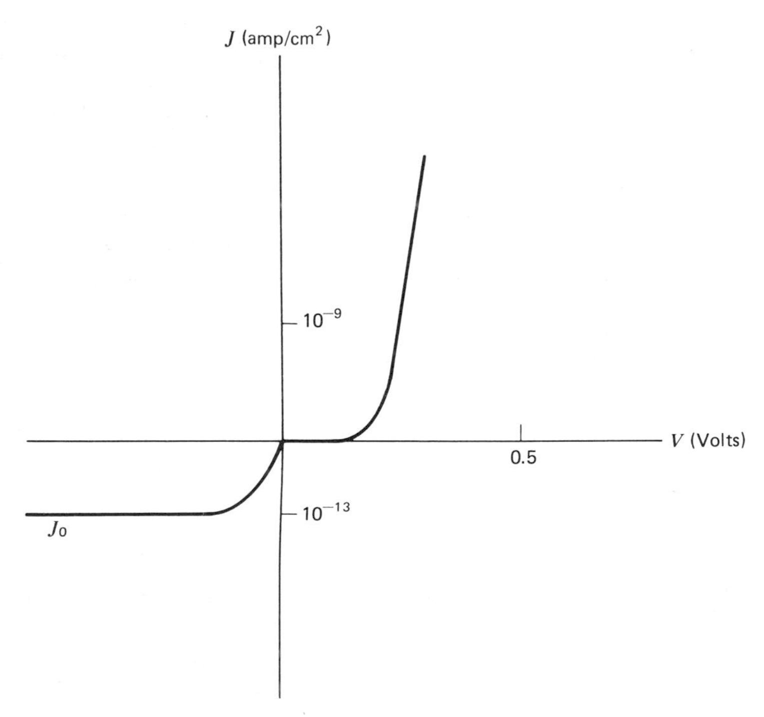

FIGURE 9.6 *The J versus V characteristic for a typical diode. The scales for reverse and forward currents are different. (The curve is a plot of the rectifier Equation 9.14.)*

current density at a temperature of 300 K. Also, find the current density when voltages of 0.5 and −0.5 V are applied to the diode.
Using Equation 9.13, we find

$$J_0 = (0.2)(300)^3 \exp[-(11{,}600)(1.12)/300]$$
$$= 8.4 \times 10^{-13}\ \text{amp/cm}^2$$

Using Equation 9.14 for $V = 0.5$ V, we find

$$J = (8.4 \times 10^{-13})(\exp[(11{,}600)(0.5)/300] - 1)$$
$$= 2.1 \times 10^{-4}\ \text{amp/cm}^2$$

For $V = -0.5$ V, Equation 9.14 gives $J = -J_0$ so that the forward current is $\sim 10^8$ times as large as the reverse current.

When a forward voltage is applied to the *p-n* diode, a forward current density flows in accordance with Equation 9.14. If, however, a *current source* (a device that supplies a fixed current) is applied to the *p-n* junction, the diode responds with a voltage across its terminals. This terminal voltage is found by solving Equation 9.14 for V in terms of J or

$$\boxed{V = kT \ln\left(\frac{J}{J_0} + 1\right)} \qquad (9.15)$$

The Junction Photovoltaic

Consider an isolated *p-n* junction fabricated of a wafer of *p*-type silicon on which a thin layer of *n*-type silicon has been deposited. The wafer is called the *base* and the deposition the *surface layer.* Electrodes are affixed to the outer surfaces of the device. The electrode for the surface layer is composed of an extremely thin metallic deposition. This electrode is essentially transparent so that sunlight reaches the surface layer with little attenuation. The surface layer is also thin so that solar radiation can reach the junction (Figure 9.7).

When solar energy is incident on the device, some of the photons create hole–electron pairs, which in effect generate a *photocurrent* flowing from the *n*-type to the *p*-type material. This photocurrent is

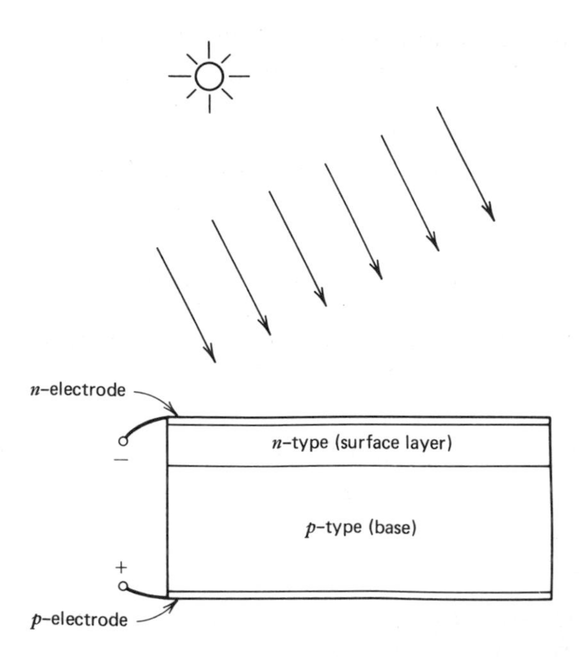

FIGURE 9.7 *A typical n- on p-photovoltaic.*

given by

$$J_p = e\dot{n}$$

where $\dot{n}$ is the rate at which the hole–electron pairs (per unit area) are being created. However, if the device is electrically isolated from external circuitry, the net steady-state current crossing the junction must be zero. This means that there must be a return current called the *junction current*, J_j, and that this current must be equal to the photocurrent. This junction current flows from the *p*-type to the *n*-type material and is therefore a *forward* current. According to Equation 9.15, this forward junction current is related to the voltage across the device by

$$\boxed{J_p = J_j = J_0(e^{V_{oc}/kT} - 1)} \tag{9.16}$$

where V_{oc} refers to the *open-circuit* voltage. A simple model for a photovoltaic cell consists of a current source in parallel with a diode as depicted in Figure 9.8. The open-circuit photovoltaic is depicted electronically in Figure 9.8*a*. Because the device is isolated, the

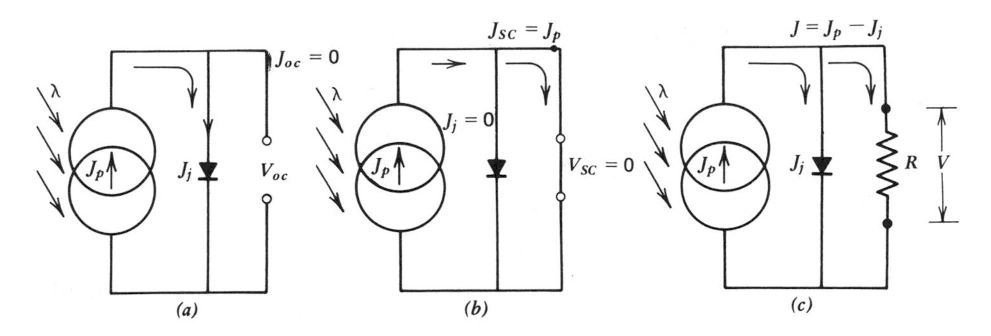

FIGURE 9.8 *A simple electronic model of a photovoltaic consisting of a photocurrent source in parallel with a diode. (a) Open-circuit operation. (b) Short-circuit operation. (c) Supplying current to an external load resistor.*

photocurrent generated by the solar radiation returns through the diode. As a result, an open-circuit voltage appears across the diode and therefore also across the terminals of the device. Solving Equations 9.16 for the open-circuit voltage, we find

$$\boxed{V_{oc} = kT \ln\left[\frac{J_p}{J_0} + 1\right]} \qquad (J_{oc} = 0) \qquad (9.17)$$

The open-circuit voltage produced by a *p-n* junction photovoltaic depends on the temperature, the reverse saturation current, and the photocurrent. As we will see, the photocurrent depends, in part, on the intensity and spectral distribution of the incident radiation.

If the terminals of the photovoltaic are short-circuited, the entire photocurrent returns via the external circuitry and the junction current vanishes. As a result, the terminal voltage is zero as expected (Figure 9.8*b*). The short-circuit current is therefore

$$\boxed{J_{sc} = J_p} \qquad (V_{sc} = 0) \qquad (9.18)$$

When a load resistor R is placed across the terminals of the photovoltaic, a fraction of the photocurrent is shunted from the diode while the rest is supplied to the load (Figure 9.8*c*). The load current is

$$J = J_p - J_j$$

or, using Equation 9.15,

$$\boxed{J = J_p - J_0(e^{V/kT} - 1)} \tag{9.19a}$$

Solving for the terminal voltage, we have

$$\boxed{V = kT \ln\left(\frac{J_p - J}{J_0} + 1\right)} \tag{9.19b}$$

From Equation 9.19b it follows that the output voltage increases with the available photocurrent but decreases as the current drawn by the load increases.

Spectral Responsivity of the Photocurrent

The current versus voltage characteristics of a *p-n* junction solar photovoltaic can be determined from Equation 9.19 once J_p is related to the radiative flux incident on the surface layer of the device. A detailed analysis of the microscopic phenomena that play a role in production of a photocurrent is beyond the level of this text. Instead, we will present a semiquantitative empirical analysis.

Imagine that the incident solar flux is represented by a spectral distribution F_λ. The spectral function represents the radiant energy per second per unit area per unit wavelength incident on the surface layer of the photovoltaic. Because a photon of wavelength λ carries an energy $E_\lambda = h\nu = hc/\lambda$, the number of photons incident per unit time per unit area per unit wavelength is

$$\dot{N}_\lambda = F_\lambda \lambda / hc \tag{9.20}$$

All the photons do not produce a photocurrent. Those with energies smaller than that of the band gap of the semiconductor do not even generate hole–electron pairs. Some of those with sufficient energy to produce such pairs are reflected and absorbed by the surface layer before they can do so. In fact, even when pairs *are* created, their tendency to recombine reduces the net photocurrent.

All these factors can be accounted for empirically using the relation

$$\dot{n}_\lambda = \beta_\lambda \dot{N}_\lambda \tag{9.21}$$

where $\dot{n}_\lambda$ is the pair creation rate (per unit area) generated by $\dot{N}_\lambda$, and

β_λ is the *quantum efficiency* for photons of wavelength λ. The parameter β_λ is zero for wavelengths larger than the cutoff wavelength λ_0. The parameter λ_0 is the wavelength of a photon whose energy is equal to that of the band gap. Generally, β_λ is largest when λ is slightly shorter than λ_0 and decreases as $\lambda/\lambda_0 \rightarrow 0$. Therefore those photons with energies slightly greater than the band gap energy are most effective in producing a photocurrent.

Using the relation $J_{p,\lambda} = e\dot{n}\lambda$ along with Equations 9.20 and 9.21, we find

$$J_{p,\lambda} = e\beta_\lambda \dot{N}_\lambda = \frac{e\beta_\lambda \lambda}{hc} F_\lambda$$

This can be conveniently expressed as

$$\boxed{J_{p,\lambda} = K_\lambda F_\lambda} \tag{9.22}$$

where $J_{p,\lambda}$ is the *spectral* photocurrent and

$$K_\lambda = \frac{e\beta_\lambda \lambda}{hc} \tag{9.23}$$

is the *spectral responsivity* of the photovoltaic. The responsivity of a typical silicon photovoltaic is shown in Figure 9.9.

The photocurrent produced by the entire spectrum is

$$J_p = \int J_{p,\lambda}\, d\lambda = \int_0^{\lambda_0} K_\lambda F_\lambda\, d\lambda = \bar{K} F \tag{9.24}$$

where

$$F = \int_0^\infty F_\lambda\, d\lambda$$

and

$$\bar{K} = \int_0^{\lambda_0} K_\lambda F_\lambda\, d\lambda / F$$

The cutoff wavelength is defined as $\lambda_0 = hc/\epsilon_g$. The parameter $\bar{K}$ is the *average responsivity* and depends on the spectral distribution of the incident radiation. It is measured in amp/watt.

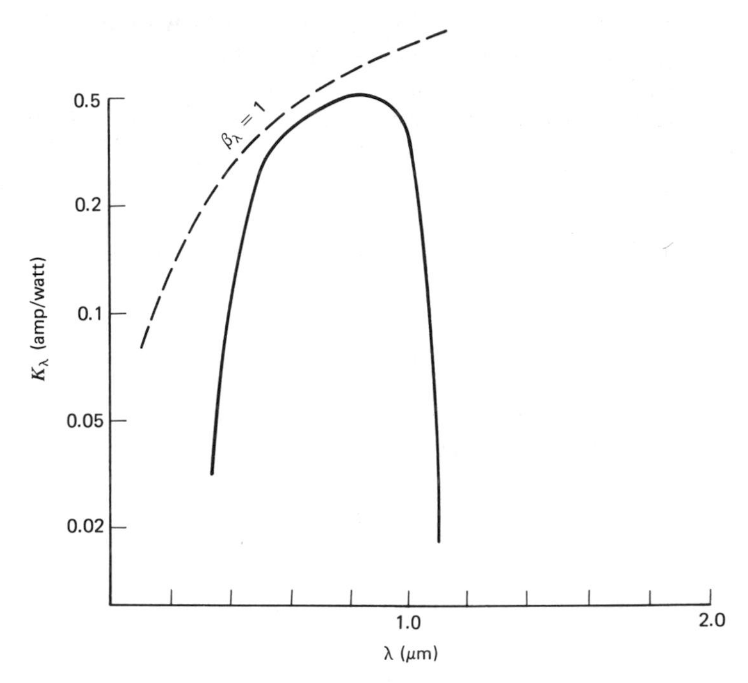

FIGURE 9.9 *The spectral responsivity of a typical silicon photovoltaic. The dashed line is the responsivity of a cell whose quantum efficiency is unity.*

Substituting Equations 9.24 into 9.19, we have

$$\boxed{J = \bar{K}F - J_0[e^{V/kT} - 1]} \tag{9.25a}$$

or

$$\boxed{V = kT \ln\left[\frac{\bar{K}F - J}{J_0} + 1\right]} \tag{9.25b}$$

Using Equation 9.20, we find the short-circuit current and open-circuit voltage to be

$$\boxed{J_{sc} = \bar{K}F} \tag{9.26a}$$

and

$$\boxed{V_{oc} = kT \ln\left(\frac{\bar{K}F}{J_0} + 1\right)} \tag{9.26b}$$

Whereas the short-circuit current is *linear* in the incident flux, the open-circuit voltage varies *logarithmically* with F (Figure 9.10). Because the average responsivity depends on the spectral distribution, it will vary with atmospheric conditions and with the position of the sun. In order to provide a basis with which to characterize the performance of a solar photovoltaic, we use a 5800 K black-body spectral distribution. It is also conventional to measure the flux in *suns* where

$$1 \text{ sun} = 1000 \text{ W/m}^2 = 100 \text{ mW/cm}^2$$

Typical electrical characteristics for a silicon photovoltaic operating at $T = 300$ K are shown in Figure 9.11. The curves were obtained from Equation 9.20 using $\bar{K} = 0.25$ mA/mW and $J_0 = 5 \times 10^{-10}$ mA/cm^2. For a fixed flux level, the output voltage decreases from its open-circuit value as the current drawn increases. Because the output power density (i.e., power per unit cell area) is given by

$$P_A = V \times J$$

the power vanishes under both open-circuit and short-circuit conditions. It reaches a maximum at some intermediate values of voltage and current, which we denote by V_{mp} and J_{mp}, respectively. These maximum power points (Figure 9.11) generally occur in the region of the "bend" of each curve.

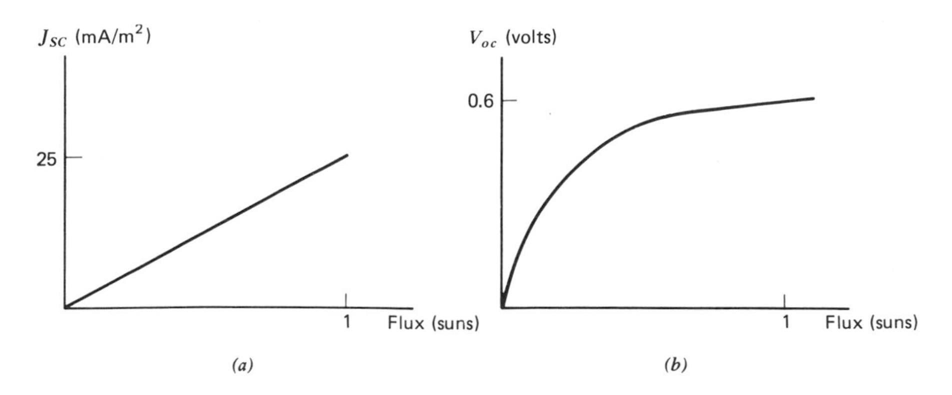

FIGURE 9.10 *(a) The short-circuit current versus the incident solar flux for a typical silicon photovoltaic. The relationship is linear. (b) The open-circuit voltage versus the incident solar flux for the same photovoltaic. The relationship is logarithmic.*

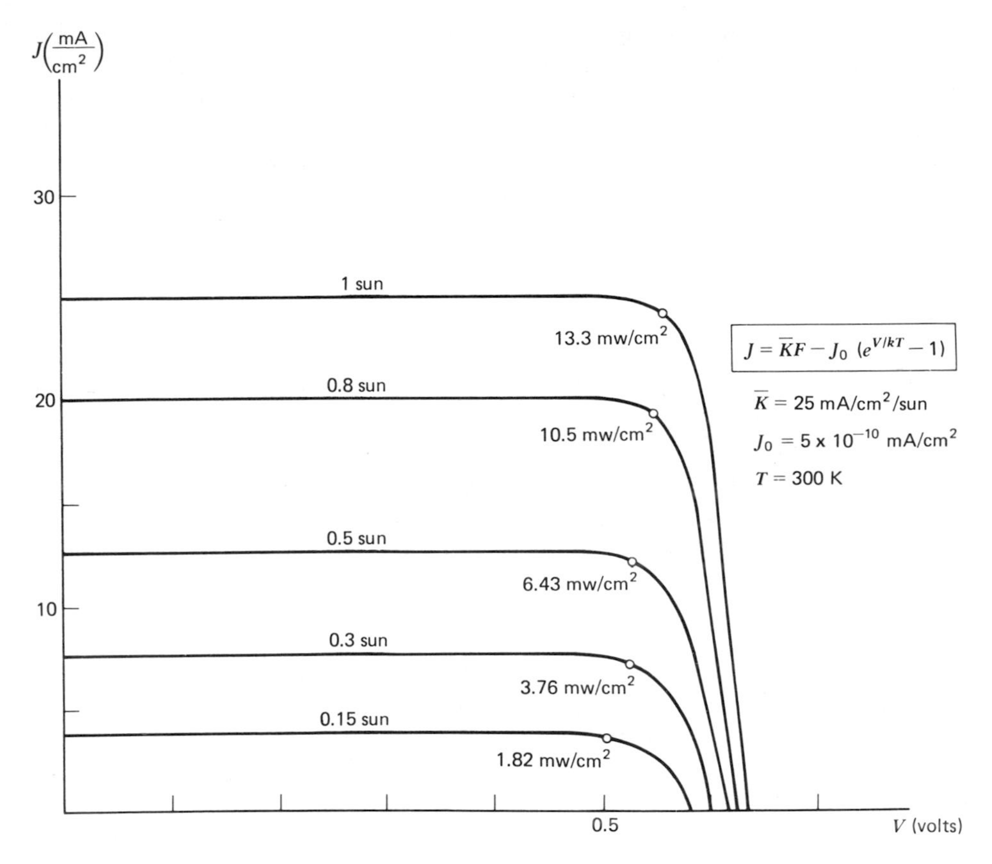

FIGURE 9.11 *The J versus V characteristics of a typical photovoltaic cell. The curve is obtained from Equation 9.25 using* $\bar{K} = 25$ *mA/cm²/sun,* $J_0 = 5 \times 10^{-10}$ *mA/cm², and* $T = 300$ *K. The maximum power points are shown for each of the insolation levels.*

EXAMPLE

The photovoltaic whose characteristic curves are shown in Figure 9.11 is operating at 1 sun ($100\ \text{mW/cm}^2$). Its area is $10\ \text{cm}^2$. Find the voltage, load resistor, power produced, and efficiency when a current density of $19\ \text{mA/cm}^2$ is drawn from the cell. Find the load resistor required to produce maximum power and determine the corresponding efficiency.

From Figure 9.11 or from Equation 9.25, we find that when $F = 1$ sun and $J = 19\ \text{mA/cm}^2$, the voltage is $V = 0.638$ V. Because the total current drawn is $I = JA = (19)(10) = 190$ mA, the required load resistor is

$$R = \frac{V}{I} = \frac{0.638\ \text{V}}{190\ \text{mA}} = 3.36\ \Omega$$

The power density and efficiency are

$$P_A = V \times J = (0.638)(19) = 11.4\ \text{mW/cm}^2$$

and

$$\eta = \frac{11.4}{100} = 11.4\%$$

From Figure 9.11, we find $J_{mp} \simeq 23.9\ \text{mA/cm}^2$ so that $V_{mp} = 0.557\ \text{V}$. The total current is $I_{mp} = J_{mp}A = 239\ \text{mA}$ so that the required load resistor is

$$R_{mp} = \frac{0.557}{239} = 2.33\ \Omega$$

The maximum power density and efficiency are

$$P_{A(max)} = V_{mp} \times J_{mp} = 13.3\ \text{mW/cm}^2$$

and

$$\eta_{(max)} = \frac{13.3}{100} = 13.3\%$$

Maximum Theoretical Efficiency as a Function of Band Gap

Note that the maximum efficiency of the silicon photovoltaic represented in Figure 9.11 is approximately 12 to 13 percent regardless of the insolation level. It is interesting to estimate the maximum theoretical efficiency of a silicon device and see how it is related to the size of the energy gap. Since V_{mp} and J_{mp} are always smaller than V_{oc} and J_{sc}, an absolute upper limit to efficiency is

$$\eta_{max} \simeq \frac{V_{oc} \times J_{sc}}{F} \tag{9.27}$$

Setting $J_{sc} = J_p$ and using Equation 9.24, we have

$$\eta_{max} \simeq V_{oc}\bar{K}$$

The average responsivity is, using Equations 9.23 and 9.24,

$$\bar{K} = \frac{e}{hc}\int_0^{\lambda_0} \beta_\lambda \lambda F_\lambda \, d\lambda \bigg/ \int_0^\infty F_\lambda d\lambda$$

Under optimum conditions, $\beta_\lambda = 1$ so that

$$\bar{K}_{max} = \frac{e}{hc}\int_0^{\lambda_0} \lambda F_\lambda \, d\lambda \bigg/ \int_0^\infty F_\lambda \, d\lambda \tag{9.28}$$

For silicon ($\epsilon_g = 1.12$ ev), the cutoff wavelength is $\lambda_c \simeq 1.1\ \mu$m. Assuming F_λ to represent a black-body spectrum at 5800 K, we find the value of $\bar{K}_{max}$ in Equation 9.28 to be

$$\bar{K}_{max}(\text{silicon}) \simeq 0.4 \text{ mA/mW}$$

The value of $\bar{K}_{max}$ is determined by the number of incident photons whose energy is smaller than that of the band gap. The smaller the band gap is, the larger $\bar{K}_{max}$ and vice versa.

According to Equation 9.26, the value of V_{oc} decreases with increasing J_0. Yet according to Equation 9.13, the reverse saturation current increases as we decrease the band gap. The net result is that V_{oc} decreases with a decrease in the band gap energy and vice versa. Thus materials with either very large or very small band gaps have low upper limits to their efficiencies. The highest theoretical limit occurs for materials whose band gap is ~1.5 ev.

With silicon, V_{oc} is ~0.6 V for typical insolation levels. Thus the maximum theoretical efficiency of a silicon photovoltaic is

$$\eta_{\text{theoretical max}} \simeq (V_{oc})(\bar{K}_{max}) = (0.6)(0.4) = 24\%$$

Hence the actual silicon photovoltaics already exceed 50 percent of the theoretical limit. Semiconductors with band gaps of ~1.5 ev have a higher value of $\eta_{\text{theoretical max}}$. However, the *actual* efficiencies obtained for such materials are well below that of silicon. It should also be noted that because efficiency decreases with increasing J_0 and J_0 increases with temperature, the efficiency of a photovoltaic generally decreases as its temperature is raised. It is therefore desirable to keep photovoltaics cool to increase both life expectancy and operating efficiency.

Photovoltaic Arrays and Systems

The output voltage of a single silicon photovoltaic $V^{(1)}$ is usually ~0.5 V and is therefore not high enough to be used in conventional electrical applications. The current output of a single cell $I^{(1)}$ is typically 1 A for a cell 3 in. in diameter. To produce larger voltages and currents, we use an array consisting of many cells. The array usually consists of n cells in series and m cells in parallel. Figure 9.12 shows an array with $n = 5$ and $m = 4$. The dotted lines are optional connections that do not affect the array's performance. They are used to ensure that a cell that becomes defective or "open" due to a faulty contact does not render an entire series branch inoperative. The output voltage and current of the array are increased to $n \times m \times P^{(1)}$ where $P^{(1)} = V^{(1)} \times I^{(1)}$ is the output power of a single cell. In theory, the efficiency of the array should be the same as that of the individual cell. However, the efficiency is somewhat reduced because of the wasted area between the cells.

Except for the fact that the V-scale and I-scale have been increased by factors n and m, respectively, the I versus V characteristics of an array are similar to those of the individual cell. Typical curves for an

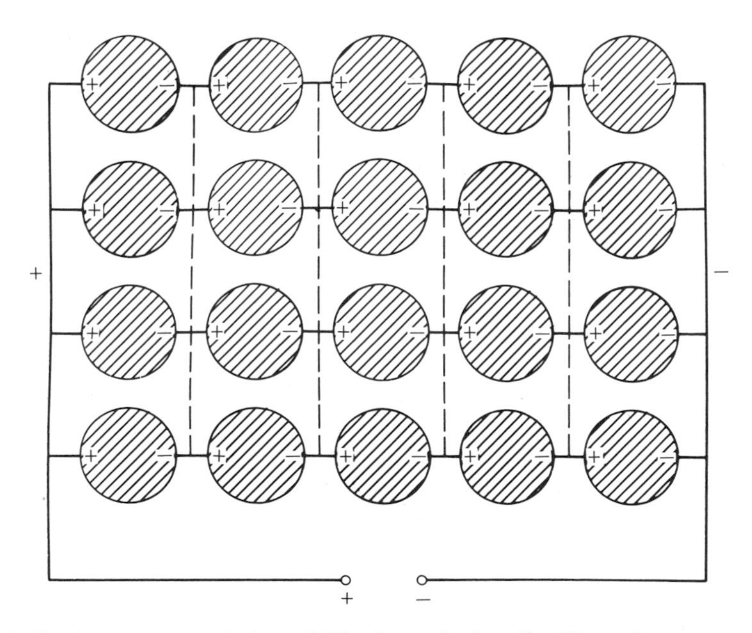

FIGURE 9.12 *An array consisting of 20 photovoltaic cells. The voltage, current, and power are, respectively, 5, 4, and 20 times as large as that of a single cell. The dashed lines are optional connections (see text).*

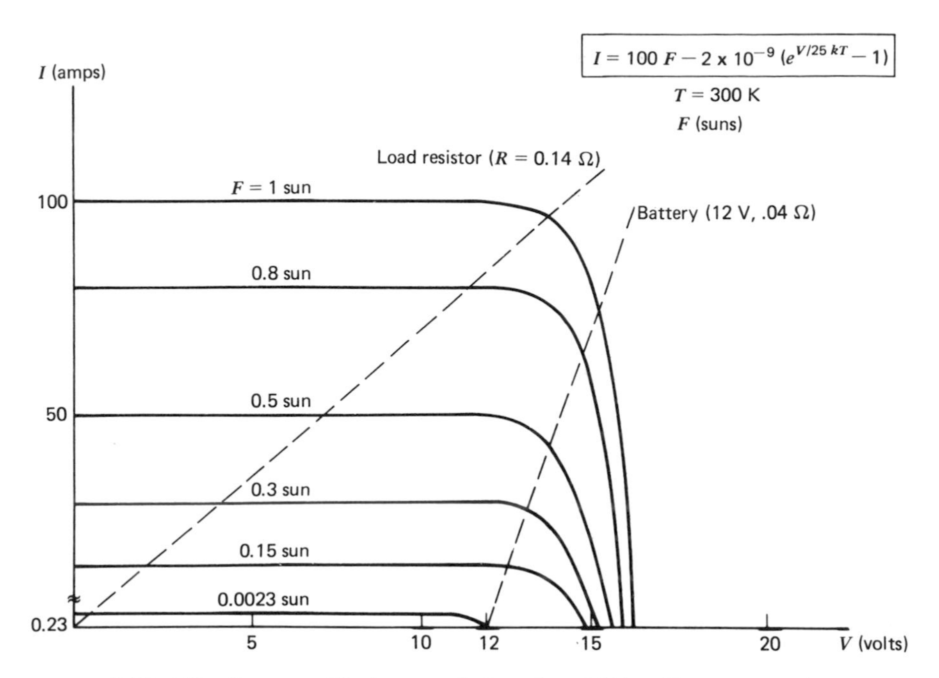

FIGURE 9.13 *The I versus V characteristics of a 1000-cell array consisting of 40 parallel strings with each string containing 25 cells in series. Each cell has an area of 100 cm² and has J versus V characteristics given in Figure 9.11. The curves are actually plots of the equation* $I = 100\,F - 2 \times 10^{-9}(e^{V/25\,kT} - 1)$ *where* $T = 300$ *K and F is expressed in suns. A resistor load line and a battery load line are shown.*

array consisting of 25 cells in series and 40 cells in parallel are shown in Figure 9.13. Each cell is assumed to have an area of $A^{(1)} = 100\ \text{cm}^2$. These curves are derived from those given for a single cell in Figure 9.11.

A photovoltaic array can be used to supply electrical energy to a load resistor or to store it in a battery. When a load resistor R is used, the operating point (i.e., the voltage and current) is determined by the intersection of a characteristic curve of the array and the *load line* of the resistor. The equation of the load line is, according to Ohm's law, $I = V/R$. The resistor load line for $R = 0.14\ \Omega$ is shown in Figure 9.13. A load line is fixed for a given load resistor. As the insolation level changes, the operating point shifts along the load line.

EXAMPLE

The photovoltaic array represented in Figure 9.13 is supplying power to a load resistor $R = 0.14\ \Omega$. Find the voltage, current, and power supplied to the load at insolation levels of 1, 0.8, and 0.5 sun.

By noting the intersection points of the load line, $I = V/0.14$, and the curves for the specified insolation levels, we find

$$V = 13.9\ \text{V}, \qquad I = 95\ \text{A}, \qquad P = VI = 1330\ \text{W}, \qquad \text{for 1 sun}$$

$$V = 11.7\ \text{V}, \qquad I = 80\ \text{A}, \qquad P = VI = 930\ \text{W}, \qquad \text{for 0.8 sun}$$

$$V = 7.3\ \text{V}, \qquad I = 50\ \text{A}, \qquad P = VI = 365\ \text{W}, \qquad \text{for 0.5 sun}$$

If the array is used to charge a battery of emf $\mathscr{E}$ and internal resistance r, the operating point is determined by the intersection of a characteristic curve of the photovoltaic array and the load line of the battery. The equation of this load line is $V = I/r - \mathscr{E}/r$; the battery load line for $\mathscr{E} = 12$ V and $r = 0.04\ \Omega$ is shown in Figure 9.13. The output power of the array is $P = VI$; this is divided into an energy storage rate $\mathscr{E}I$ and into a heat dissipation rate I^2r within the battery. Ideally, the internal resistance should be as small as possible to minimize dissipation. The load line for a battery with negligible internal resistance is almost vertical. If the insolation falls to a level for which $V_{oc} < \mathscr{E}$, the battery will begin to discharge through the array. To prevent this, we can use a *blocking* diode, as shown in Figure 9.14.

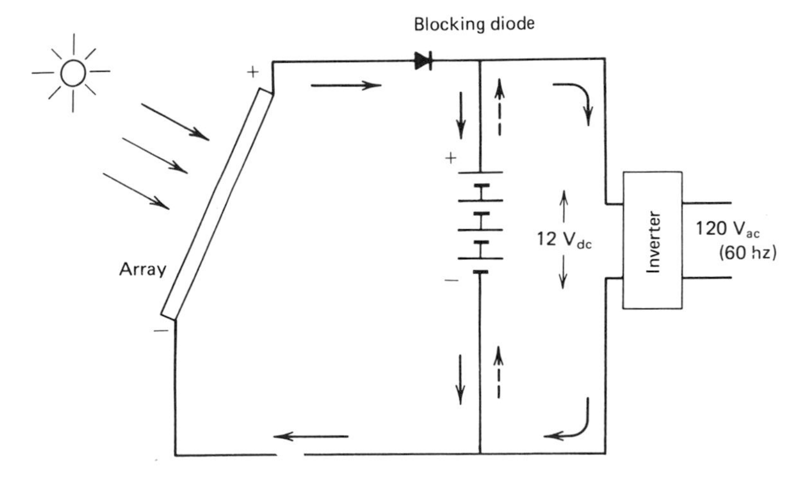

FIGURE 9.14 *A schematic of a photovoltaic array used to supply electricity at* 120 V *ac* (60 *hz*). *Batteries store excess solar energy for use during sunless periods.*

EXAMPLE

The array represented in Figure 9.13 is used to charge a battery with $\mathscr{E} = 12$ V and $r = 0.04\ \Omega$. Find the charging current, the useful charging power, and the dissipation rate at insolation levels of 1, 0.8, and 0.5 sun. Determine the minimum insolation level for which charging is possible.

Applying the load line to the appropriate curves, we find

$$I = 75\text{ A}, \quad \mathscr{E}I = 900\text{ W}, \quad I^2r = 225\text{ W}, \quad \text{for 1 sun}$$

$$I = 66\text{ A}, \quad \mathscr{E}I = 792\text{ W}, \quad I^2r = 174\text{ W}, \quad \text{for 0.8 sun}$$

$$I = 46\text{ A}, \quad \mathscr{E}I = 552\text{ W}, \quad I^2r = 85\text{ W}, \quad \text{for 0.5 sun}$$

The charging stops at an insolation level that results in an open-circuit voltage of $V_{oc} = \mathscr{E} = 12$ V. From Figure 9.13 this value is $F = 0.0023$ sun (~ 2.3 W/m^2).

In a typical photovoltaic system, 12 V batteries can be used to store energy from an array whose output voltage is greater than 12 V. However, conventional electrical appliances require 120 V-60 hz ac. Consequently, an inverter is required to convert the low voltage dc battery output. An inverter consists of a multivibrator that chops the dc voltage producing a 60 hz-12 V peak-to-peak square wave. This oscillating voltage can be boosted to 120 V ac using a step-up transformer. When insolation is available and demand small, the array will charge the batteries. When insolation is low and demand substantial, the batteries will supply the required energy (Figure 9.14).

A photovoltaic array that is 10 percent efficient could supply electricity at a rate of 100 W/m^2-sun. For a daily insolation level of 0.25 sun, a 20 m^2 array would supply an average power of 0.5 kw. This represents a substantial fraction of the power requirements of a small home.

Photovoltaics have some very desirable features. They are unobtrusive and have no moving parts. Their efficiency is not appreciably affected by changes in ambient temperatures. They are also relatively simple to install and interface easily with existing systems. The major problem with photovoltaics is their very high cost. At retail, the current cost is between $2000 and 4000 per square meter of cell area. Even with the lower figure, the cost of a 20 m^2 array would be in excess of $40,000, excluding peripheral equipment. To appreciate why costs are so high, we consider the fabrication of silicon cells.

Fabrication of Silicon Photovoltaics

Silicon photovoltaics are among the most efficient and best understood solar cells in use today. Their fabrication begins with the purest grade (solar grade) of silicon available. This silicon is derived from a lower grade (semiconductor grade) used in electronic components. The final purification stage is called *zone melting* in which zones of molten silicon are drawn through the bulk material carrying with them residual minute impurities. The solar grade left in the bulk is 99.999 percent pure.

The pure silicon is kept molten in a crucible using radio frequency waves similar to those used in a microwave oven. This keeps the material uniformly heated. The melt is kept in an inert gas environment. Doping is carefully added to the melt to produce the desired *n*- or *p*-type material. The melt is then crystallized. The method of crystallization most commonly used is the *Czochralski* method. A small seed crystal attached to a special holder is introduced to the melt. As the seed is withdrawn, a cylindrical ingot of crystalline silicon is formed and slowly drawn from the melt. The diameter of the ingot depends on the withdrawal rate; the slower the rate is, the larger the diameter. Typical rates vary from inches to fractions of an inch per hour producing ingots as large as 4″ in diameter.

The ingots of solar grade (*n*- or *p*-type) silicon are then sliced into circular wafers ~0.03 in thick. These wafers form the base of the cell. If the base material is *p*-type, then a thin layer of *n*-type material is formed on the wafer surface by diffusing appropriate impurities. The surface layer of the photovoltaic is typically a few microns (1 micron = 10^{-4} cm) thick. A metal electrode is affixed to the underside of the base and a network of fine grid electrodes is affixed to the surface layer. The surface layer is commonly coated with an antireflective film to reduce the reflection of radiation at those wavelengths at which the cell is most sensitive. The fabrication process requires that the base thickness be thin enough to reduce series resistance in the cell, yet thick enough to provide structural support. The surface layer must be thin enough to allow radiation to reach the junction yet thick enough so that the electron–hole production within the bulk is appreciable. Care must be taken to see that good contact is made with the metallic electrodes.

Fabrication time is long and costs are high even though the raw material, sand (SiO_2), is abundant and very inexpensive. Recent research into thin films and amorphous (noncrystalline) semiconductors has shown that it might be possible to produce photovoltaics at a lower cost. Another technique being considered involves the use of plastic prism concentrators that allow large flux levels to fall on cells of smaller area. If the cost of cells could be reduced by a factor of 100, say, to $40 per square meter, even at only half the current efficiency, the net gain would be a factor of 50. This would make photovoltaics very competitive with conventional electrical energy, especially as the cost of fossil fuels continues to rise.

PROBLEMS

9-1. (a) Find the radiation pressure exerted on a stationary reflector that is intercepting 1000 W/m^2 of radiative flux.
(b) What is the corresponding pressure on a perfect absorber?
(c) Using Equation 9.6, find the conversion efficiency if the mirror is moving at 1000 mph.

9-2. Consider a semiconductor whose parameters are $A = 9.5 \times 10^5$ cm^{-3}-$K^{-3/2}$ and $\epsilon_g = 0.75$ ev.
(a) Using Equation 9.7, find the intrinsic carrier concentration at 300 K.
(b) If the electron and hole mobilities are $\mu_n = 3800$ cm^2/V and $\mu_p = 1800$ cm^2/V, respectively, find the intrinsic resistivity at 300 K.
(c) If a donor impurity is added where $N_d = 2 \times 10^{17}$ part/cm^3, find the majority (electron) and minority (hole) concentrations and estimate the extrinsic resistivity.

9-3. A *p-n* diode made of a semiconductor whose gap is $\epsilon_g = 0.75$ ev is operating at $T = 300$ K. The junction parameter is $D =$ 0.15 amp/cm^2-K^3.
(a) Using Equation 9.13, find the reverse saturation current density of the diode.
(b) Find the forward and reverse current densities when the applied voltage is 0.7 V.
(c) Find the voltages across the diode when forward currents of $J = 10$ mA/cm^2 and 50 mA/cm^2 are flowing through the junction.

9-4. A *p-n* junction photovoltaic operating at 300 K has an area of 100 cm^2 and a reverse saturation current of $J_0 = 6 \times 10^{-10}$ mA/cm^2 at 300 K. An incident flux of 1 sun produces a short-circuit current of 2 amps.
(a) Find the average responsivity $\bar{K}$ of the photovoltaic.
(b) Find the open-circuit voltage of the cell at 1 sun.

(c) Find the voltage, current, and power supplied to a $R = 0.32\ \Omega$ load resistor at 1 sun.

9-5. A monochromatic beam of photons of wavelength $\lambda = 0.5\ \mu m$ has a flux of $400\ W/m^2$. Find the energy of each photon and determine the photon flux, that is, the number of photons falling on unit area per unit time.

9-6. (a) Find the cutoff wavelength λ_0 for silicon ($\epsilon_g \simeq 1.2$ ev) and for germanium ($\epsilon_g \simeq 0.75$ ev).
(b) Determine the maximum fraction of a 5760 K spectrum that can possibly be harnessed by each semiconductor. (*Hint.* Use Table 1.1.)

9-7. Consider a photovoltaic whose spectral responsivity is constant and equal to $K_\lambda = 0.3$ mA/mW for $1\ \mu m \leq \lambda \leq 2\ \mu m$ and $K_\lambda = 0$ otherwise. Find the average responsivity for a 5760 K spectrum and determine the photocurrent at a flux of 1 sun. (*Hint.* Use Table 1.1.)

9-8. A *p-n* junction photovoltaic is made of a semiconductor whose band gap is $\epsilon_g = 1.2$ ev. It has a junction parameter $D = 0.2$ amp/cm^2-K^3 and an average responsivity of $\bar{K} = 0.25$ amp/W.
(a) Using Equation 9.13, find the reverse saturation currents at both $T = 15$ and 40°C.
(b) Find the open-circuit voltages at these temperatures when the intercepted flux is 1 sun.

9-9. (a) Using the J versus V characteristics of a silicon cell as given in Figure 9.11, plot the I versus V curves for an array of 200 cells. The array has 10 parallel strings, each of which has 20 cells in series. Each cell is circular with a diameter of 3″.
(b) Find the open-circuit voltage of the array at 1 sun.
(c) Find the short-circuit current at 1 sun.

9-10. Referring to the array in Problem 9-9,
(a) Determine the minimum emf of a battery that can be charged at 0.15 sun.
(b) Find the current, voltage, and power supplied to a $1.17\ \Omega$ load resistor at 1 sun.
(c) Find the current, voltage, and power supplied at 1 sun to a battery whose emf is 10 V and whose internal resistance is $0.15\ \Omega$. Determine the fraction of this power being stored in the battery.

9-11. (a) Estimate the cost per kw of the silicon array of Problem 9-10 at 1 sun. Assume that the array is operating at 12 percent efficiency and that each cell costs $10.

(b) At an average daily insolation of 0.25 sun, find the number of years it will take for the array to pay for itself in terms of the current cost of electricity ~10¢ per kwhr. Neglect any maintenance costs.

REFERENCES

1. Angrist, S. W., *Direct Energy Conversion*, 3rd ed., Allyn & Bacon, Boston (1976), Chapters 3 and 5.
2. Brinkworth, B. J., *Solar Energy for Man*, Wiley, New York (1972), Chapter 8.
3. Kittel, C., *Elementary Solid State Physics*, Wiley, New York (1962), Chapters 6 and 7.
4. Meinel, A. B. and M. P. Meinel, *Applied Solar Energy*, Addison-Wesley, Reading, Mass. (1976), Chapter 15.
5. Merrigan, J. A., *Sunlight to Electricity*, MIT Press, Cambridge, Mass. (1975).
6. Millman, J., *Vacuum Tube and Semiconductor Electronics*, McGraw-Hill, New York (1958), Chapters 3 and 5.
7. Pulfrey, D. L., *Photovoltaic Power Generation*, Van Nostrand Reinhold, New York (1978).
8. RCA, *Electro-Optics Handbook*, RCA/Commercial Engineering, Harrison, N.J. (1968), Section 10.

APPENDIX 1

Equations for Solar Coordinates

1. To find solar noon:
   solar noon (in local standard time) =
   $12{:}00 - 4(\text{Long}_{st} - \text{Long}_{loc}) - \text{EOT}$ where
   Long_{st} = standard meridian for observer's time zone
   Long_{loc} = local meridian of observer
   EOT = equation of time
2. To convert local standard time to solar time:
   solar time = local standard time + $4(\text{Long}_{st} - \text{Long}_{loc})$ + EOT
3. To find the codeclination of the sun on a given day of the year:

$$\cos D' = [\sin(23.5°)]\left[\sin\left(\frac{360° \times n}{365.25 \text{ days}}\right)\right]$$

   where n is the number of days after the vernal equinox. The declination is $D = 90° - D'$.
4. To find the hour angle of the sun H:

$$H = \pm\frac{360°}{24 \text{ hr}}t$$

   where t is the number of hours before (negative sign) or after (positive sign) solar noon.

5. To find the solar zenith angle Z:

$$\cos Z = \cos D' \cos L' + \sin D' \sin L' \cos H$$

where

D' = solar codeclination
L' = observer's colatitude (90° − latitude)
H = solar hour angle

6. To find the solar azimuth angle A:

$$\tan A = \sin D' \sin H / (\sin D' \cos L' + \sin D' \sin L' \cos H)$$

7. To find the noontime zenith angle Z_{noon}:

$$Z_{noon} = |D' - L'|$$

8. To find the sunset (sunrise) hour angle H_s and the number and daylight hours $T_{daylight}$:

$$H_s = \pm \cos^{-1}(-\cot D' \cot L')$$

where the negative sign refers to sunrise.

$$T_{daylight} = \frac{24 \text{ hr}}{180°} H_s$$

9. To find the sunset (sunrise) azimuth:

$$A_s = \pm \cos^{-1}(-\cos D' / \sin L')$$

where the negative sign refers to sunrise.

10. To find the obliquity angle θ of the sun's rays to the surface whose normal is tilted Δ from the vertical and rotated through an azimuth angle ψ:

$$\cos \theta = \cos Z \cos \Delta + \sin Z \sin \Delta \cos(A - \psi)$$

For a southerly tilt ($\psi = 0$):

$$\cos \theta = \cos(L' + \Delta) \cos D' + \sin(L' + \Delta) \sin D' \cos H$$

APPENDIX 2

Approximate Equations for Solar Flux

The equation of radiative transfer for a plane-stratified atmosphere can be written (see Ref. 1, p. 15):

$$\mu \frac{\partial I(t, \mu)}{\partial t} = -I(t, \mu) + \frac{\tilde{\omega}_0}{2} \int_{-1}^{1} d\mu' \bar{P}(\mu, \mu') I(t, \mu') \qquad (A.2.1)$$

where $I(t, \mu)$ is the intensity of a ray traveling along the direction $\mu = \cos\theta$ at an optical depth t (measured from the top of the atmosphere). At the top of the atmosphere $t = 0$, whereas at ground $t = \tau$. The scattering is represented by the *single scattering albedo* $\tilde{\omega}_0 = \tau^s/\tau$ and by the *phase function* $\bar{P} = \bar{P}(\mu, \mu')$. The term on the left of Equation A.2.1 represents the change of the intensity of a ray at the point t as it travels along the direction defined by μ. The obliquity μ is positive for the downward hemisphere and negative for the upward hemisphere. The first term on the right gives the decrease in I due to attenuation. The second term gives the increase due to the radiation rescattered from all other directions μ' into the direction μ.

To convert Equation A.2.1 to simpler flux equations, we integrate both sides over upward and downward hemispheres and find

$$\frac{dF^{\downarrow}}{dt} = -J^{\downarrow} + \frac{\tilde{\omega}_0}{2}(J^{\downarrow} + J^{\uparrow})$$

$$= -\left(1 - \frac{\tilde{\omega}_0}{2}\right)J^{\downarrow} + \frac{\tilde{\omega}_0}{2}J^{\uparrow} \qquad \text{(A.2.2a)}$$

and

$$\frac{dF^{\uparrow}}{dt} = +J^{\uparrow} - \frac{\tilde{\omega}_0}{2}(J^{\uparrow} + J^{\downarrow})$$

$$= \left(1 - \frac{\tilde{\omega}_0}{2}\right)J^{\uparrow} - \frac{\tilde{\omega}_0}{2}J^{\downarrow} \qquad \text{(A.2.2b)}$$

where

$$F^{\downarrow}(t) = \int_0^1 \mu I(t, \mu)\, d\mu \qquad F^{\uparrow}(t) = -\int_{-1}^0 \mu I(t, \mu)\, d\mu$$

and

$$J^{\downarrow}(t) = \int_0^1 I(t, \mu)\, d\mu \qquad J^{\uparrow}(t) = \int_{-1}^0 I(t, \mu)\, d\mu$$

The functions $F^{\uparrow\downarrow}$ and $J^{\uparrow\downarrow}$ are the upward and downward *fluxes* and *net intensities*, respectively. In deriving Equation A.2.2, we interchanged the derivative and integral on the left; on the right we used the relation

$$\int_0^1 \bar{P}(\mu, \mu')\, d\mu = \int_{-1}^0 \bar{P}(\mu, \mu')\, d\mu = 1$$

which is valid for any scattering process that is symmetric with respect to forward and back scattering (e.g., Rayleigh scattering).

It is convenient to define the downward and upward obliquity functions as

$$\mu^{\downarrow}(t) = \frac{F^{\downarrow}(t)}{J^{\downarrow}(t)} \qquad \text{and} \qquad \mu^{\uparrow}(t) = \frac{F^{\uparrow}(t)}{J^{\uparrow}(t)}$$

so that Equation A.2.2 can be written

$$\frac{dF^{\downarrow}}{dt} = -\left(\frac{2 - \tilde{\omega}_0}{2\mu^{\downarrow}}\right)F^{\downarrow} + \frac{\tilde{\omega}_0}{2\mu^{\uparrow}}F^{\uparrow} \qquad \text{(A.2.3)}$$

and

$$\frac{dF^{\uparrow}}{dt} = \left(\frac{2-\tilde{\omega}_0}{2\mu^{\uparrow}}\right)F^{\uparrow} - \frac{\tilde{\omega}_0}{2\mu^{\downarrow}}F^{\downarrow}$$

Flux equations of this type were originally developed by Kuznetzov (1942) and applied to special cases by Yudin and Kagan (1956) and by Kondratyev (1956) (see Ref. 2, p. 203).

If the downward solar flux is primarily monodirectional and the upward flux isotropic, we may set

$$\mu^{\downarrow}(t) = \mu_0 = \cos Z \qquad \text{and} \qquad \mu^{\uparrow}(t) = \tfrac{1}{2}$$

where Z is the solar zenith angle. Equations A.2.3 become

$$\frac{dF^{\downarrow}}{dt} = -AF^{\downarrow} + BF^{\uparrow} \tag{A.2.4}$$

and

$$\frac{dF^{\uparrow}}{dt} = CF^{\uparrow} - DF^{\downarrow}$$

where the constants are

$$A = (2-\tilde{\omega}_0)/2\mu_0 \qquad B = \tilde{\omega}_0$$
$$C = (2-\tilde{\omega}_0) \qquad D = \tilde{\omega}_0/2\mu_0$$

Equations A.2.4 are a pair of coupled first-order linear differential equations (with constant coefficients) for the upward and downward fluxes. By differentiating one and substituting into the other, we find that both $F^{\uparrow}$ and $F^{\downarrow}$ satisfy the same equation, namely,

$$\frac{d^2F^{\downarrow\uparrow}}{dt^2} + (A-C)\frac{dF^{\downarrow\uparrow}}{dt} + (BD-AC)F^{\downarrow\uparrow} = 0$$

The solutions are

$$F^{\downarrow}(t) = ae^{\gamma^+ t} + be^{\gamma^- t}$$
$$F^{\uparrow}(t) = a'e^{\gamma^+ t} + b'e^{\gamma^- t} \tag{A.2.5}$$

where

$$\gamma^{\pm} = \tfrac{1}{2}(C-A) \pm \tfrac{1}{2}[(C+A) - 4BD]^{1/2}$$

Substituting Equation A.2.5 back into Equation A.2.4, we find

$$a' = \left(\frac{\gamma^+ - A}{B}\right)a \qquad \text{and} \qquad b' = \left(\frac{\gamma^- + A}{B}\right)b$$

so that Equation A.2.5 becomes

$$F^{\downarrow}(t) = ae^{\gamma^+ t} + be^{\gamma^- t}$$

and

$$F^{\uparrow}(t) = \frac{1}{B}[(\gamma^+ + A)ae^{\gamma^+ t} + (\gamma^- + A)be^{\gamma^- t}] \qquad \text{(A.2.6)}$$

The remaining constants, a and b, are obtained from the boundary conditions. At the top of the atmosphere ($t = 0$), the downward flux is due to the solar constant and we write

$$F^{\downarrow}(0) = \mu_0 S$$

At ground ($t = \tau$) the two fluxes are related by

$$F^{\uparrow}(\tau) = RF^{\downarrow}(\tau)$$

where R is the ground reflectivity. Using these conditions, we can determine a and b and the downward flux at ground can be shown to be

$$F^{\downarrow}(\tau) = \mu_0 S\left[\left(\frac{G}{1+G}\right)e^{\gamma^+\tau} + \left(\frac{1}{1+G}\right)e^{\gamma^-\tau}\right] \qquad \text{(A.2.7)}$$

The diffuse downward component is obtained by subtracting the downward direct flux, which is $F^{(\text{dir})} = \mu_0 Se^{-\tau/\mu_0}$, so that

$$\boxed{F^{\downarrow(\text{diff})} = \mu_0 S\left[\left(\frac{G}{1+G}\right)e^{\gamma^+\tau} + \left(\frac{1}{1+G}\right)e^{\gamma^-\tau} - e^{-\tau/\mu_0}\right]} \qquad \text{(A.2.8a)}$$

The upward flux is assumed to be diffuse so that

$$F^{\uparrow(\text{diff})} = F^{\uparrow} = RF^{\downarrow}$$

or

$$F^{\uparrow(\text{diff})} = R\mu_0 S\left[\left(\frac{G}{1+G}\right)e^{\gamma^+\tau} + \left(\frac{1}{1+G}\right)e^{\gamma^-\tau}\right] \tag{A.2.8b}$$

Equations A.2.8 are given in Equation 3.16.

REFERENCES

1. Chandrasekhar, S., *Radiative Transfer*, Dover, New York (1960), Chapter I.
2. Kondratyev, K. Y., *Radiation in the Atmosphere*, Academic, New York (1969), Chapters 1–5.
3. Menzel, D. H., *Selected Papers on the Transfer of Radiation*, Dover, New York (1966).
4. Schuster, A., Radiation Through a Foggy Atmosphere, *Astrophys. J.* 21, 1 (1905), p. 1.
5. Schwarzchild, K., Uber Diffusion und Absorption in der Sonnenatmosphäre, *Sitzungs-Berichte d. Preuss*, Akad. Berlin (1914), p. 1183.

תושלב״ע.

INDEX